STEREO-DIFFERENTIATING REACTIONS

STEREO-
DIFFERENTIATING
REACTIONS

STEREO-
DIFFERENTIATING
REACTIONS

the nature of asymmetric reactions

Yoshiharu IZUMI and Akira TAI

Institute for Protein Research, Osaka University, Suita-shi, Osaka 565, Japan

KODANSHA LTD.
TOKYO

Ⓐ︎Ⓟ︎
ACADEMIC PRESS
NEW YORK   SAN FRANCISCO   LONDON
A Subsidiary of Hartcourt Brace Jovanovich, Publishers

Ⓢ KODANSHA SCIENTIFIC BOOKS

Copyright © 1977 by Kodansha Ltd.

This book is essentially a translation of the authors' book, *Rittai Kubetsuhanno* (Japanese), Kodansha, 1975, but includes several rearrangements of items and detailed modifications and additions.

I.S.B.N. 0-12-377850-6
Library of Congress Catalog Card Numer 76-51533

Co-published by

KODANSHA LTD.
12-21 Otowa 2-chome, Bunkyo-ku, Tokyo 112, Japan

and

ACADEMIC PRESS, INC.
111 Fifth Avenue, New York, N.Y. 10003

ACADEMIC PRESS, (LONDON) LTD.
24/28 Oval Road, London NW1

PRINTED IN JAPAN

Preface

About a century has passed since Fischer first discovered the asymmetric reaction. Since that time, studies on asymmetric reactions have developed remarkably and represent a significant part of organic chemistry today. However, the chemistry of these reactions has been considered simply as a part of stereochemistry, and no systematic discussions have emerged. In fact, asymmetric reactions have considerably different features from general stereochemical reactions; of course, differentiation of the substrate by the reagent or catalyst is the most important feature. Accordingly, a special approach including a knowledge of the topology involved, is necessary to understand these reactions.

Although many books dealing with asymmetric reactions have been published, they are not always suitable as introductory texts, since they usually consist of reviews of individual reactions and do not treat the fundamental principles of asymmetric reactions. To understand such reactions theoretically, a book which contains the basic mathematics and physics appropriate to asymmetric reactions is desirable.

In 1971, one of the authors briefly introduced a classification of asymmetric reactions based on a new concept in *Angew. Chem. Intern. Ed. (Engl.)*. Subsequently, in 1974, we and several others proposed a full system of classification based on the new concept of "differentiation", and reviewed reactions involving differentiation based on the new system in *"Chemistry of Asymmetric Reactions"* (*Kagaku Sosetsu*), vol. 4, edited by the Chemical Society of Japan in Japanese.

The presentation of the new concept is not only intended to classify asymmetric reactions under a logical system, but also to provide a means for understanding the reaction processes of asymmetric reactions more fully. The publishing of *"Chemistry of Asymmetric Reactions"* and the successful application of the new concept in our research on asymmetrically modified Raney nickel catalysts prompted us to publish the present book. For these reasons, this book is entitled *"Stereo-differentiating Reactions: the Nature of Asymmetric Reactions"*.

In the book, synthetic organic chemistry, stereochemistry, group theory, the theory of optical rotation, experimental methods, etc., which are all basic to the study of stereo-differentiating reactions, are brought together to form a unified approach based on the new concept of "differentiation".

The authors hope that the value of the new concept, which is rather more complex than conventional treatments of asymmetric reactions, will become clear in the present book. This new concept should be useful in many fields of study, not only the development of stereo-differentiating reactions, but also in the study of general reaction mechanisms in organic chemistry. The consistent overall treatment of asymmetric reactions in this book has been continuously developed by our research group since Dr. Akabori led the group, and we wish to dedicate this book to Dr. Akabori and to co-workers who have participated in these studies, as listed below:

A. Akamatsu	M. Imaida	K. Okubo
K. K. Babievsky	R. Imamura	K. Okuda
T. Chihara	H. Koizumi	H. Ozaki
Y. Fujii	S. Komatsu	K. Ri
H. Fukawa	T. Masuda	Y. Saito
Y. Fukuda	K. Matsunaga	S. Sakurai
T. Furuta	K. Miyamoto	K. Shimizu
T. Harada	K. Morihara	H. Takizawa
F. Higashi	J. Muraoka	T. Tanabe
Y. Hiraki	K. Nakagawa	S. Tatsumi
Y. Hironaka	T. Ninomiya	K. Toi
K. Hirota	K. Okada	T. Yajima
Y. Imai	H. Okamura	

Also, the authors wish to express their sincere thanks to the following persons: Professor H. Matsuda, College of General Education, Osaka University, for his helpful advice in the field of electromagnetism; Professor T. Miyazawa, Faculty Science, Tokyo University, Assistant Professor T. Takagi, Drs. T. Kitagawa and H. Akutsu, Institute for Protein Research, Osaka University, for their helpful advice in the field of optics; Drs. T. Harada and H. Ozaki for their sincere discussions during the preparation of this book; Miss K. Mimaki, Mrs. K. Hara, Mrs. M. Fukuda, Mrs. N. Okuhara, Mrs. M. Yamamoto, Mrs. T. Kobayashi and Mrs. A. Kobatake for their excellent technical assistance in performing experimentation. Mr. W. R. S. Steele and the staff of Kodansha for their linguistic and editorial assistance in preparing the final manuscript for publication.

October, 1976   Yoshiharu IZUMI
Osaka, Japan   Akira TAI

Contents

CHAPTER 1

Historical Background

All discoveries develop from a historical background of related events, and the identification of asymmetric reactions is no exception. The origin of studies on asymmetric reactions perhaps lies in the discovery of polarized light, specifically in the discovery of the double-refractive properties of calcite. It was in 1669 that Bartholinus first concluded that the double images seen through Iceland spar, a form of calcite, were due to the phenomenon of double refraction. Soon afterwards, in 1678, Huygens deduced the existence of a relation between the orientation of the crystal and the deflection of light rays after observing that two of the four images seen through two crystals of calcite disappear when the upper crystal is rotated until it is parallel to the lower one. To explain this phenomenon he supposed that light must possess transverse wave properties, but at that time the longitudinal wave theory of light was almost universally accepted. Support for a transverse wave theory of light gradually developed, but it was not until 19th century dynamics were superseded by Maxwell's unification of optics and electromagnetism at the beginning of the 20th century that a sound theoretical basis was developed for the transverse wave properties of light.

However, experimental work had continued. One evening in 1808, Malus was watching the image of the setting sun reflected on a window of the palace of Luxembourg through a piece of Iceland spar, and he observed that the intensity of the double image changed as the piece of Iceland spar was rotated. Subsequently, he found that the light reflected from a polished surface of the transparent substance and the light passing through the Iceland spar had the same physical properties. Thus, when two plates were placed facing each other, one facing north and the other south, the image of an object could be seen by reflection from the two plates as shown in Fig. 1.1. However, when the upper plate was rotated 90° to the east or west,

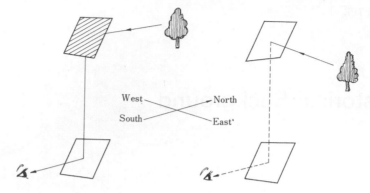

Fig. 1.1.   Diagrammatic representation of Malus' experiment.

the image could no longer be seen. Malus referred to these plates as poles, saying "Giving to these sides the names of poles, I will describe as polarized the modification which gives to light its properties relative to those poles". Thus, he was the first to introduce the term "polarized light". Malus tried to explain his results by supposing that light has two different polarities, like a magnet. These polarities were supposed to be at right angles to each other, and Iceland spar had the property of separating the two.

Although Malus's experiments were terminated by his death in 1812, studies on these phenomena were continued by Biot, who made a polarimeter based on the apparatus used by Malus. Biot showed that a plate of quartz cut along the crystal axis rotates the plane of incident polarized light and he found that right and left forms of quartz crystals rotated the plane of polarized light in different directions. Herschel later demonstrated the relationship between the crystallographic and optical properties of quartz. Biot was also the first to show that many organic compounds could rotate the plane of polarized light. Since this occurred with liquids or in solutions, as well as in the solid state, the effect was clearly attributable to the structures of the molecules themselves. Tartaric acid obtained from wine lees during the manufacture of wine was shown by Biot to be dextrorotatory. On the other hand, racemic acid, which was isolated during the purification of tartaric acid in the 1820's, was found by Biot to be optically inactive. By 1831, however, Berzelius found that tartaric acid and racemic acid both had the same molecular formula, $C_4H_6O_6$, and he introduced the term *isomerism* to describe different substances having the same molecular formula. Subsequently, Mitschelrich studied the crystal structures of the ammonium salts of tartaric and racemic acids, and reported in 1844 that the crystals are isomorphous and have the same physical properties. In 1846, Pasteur observed that all the crystals of dextrorotatory tartaric acid that he obtained had hemihedral faces with the same orientation. On the basis of Herschel's study of quartz, Pasteur assumed intuitively that the hemihedral structure of tartaric acid salts must be related to their optical rotatory power. He therefore expected that crystals of racemic acid would not have hemihedral faces, in conflict with Mitschelrich's earlier observations, but when he prepared crystals of sodium ammonium racemate he found that they did indeed have hemihedral faces. However, he made an important observation which Mitschelrich did not: the crystals of the racemate were a conglomerate of left- and right-handed hemihedral faces. In 1848, Pasteur succeeded in separating the two types of crystals from sodium ammonium racemate by hand, using a microscope. He found that the crystals resembling those of sodium ammonium tartrate were in fact dextrorotatory and were identical with the tartaric acid salt. The remaining crystals rotated the plane of polarized light by the same amount, but were levorotatory. He confirmed that a solution of an equal mixture of the two types of separated crystals was optically inactive, and described such a mixture as a *racemate*, from the name racemic acid.

Pasteur was fortunate in his choice of crystallization conditions, since sodium ammonium racemate can be crystallized as a conglomerate below 28°C, and in any event, not many compounds could be resolved by this method. In 1852, he found that racemic acid could be resolved by the use of optically active natural bases such as quinine and brucine, and in 1858 he developed a method for resolving tartaric acid by using *Penicillium* sp., yielding *levo*-tartaric acid. He thus established the basis for all techniques now used for optical resolution. Throughout this work, Pasteur assumed that the two forms of tartaric acid corresponded to "real" and "mirror" images in structure, as described in his lecture in 1860.

Two years earlier, in 1858, Kekulé published his theory of molecular structure, and in 1873, based on this theory, Wislicenus recognized *dextro*- and *levo*-lactic acids to have the same structure, saying "If it is once granted that molecules can be structurally identical and yet possess dissimilar properties, this can only be explained on the grounds that the difference is due to a different arrangement of these atoms in space". In 1874, van't Hoff and Le Bel simultaneously proposed the concept of the spatial arrangement of four substituents around a carbon atom: van't Hoff developed his ideas from those of Kekulé and Wislicenus, while Le Bel's proposal was based on Pasteur's work. This provided the basis for modern views on structure, and the relationship between the configuration and optical isomerism of organic compounds was settled. However, the formation of optically active substances in living systems still remained a mystery, the phenomenon being explained in terms of an unscientific concept, "vital force".[1]

In 1894, the study of asymmetric reactions began with Fischer's work[2] on the reaction of hydrogen cyanide with sugars, yielding cyanhydrin epimers in different proportions. Fischer was interested by this result, especially in relation to the mechanisms of synthesis of optically active sugars in plant cells. However, his results were still considered by many to be attributable to a "vital force" remaining within sugars derived from living systems. In 1912, Bredig and Fiske[3] successfully achieved a catalytic asymmetric reaction, viz. benzcyanhydrin synthesis with quinine as a catalyst.

Gradually examples of asymmetric reactions were accumulated, and the experimental results were organized by Marckwald in 1904.[4] His definition of the asymmetric reaction, which clarified its essential features, permitted studies of asymmetric reactions to be put on an organic chemical basis. In order to explain the reaction mechanism of asymmetric reactions, it was proposed that they proceeded by an induced asymmetric force[5] acting through the chemical bond from the asymmetric center to the reaction center. This hypothesis gained wide acceptance by the 1940's, and was supported by McKenzie, Ritchie et al.[6] However, in the 1950's, a reassessment was made as studies on the mechanism of steric hindrance grew.[7,8] In 1950, Doering et al.[9] and Jackman et al.[10] interpreted the reaction mechanism of the Meerwein-Ponndorf-Verley reduction, which yields optically active alcohols, by postulating the existence of a transition intermediate. This was the first modern explanation of an asymmetric reaction mechanism. Empirical rules for asymmetric reduction were proposed by Cram et al. in 1952[11] and Prelog et al. in 1953,[12] so permitting a predictive treatment of asymmetric reactions.

In Japan, studies centered on the synthesis of amino acids after World War II, and catalytic asymmetric reactions soon emerged as preferred methods. In 1956, the present authors and co-workers[13] achieved asymmetric reduction with a heterogeneous silk-Pd catalyst. Studies of this catalyst then led to work on modified Raney nickel, and later to systematic investigations of asymmetric modified Raney nickel catalysts.[14] Gradually the importance of asymmetric reactions from a practical viewpoint was recognized. For instance, in the late 1960's it was found that L-3,4-dioxyphenylalanine (L-DOPA) could be used to treat Parkinson's disease and, in 1968, Knowles et al[15] succeeded in synthesizing L-DOPA with almost 100% optical purity using the asymmetric Wilkinson complex catalyst. At this time the main focus of interest in relation to asymmetric synthesis was the use of asymmetric catalysts.

Concurrent with the development of asymmetric catalytic reactions, progress has also been made in the last decade with asymmetric reactions using reagents. Research on the chemistry of boranes by Brown et al. in the 1960's led to the development of many asymmetric hydrogenation reactions involving boranes.[16] While asymmetric syntheses with almost 100% optical purity were developed by Horeau *et al.* in 1968.[17]

However, in spite of the great advances in studies of asymmetric reaction procedures, no remarkable progress was made in the theoretical treatment of asymmetric reactions. In order to systematize the classification of asymmetric reactions, the authors[18] proposed a new definition and classification of asymmetric reactions in 1971, and restated them in detail in 1974.[19]

REFERENCES WORKS

a. M. Born and E. Wolf, *Principles of Optics,* XXI, Pergamon, 1970.
b. R. E. Lyle and G. G. Lyle, *J. Chem. Educ.,* **41,** 309 (1964).
c. J. B. Cohen, *Organic Chemistry for Advanced Students,* Edward Arnold, 1931.
d. M. Nakazaki, *Bunshi no Katachi to Taisho* (Japanese), Nankodo, 1972.
e. *Asimov's Biographical Encyclopedia of Science and Technology,* Doubleday, 1964.
f. *World Who's Who in Science,* Marquis, 1968.
g. Ed. T. Kaneko, Y. Izumi, I. Chibata and T. Itoh, *Synthetic Production and Utilization of Amino Acids,* Kodansha-Wiley, 1974.

REFERENCES

1. F. R. Japp, *Nature,* **58,** 452 (1898).
2. E. Fischer, *Ber.,* **23,** 2611 (1890).
3. G. Bredig and P. S. Fiske, *Biochem. Z.,* **46,** 7 (1912).
4. W. Marckwald, *Ber.,* **37,** 1368 (1904).
5. T. M. Lowry and E. E. Walker, *Nature,* **113,** 565 (1924).
6. P. D. Ritchie, *Advan. Enzymol.,* **7,** 65 (1947).
7. D. M. Bovey, J. A. Reid and E. F. Turner, *J. Chem. Soc.,* **1951,** 3227.
8. W. A. Bonner, *J. Org. Chem.,* **26,** 2194 (1961).
9. W. von E. Doering and R. W. Young, *J. Am. Chem. Soc.,* **72,** 631 (1950).
10. L. M. Jackman, J. A. Mills and J. S. Shannon, *ibid.,* **72,** 4814 (1950).
11. D. J. Cram and F. A. Abd Elhatez, *ibid.,* **74,** 5828 (1952).
12. V. Prelog, *Helv. Chim. Acta.,* **38,** 308 (1953).
13. S. Akabori, Y. Izumi, Y. Fujii and S. Sakurai, *Nature,* **178,** 323 (1956).
14. Y. Izumi, M. Imaida, H. Fukawa and S. Akabori, *Bull. Chem. Soc. Japan,* **36,** 21 (1963).
15. W. S. Knowles and M. J. Sabacky, *Chem. Commun.,* **1968,** 1445.
16. H. C. Brown and G. Zweifel, *J. Am. Chem. Soc.,* **83,** 2544 (1961).
17. J. P. Vigneron, H. Kagan and A. Horeau, *Tetr. Lett.,* **1968,** 750.

18. Y. Izumi, *Angew. Chem. Intern. Ed. Engl.*, **10**, 871 (1971).
19. Y. Izumi *et al.*, *Kagaku Sosetsu* (Japanese) (ed. Chem. Soc. Japan), vol. 4, Tokyo University Press, 1974.

CHAPTER 2

Molecular Symmetry and Chirality

2.1 STEREOISOMERISM

Compounds which have the same rational formula but different stereo systems are called stereoisomers, i.e., stereoisomers cannot be superposed upon each other spatially. Stereoisomerism has never been defined consistently, only on a case-by-case basis, so there is no unity of approach. However, if stereoisomerism is considered on the basis of a definite physical or mathematical treatment, certain relationships should emerge. This book will seek to order stereoisomers from the viewpoints of the internal energy and symmetry of the molecules.

2.1.1 Classification according to internal energy

Stereoisomers can be grouped as shown in Fig. 2.1 on the basis of internal energy. The upper row shows stereoisomers which have the same internal energy, though they cannot be superposed spatially. Such compounds are named enantiomers and have identical physical properties, except for properties related to optical rotation, and identical chemical properties.

The stereoisomers listed in the lower row of Fig. 2.1 have different internal energies, and therefore have different physical and chemical properties. These are called diastereomers. This category can be further subdivided into diastereomers in the strict sense, i.e., those containing two or more asymmetric carbon atoms in the molecule, and molecules where the isomerism arises from geometric factors (for instance, *cis-trans* isomerism involving a carbon-carbon double bond).[†]

Stereoisomers can be also divided in a different way into two classes. Of these, one consists of configurational stereoisomers arising from differences of configuration, and the other consists of conformational

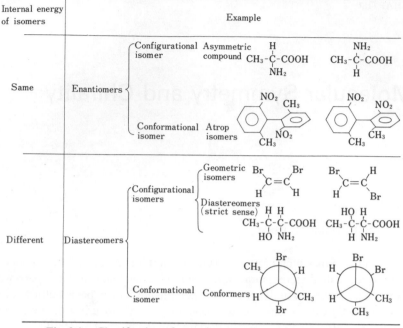

Fig. 2.1.   Classification of stereoisomers.

stereoisomers resulting from differences of steric conformation due to the rotation or twisting of atom(s) or atomic group(s) bonded to carbon atoms along a certain bond taken as the axis of rotation. An energy barrier of more than 16–20 kcal/mole between such conformational stereoisomers is necessary if the separate forms are to be stable at room temperature. Atrop

†   In the strict sense compounds are said to be diastereomers if a single asymmetric center is of different configuration in the two compounds while the remaining asymmetric centers) have the same configuration. The practical terminology for diastereomers composed of two asymmetric centers can be illustrated by reference to threose and erythrose.

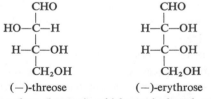

       (−)-threose   (−)-erythrose

A diastereomer such as threose in which two hydroxyl groups bound to asymmetric centers appear on different sides of the molecule on a Fischer projection is referred to as *threo*, while a diastereomer such as erythrose, in which the two hydroxyls are on the same side in the projection is referred to as *erythro*. These terms were extended to apply to the relations between an amino and a hydroxyl group in oxyamino acids.

isomers, which are simply rotational energy barrier isomers, are classified as enantiomeric from the viewpoint of internal energy, since the internal energies are the same.

2.1.2   Classification according to molecular symmetry

The chemistry of optically active compounds was originally simply the chemistry of compounds containing asymmetric centers, but in 1955 McCasland[1] synthesized 3,4,3',4'-tetramethyl-spiro-(1,1')-bipyrrolidium salts and confirmed that their isomers are optically active.

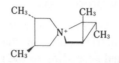

3,4,3',4'-tetramethyl-spiro-(1,1')-bipyrrolidium ion

Since 1955, many optically active substances which do not contain any asymmetric centers have been synthesized. Thus it was shown that the only requirement for optical activity was that the real image and mirror image of the substance should not be superposable. These optically active compounds without asymmetric centers are referred to as dissymmetric.

However, the terms asymmetric and dissymmetric have essentially the same meaning, i.e., the opposite of symmetric, and in addition the unwieldly term nondissymmetric had to be introduced for structures having no enantiomers. In order to avoid confusion, Cahn, Ingold and Prelog[2] suggested in 1966 that the term *chiral* should be used for enantiomers, based on the Greek word *cheiro* used in the same connection by Kelvin[3] in 1904. Thus, a new classification system based on molecular symmetry was introduced. The term chirality replaced both asymmetry and dissymmetry, while achirality replaced nondissymmetry.

2.2   MOLECULAR SYMMETRY AND CHIRALITY

2.2.1   Method of classification

A.   Symmetry operations and symmetry elements

The requirement for a molecule to have optical activity is that the structure of the molecule should be chiral. Chirality exists if the real image of the compound cannot be superposed upon the mirror image, but this is still not very useful if the decision must be made intuitively. We shall therefore consider the use of symmetry to classify molecular structures, since the concept of symmetry is well formulated mathematically. In this way we can decide what differences of symmetry exist between chiral and achiral compounds.

By considering molecular structures as arrays of geometrical points, we can classify stereo structure in terms of the type and number of symmetric elements. As geometrical symmetry elements we can consider two basic elements: the proper axis and the improper axis. A stereo structure is said to have a proper or improper axis of symmetry if the operation of proper rotation or improper rotation produces a structure equivalent to the initial structure.[†]

B. Proper rotation and the proper axis

Proper rotation is the operation of rotating a shape about the axis passing through the geometric center of the shape. The operations of proper rotation are classified according to the rotational angle, and proper rotation through an angle of $2\pi/n$ is referred to as C_n operation. When a shape is transformed into an equivalent shape by a rotation of $2\pi/n$ about the axis passing through its center (C_n operation is carried out), it is said to have an n-fold proper axis (C_n axis) as a symmetric element. Successive application of the C_n operation, such as C_n^2 or C_n^3 (i.e., two or three successive C_n operations) will naturally produce equivalent shapes, and any operation which produces a shape identical with the original one is referred to as an identity operation (E).

Examples of C_n operations are given in Fig. 2.2, which shows compounds with C_1, C_2 and C_3 symmetry elements. Compounds with a C_1 axis of symmetry, such as compound I, are generally said to have nonsymmetric structures. Compound II has a 2-fold proper axis of symmetry (C_2). Application of the C_2^1 operation yields the equivalent molecule IIa, as does C_2^2, which yields IIb. However, if the methyl groups could be tagged as shown, C_2^2 produces a molecule which is not only equivalent but also identical and thus $C_2^2 = E$. In the case of compound III in the same figure, there is a 3-fold proper axis, and it can be seen that $C_3^3 = E$. In these examples, the only symmetric element is C_n.

Fig. 2.3 shows the same molecules with their mirror images. It is clear that the real and mirror images cannot be superposed. Fig. 2.4 shows two more complex molecules, the top one possessing three 2-fold proper axes (C_2) perpendicular to each other, and the bottom one possessing a 3-fold proper axis (C_3) and three 2-fold proper axes perpendicular to the C_3 axis. For these compounds too, the real and mirror images are not superposable, so that mirror stereoisomers exist. It is clear that the structure of a chiral molecule requires only a proper axis of symmetry as a symmetric element.

[†] Structures are said to be equivalent if they cannot be distinguished by means of a symmetry operation.

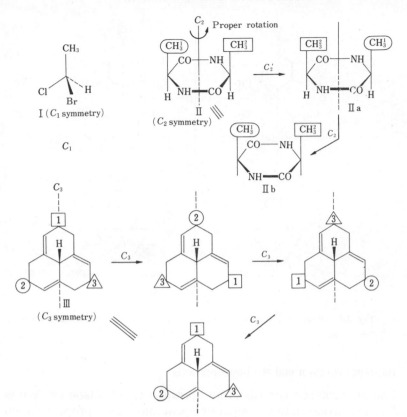

Fig. 2.2. Molecules with C_1, C_2 and C_3 symmetry elements.

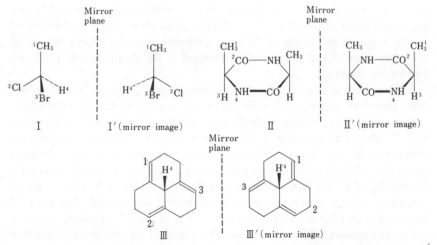

Fig. 2.3. Real and mirror images of the compounds shown in Fig. 2.2.

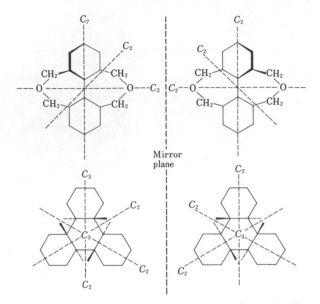

Fig. 2.4. Real and mirror images of more complex molecules, showing various axes of symmetry.

C.   Improper rotation and the improper axis

Improper rotation consists of an operation called rotation-reflection which can be divided into two separate operations: first a proper rotation and then a reflection through a plane perpendicular to the axis of rotation and containing the geometric center. Improper rotation is specified by the rotational angle $(2\pi/n)$ of the first rotation, and is specified by the symbol S_n for a rotational angle of $2\pi/n$. A structure is said to have an n-fold improper axis (S_n) as a symmetric element if the S_n operation yields an equivalent structure.

First we will consider S_1, which consists of C_1 operation and reflection. However, C_1 is an identity operation, and so S_1 simply produces a mirror image by reflection through a plane including the geometric center. Fig. 2.5 shows an example of a molecule with an S_1 symmetric element. It can be seen that S_1 operation produces an equivalent but not identical structure, while two S_1 operations, i.e., S_1^2, yield an identical structure, so that $S_1^2 = E$. We can see intuitively that the molecule has the plane of symmetry shown, which includes the geometric center, and thus a 1-fold improper axis (S_1) is exactly the same as a plane of symmetry (σ) as a symmetric element. We will use the term plane of symmetry (σ), since it is easier to understand intuitively.

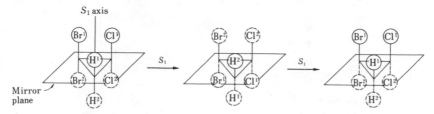

Fig. 2.5. Example of a molecule with an S_1 symmetric element.

Next, S_2 is the operation of rotation by $2\pi/2$ followed by reflection, and is thus equivalent to the inversion of a center of symmetry. Fig. 2.6 shows an example of a molecule with S_2 as a symmetric element. In general, a 2-fold improper axis of symmetry (S_2) is equivalent to an inversion center (i) as a symmetry element, and as can be seen from the figure, $S_2^2 = E$.

S_3 is the operation of rotation by $2\pi/3$ followed by reflection. Fig. 2.7 shows an example of a compound having S_3 as a symmetry element. This is equivalent to a C_3 axis with a symmetry plane (σ) perpendicular to it, and Fig. 2.7(b) shows that $S_3^6 = E$.

Fig. 2.8 shows a molecule with an S_4 element of symmetry. S_4 involves a proper rotation of $2\pi/4$ followed by reflection through a plane perpendicular to the axis of rotation. Fig. 2.8(b) shows the successive application of S_4 operations to a schematic diagram of the molecule shown in (a).

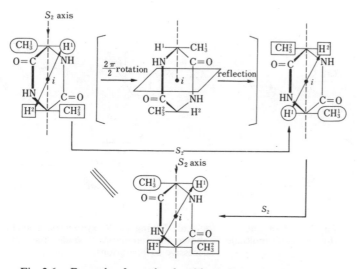

Fig. 2.6. Example of a molecule with an S_2 symmetric element.

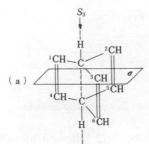

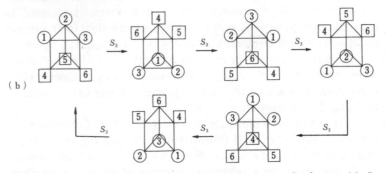

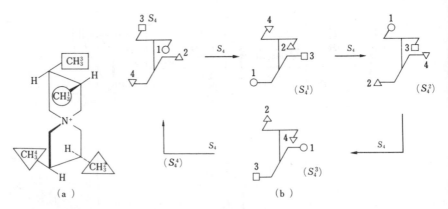

Fig. 2.7. Example of a molecule having an S_3 symmetric element (a). In (b) the same molecule is shown diagrammatically under the influence of successive S_3 operations.

Fig. 2.8. Example of a molecule having an S_4 symmetric element (a). In (b) the same molecule is shown diagrammatically under the influence of successive S_4 operations.

It is clear that $S_4^4 = E$, but in addition, the result of S_4^2 operation is the same as that which would be obtained by C_2 operation, so that the presence of S_4 symmetry automatically implies the presence of a C_2 axis of symmetry.

We may note here that S_1, S_2 and S_3 all contain symmetric elements which can be expressed in terms of σ and i, whereas S_4 is a different and independent symmetry element. These features are summarized in Table 2.1.

TABLE 2.1
Characteristics of improper axes

S_{2p-1}		Symmetry plane (σ) and proper axis (C_{2p-1})	$S_{2p-1}^{2p-1} = \sigma$	$S_{2p-1}^{2(2p-1)} = E$
	$S_1(p = 1)$	Symmetry plane (σ)	$S_1^1 = \sigma$	$S_1^2 = E$
	$S_3(p = 2)$	Symmetry plane (σ) and 3-fold proper axis (C_3)	$S_3^3 = \sigma$	$S_3^6 = E$
S_{4p-2}		Inversion center (i) and proper axis (C_{2p-1})	$S_{4p-2}^{2p-1} = i$	$S_{4p-2}^{4p-2} = E$
	$S_2(p = 1)$	Inversion center (i)		$S_2^2 = E$
	$S_6(p = 2)$	Inversion center (i) and 3-fold proper axis (C_3)	$S_6^3 = i$	$S_6^6 = E$
S_{4p}		No σ or i, but S_{4p} and C_{2p}	$S_{4p}^{2p} = C_{2p}$	$S_{4p}^{4p} = E$
	$S_4(p = 1)$	S_4 and C_2	$S_4^2 = C_2$	$S_4^4 = E$

Next, we will consider the relation between the real and mirror images of shapes with an improper axis of symmetry as a symmetric element. Fig. 2.9 shows the real and mirror images of some of the compounds we have already looked at. In this figure, compound I can clearly be superposed on its mirror image I'. When the mirror image of compound II is rotated by 180° about an axis perpendicular to the plane of reflection, II'' is obtained, and this can be superposed upon II, so that compounds II and II' are also equivalent. Compound III and its mirror image are superposable. Similarly, if the mirror image of IV is rotated by 180° about an axis perpendicular to the mirror plane (IV'') and further rotated by 90° about an axis parallel to the mirror plane, IV''' is obtained. This is superposable upon IV, and thus IV and its mirror image IV' are equivalent. As illustrated by these examples of S_1 through S_4, the real and mirror images of shapes having an improper axis S_n as a symmetric element are generally superposable, so that such compounds are achiral.

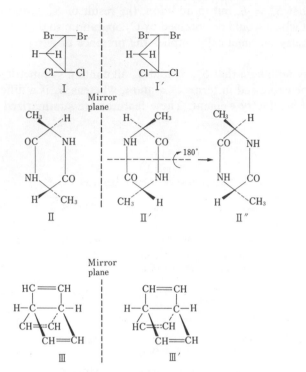

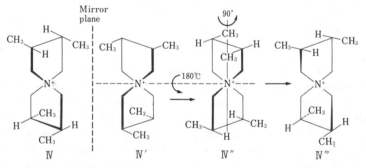

Fig. 2.9. Relationship between the real and mirror images of compounds
with improper axes of symmetry as symmetric elements.

D. Expression of symmetry operations in Cartesian coordinates

In the previous sections, we have looked at concrete examples of symmetric elements and operations in relation to the shapes of molecules, and now we will consider how to express these in Cartesian coordinates.

Having identified the symmetry elements which are present, we choose Cartesian coordinates as follows:
1) Take the geometrical center as the origin of Cartesian coordinates.
2) Decide the z axis by means of the following rules:
    (a) If the molecule has only one proper axis, take this as the z axis.
    (b) If there are several kinds of proper axis, take the axis with the largest value of n as the z axis. If there are several axes with the same value of n, choose the axis which passes through the largest number of atoms.
    (c) If the molecule has no proper axis, choose an S_n axis as the z axis on the same basis.
3) Choose the x axis in the plane perpendicular to the z axis such that it passes through the largest possible number of atoms. The y axis is then automatically fixed.
4) Use the right-handed system for the x, y, z axes.

Fig. 2.10 shows the Cartesian coordinates fixed for three compounds by means of these rules.

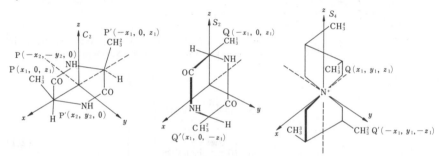

The z axis is fixed as the C_2 axis then the x axis is fixed in the plane including CH_3 and H.

The z axis is fixed as the S_2 axis then the x axis is fixed in the plane of the 6-membered ring.

The z axis is fixed as the $C_2(S_4)$ axis then the x axis is fixed in the plan of one of the 5-membered rings.

Fig. 2.10. Cartesian molecular coordinates determined as described in the text.

We shall now consider the relation between Cartesian coordinates chosen in this way and symmetry operations. Fig. 2.11(a) shows the relation between $P(x, y, z)$ and $P'(x', y', z')$ obtained by C_n operation on $P(x, y, z)$. Fig. 2.11(b) shows a projection chart of P, P' on the xy plane. The relationship between $P(x, y, z)$ and $P'(x', y', z')$ can clearly be formulated as follows.

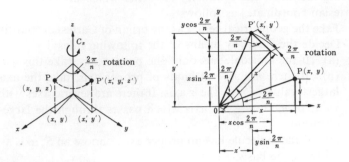

Fig. 2.11. Transformation of points expressed in Cartesian coordinates under C_n operation.

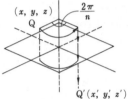

Fig. 2.12. Transformation of points expressed in Cartesian coordinates under S_n operation.

$$\left.\begin{aligned} x' &= x\cos\frac{2\pi}{n} - y\sin\frac{2\pi}{n} \\[2mm] y' &= x\sin\frac{2\pi}{n} + y\cos\frac{2\pi}{n} \\[2mm] z' &= z \end{aligned}\right\} \qquad (2.1)$$

The situation is very similar for improper operations S_n, as shown in Fig. 2.12, where the point $Q(x, y, z)$ is transformed to $Q'(x', y', z')$ by S_n operation. The only difference from a C_n operation is the reflection through the xy plane, so that the z coordinate is changed. Thus Eq. 2.2 gives the relation between $Q(x, y, z)$ and $Q'(x', y', z')$.

$$\left.\begin{aligned} x' &= x\cos\frac{2\pi}{n} - y\sin\frac{2\pi}{n} \\[2mm] y' &= x\sin\frac{2\pi}{n} + y\cos\frac{2\pi}{n} \\[2mm] z' &= -z \end{aligned}\right\} \qquad (2.2)$$

These equations can be restated in matrix form as follows.

For C_n operation
$$\begin{pmatrix} x' \\ y' \\ z' \end{pmatrix} = \begin{pmatrix} \cos\dfrac{2\pi}{n} & -\sin\dfrac{2\pi}{n} & 0 \\ \sin\dfrac{2\pi}{n} & \cos\dfrac{2\pi}{n} & 0 \\ 0 & 0 & 1 \end{pmatrix} \begin{pmatrix} x \\ y \\ z \end{pmatrix} \tag{2.3}$$

For S_n operation
$$\begin{pmatrix} x' \\ y' \\ z' \end{pmatrix} = \begin{pmatrix} \cos\dfrac{2\pi}{n} & -\sin\dfrac{2\pi}{n} & 0 \\ \sin\dfrac{2\pi}{n} & \cos\dfrac{2\pi}{n} & 0 \\ 0 & 0 & -1 \end{pmatrix} \begin{pmatrix} x \\ y \\ z \end{pmatrix} \tag{2.4}$$

When a molecule with symmetry element C_n is placed in Cartesian coordinates as described above, if an atom A is at the point P with coordinates (x, y, z) then the same kind of atom must also be found at the point P′ with coordinates (x', y', z'), and similar relations exist for all other atoms in the molecule. Thus the configuration of the molecule can be expressed as a set of points satisfying the relation given by Eq. 2.3. Similarly, the configuration of a molecule with symmetry element S_n can be expressed as a set of points satisfying the relation given by Eq. 2.4.

E. Chirality and the point group

If we regard a molecule as a set of equivalent points under symmetry operations, the configuration can be classified in terms of the group of symmetry operations that is sufficient to express the positional relations of the points. Such a group of symmetry operations is called a point group (see Appendix A). The symmetry of a molecule can be expressed necessarily and sufficiently by a particular point group. C_n and S_n are operations which organize a point group, and are defined by means of axes which pass through the geometric center of the molecule. A point group can therefore express the positional relationships of points in a molecule without any movement of the geometric center of the molecule within the coordinates.

Table 2.2 shows the kinds of point groups and their organizing operations. By applying various kinds of symmetry operations to the molecule and noting which produce equivalent structures, the symmetry elements of a molecule can be identified, and the resulting group of symmetry operations will belong to one of the point groups shown in the table.

As mentioned in section B, symmetry operations of chiral molecules

TABLE 2.2
Point groups

Notation	Symmetry elements	Example	Element	C_n axis	S_n axis	
C_n	$\{C_n\}$	C_1 C_2 C_3	E E, C_2 E, C_3^1, C_3^2			Chiral point groups
D_n T O	$\{C_n, C_2^{xy}\}$ $\{C_2^z, C_3^{xyz}\}$ $\{C_4^z, C_3^{xyz}\}$	D_2 D_3 T O	E, C_2^x, C_2^y, C_2^z $E, 2 \times C_3, 3 \times C_2$ $E, 4 \times C_3, 4 \times C_3^2, 3 \times C_2$ $E, 8 \times C_3, 3 \times C_2, 6 \times C_2,$ $6 \times C_4$	O	×	
C_s C_i		$C_s(= S_1)$ $C_i(= S_2)$	E, σ_h E, i	×	O	
C_{nh}	$\{C_n, \sigma_h\}$	C_{2h} C_{3h} C_{4h}	E, C_2, i, σ_h $E, C_3, C_3^2, \sigma_h, S_3, S_3^5$ $E, C_4, C_2, C_4^3, i, S_4^3, \sigma_h, S_4$			Achiral point groups
C_{nv}	$\{C_n, \sigma_v\}$	C_{2v} C_{3v} C_{4v}	$E, C_2, \sigma^{xy}, \sigma^{yz}$ $E, 2 \times C_3, 3 \times \sigma_v$ $E, 2 \times C_4, C_2, 2 \times \sigma_v,$ $2 \times \sigma_d$			
S_n	$\{S_n\}$	S_4 S_6	E, S_4, C_2, S_4^3 $E, C_3, C_3^2, i, S_6^5, S_6$			
D_{nh}	$\{C_n, C_2', \sigma_n\}$	D_{2h} D_{3h}	$E, C_2^x, C_2^y, C_2^z, i,$ $\sigma^{xy}, \sigma^{xz}, \sigma^{yz}$ $E, 2 \times C_3, 3 \times C_2, \sigma_h,$ $2 \times S_3, 3 \times \sigma_v$	O	O	
D_{nd}	$\{C_n, C_2, \sigma_d\}$	D_{2d} D_{3d}	$E, 2 \times S_4, C_2, 2 \times C_2',$ $2 \times \sigma_d$ $E, 2 \times C_3, 3 \times C_2, i,$ $2 \times S_6, 3 \times \sigma_d$			
T_d	$\{S_4^z, C_3^{xyz}\}$	T_d	$E, 8 \times C_3, 3 \times C_2, 6 \times S_4,$ $6 \times \sigma_d$			
O_h	$\{C_4^z, C_3^{xyz}\}$	O_h	$E, 8 \times C_3, 6 \times C_2, 6 \times C_4,$ $3 \times C_2, i,$ $6 \times S_4, 8 \times S_6, 6 \times \sigma_h,$ $3 \times \sigma_d$			
$C_{\infty v}^*$ $D_{\infty h}$		$C_{\infty v}$ $D_{\infty h}$	$E, 2 \times C_\infty, \infty \times \sigma_v$ $E, 2 \times C_\infty, \infty \times \sigma_v, i, 2S_\infty,$ ∞C_2			

$C_n = n$-fold proper axis
$C_2' = $ 2-fold axis normal to the principal C_n axis
$\sigma_v = $ reflection plane including the principal axis
$\sigma_n = $ reflection plane normal to the principal axis
$\sigma_d = $ reflection plane including the principal axis and bisecting two C_2' axes
$C_n^z = C_n$ axis directed along the z coordinate; $C_2^{xy} = C_2$ axis in the xy plane
$C_3^{xyz} = C_3$ axis determined by the coordinates x, y, z
$* = $ continuous group

belong to the C_n group and do not include S_n operations. From Table 2.2 it can be seen that only four kinds of point group satisfy this condition, i.e., C_n, D_n, T and O. However, no real molecules exist belonging to the T and O point groups. Therefore a necessary condition for a compound to be optically active is that the molecular symmetry must be C_n or D_n.

2.2.2 Characteristics of chiral molecules and counting stereoisomers

In section 2.2.1 we have seen that the elements of symmetry of a molecule can be identified and a point group assigned to the molecule. It is then possible to decide whether the molecular structure is chiral or not. In this section we shall consider the next problem, i.e., deciding what kinds of stereoisomers of a given molecule exist, how many there are, and which of them are chiral molecules.

A. Constructing molecular skeletons

We shall consider a molecule as being made up of a skeleton together with various ligands attached to the positions on the skeleton. The only restriction on the choice of skeleton (assuming that a molecule with the chosen skeleton can exist, at least in principle) is that the structure of the skeleton itself must not be chiral, i.e., if ligands of a single type are attached to all the positions on the skeleton, the resulting molecule will be achiral.

If molecules are constructed in this way, all molecules with the same skeleton are said to belong to the same molecular class. If molecules in the same molecular class and with the same ligands have different distributions of the ligands on the skeleton, they correspond to stereoisomers as defined in section 2.1. Enantiomers are then stereoisomers in which the distribution of ligands produces a chiral molecule.

In order to classify isomers, a reference molecule is chosen for each molecular class and is taken as having n positions on the skeleton to which ligands may be attached. These positions are numbered $d_1, d_2 \ldots d_n$ (these are known as skeletal numbers; see section 2.2.3 for the rules determining the locations of the numbers), and ligands $l_1, l_2 \ldots l_n$ are added until all the positions on the skeleton are filled, i.e., according to skeletal number and so that the skeletal numbers and ligand numbers coincide (Fig. 2.13). The reference molecule thus constructed can be expressed by Eq. 2.5.

$$L = \begin{pmatrix} l_1 \\ l_2 \\ \vdots \\ l_n \end{pmatrix} \tag{2.5}$$

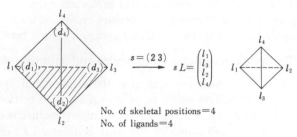

No. of skeletal positions $=4$
No. of ligands $=4$

Fig. 2.13. Numbering of the ligands and molecular skeleton, showing the operation of a permutation of the ligand numbers on the reference molecule.

If a ligand i is the same as another ligand k, then $l_i = l_k$. Thus each isomer of the same molecular class can be expressed unambiguously in terms of the distribution of ligands l_i on the fixed skeletal numbers d_j of the reference molecule. Since the distribution of the ligand numbers $(1, 2, \ldots i, \ldots n)$ on the fixed skeletal numbers $(1, 2, \ldots j, \ldots n)$ corresponds to a one-to-one mapping of a set of n numbers $(1, 2, \ldots n)$ onto itself, the distribution of ligands in a certain isomer can be derived by the operation of an element s belonging to the symmetry permutation group $\mathscr{S}_n$ on the set of ligand numbers of the reference molecule (see Fig. 2.13). The same isomer is also obtainable by rearrangement of the skeletal numbers of the reference molecule if the ligand numbers are fixed, and in this case the element t which operates to derive the isomer from the reference molecule also belongs to the symmetry permutation group $\mathscr{S}_n$, and the relation $s^{-1} = t$ exists between s and t (see Appendix). In this section we will use the $\mathscr{S}$ permutation and describe isomers in terms of sL ($s \in \mathscr{S}_n$).

B. Stereoisomers of molecules in which all the ligands are different

When the molecular skeleton has been decided, the point group ($\mathscr{G}$) to which the skelton belongs can be determined. Since the isomer obtained by the operation $g \in \mathscr{G}$ on the skeleton of the reference molecule corresponds to that obtained by the operation $s \in \mathscr{S}_n$ on the ligands of the reference molecule, the group $\mathscr{G}$ is homomorphic with the subgroup $\mathscr{S}$ in the group $\mathscr{S}_n$. The rotational subgroup $\mathscr{D}$ in the group $\mathscr{G}$ is also mapped onto the group $\mathscr{N}$, which is a subgroup $\mathscr{S}$ (see Appendix A).

$$\mathscr{G} \to \mathscr{S}$$
$$\mathscr{D} \to \mathscr{N}$$

$$(2.6)$$

To distinguish isomers belonging to the same molecular class, only the s permutation is required since s and t belong to the same permutation group, $\mathscr{S}_n$. Therefore we can disregard skeletal numbers in considering the structures of isomers, and molecules sL become equivalent if s belongs to $\mathscr{N}$ (equivalent to rotation of the molecule in space, the operation $s \in \mathscr{N}$ on the ligands gives an indistinguishable molecule if skeletal numbers are disregarded). The relations among these groups and subgroups are as follows.

$$\mathscr{D} \subset \mathscr{G}, \mathscr{N} \subset \mathscr{S} \subset \mathscr{S}_n \qquad (2.7)$$

Table 2.3 shows the significance of Eqs. 2.6 and 2.7 for molecules generated from four ligands (l_1, l_2, l_3, l_4) in four positions (i.e., the symmetry group of the skeleton is D_{2d}, C_{4v}, D_{4h} or T_d). The condition $\mathscr{D} = \mathscr{G}$ corresponds to the skeleton itself being chiral (elements of the point group consist only of rotational operations). In order to choose an achiral skeleton, the condition $\mathscr{D} \subset \mathscr{G}$ must hold in all cases. Subject to $\mathscr{D} \subset \mathscr{G}$, $\mathscr{N} = \mathscr{S}$ means that $\mathscr{G}$ is homomorphic with $\mathscr{S}$, as in the case where the skeleton belongs to D_{4h} in Table 2.3. In this case a symmetry operation by an element s of $\mathscr{S}$ corresponds to both skeletal rotation and an operation other than rotation (including reflection). Molecules L and sL ($s \in \mathscr{N}$) are then indistinguishable, if the skeletal number is disregarded, and so all molecules obtained from the skeleton which satisfy $\mathscr{N} = \mathscr{S}$ are achiral.

If $\mathscr{N} \subset \mathscr{S}$, its coset $\mathscr{N}$ and $\mathscr{N}'$ exist (see Appendix A). If a representative element of the coset $\mathscr{N}'$ is denoted by s', all the elements of $\mathscr{N}'$ are given by ss' ($s \in \mathscr{N}$), so the molecule given by the operation of an element of $\mathscr{N}'$ is $\{ss'L\}$ ($s \in \mathscr{N}$). Since s' includes a reflectional operation, $\{ss'L\}$ corresponds to the mirror image of $\{sL\}$.

If the skeletal number is disregarded, $\{ss'L\}$ ($s \in \mathscr{N}$) are all the same, so a real image/mirror image relationship exists between a pair of molecules, $\{ss'L\}$ ($s \in \mathscr{N}$) and $\{sL\}$ ($s \in \mathscr{N}$). In such a case, $\mathscr{N}$ is a regular subgroup of index 2 with respect to $\mathscr{S}$. Because $\mathscr{D} \subset \mathscr{G}$ and $\mathscr{N} \subset \mathscr{S}$ are satisfied when the skeletal point group $\mathscr{G}$ is D_{2d}, C_{4v} or T_d, chiral isomers can exist in molecular classes with these skeletons. In a molecule whose ligands are all different, the following relations hold.

$$\mathscr{N} = \mathscr{S} \quad \text{(achiral molecule)}$$
$$\mathscr{N} \subsetneqq \mathscr{S} \quad \text{(chiral molecule)} \qquad (2.8)$$

We will now consider a concrete example of a molecule whose skeleton belongs to the D_{2d} point group and in which all the ligands are different. All the molecules obtained by applying s (s_1, s_2, s_3, s_4), an element of $\mathscr{N}$, to the reference skeleton are indistinguishable if the

TABLE 2.3

Significance of Eqs. 2.6 and 2.7 for molecules generated from four ligands in four skeletal positions

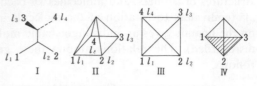

$\mathscr{S}_4$ Element	Distribution of ligand numbers†	Permutation	Class	Skeleton I $\mathscr{S}_4$ D_{2d}	Skeleton II $\mathscr{S}_4$ C_{4v}	Skeleton III $\mathscr{S}_4$ D_{4h}	Skeleton IV $\mathscr{S}_4$ T_d
$s_1 = (1\ 2\ 3\ 4) = e$			1^4	$(s_1)\ \boxed{E}$	$(s_1)\ \boxed{E}$	$(s_1)\ \boxed{E}\ \sigma_h$	$(s_1)\ \boxed{E}$
$s_2 = (2\ 1\ 4\ 3) = (1\ 2)(3\ 4)$				$(s_2)\ C_2$	$(s_2)\ \overline{\sigma_d}$	$(s_2)\ C_2\ \sigma_d$	$(s_2)\ C_2$
$s_3 = (3\ 4\ 1\ 2) = (1\ 3)(2\ 4)$			2^2	$(s_3)\ C_2'$	$(s_3)\ \boxed{C_2}$	$(s_3)\ C_2\ i$	$(s_3)\ C_2$
$s_4 = (4\ 3\ 2\ 1) = (1\ 4)(2\ 3)$				$(s_4)\ C_2'$	$(s_4)\ \overline{\sigma_d}$	$(s_4)\ C_2\ \sigma_d$	$(s_4)\ C_2$	⎫ $3C_3$
$s_5 = (1\ 3\ 4\ 2) = (2\ 3\ 4)$				s_5	s_5	s_5	$(s_5)\ C_3$
$s_6 = (4\ 2\ 1\ 3) = (1\ 4\ 3)$				s_6	s_6	s_6	$(s_6)\ C_3$
$s_7 = (2\ 4\ 3\ 1) = (1\ 2\ 4)$				s_7	s_7	s_7	$(s_7)\ C_3$
$s_8 = (3\ 1\ 2\ 4) = (1\ 3\ 2)$			$1\cdot3$	s_8	s_8	s_8	$(s_8)\ C_3$
$s_9 = (1\ 4\ 2\ 3) = (2\ 4\ 3)$				s_9	s_9	s_9	$(s_9)\ C_3$	⎫ $8C_3$
$s_{10} = (3\ 2\ 4\ 1) = (1\ 3\ 4)$				s_{10}	s_{10}	s_{10}	$(s_{10})\ C_3$
$s_{11} = (4\ 1\ 3\ 2) = (1\ 4\ 2)$				s_{11}	s_{11}	s_{11}	$(s_{11})\ C_3$
$s_{12} = (2\ 3\ 1\ 4) = (1\ 2\ 3)$				s_{12}	s_{12}	s_{12}	$(s_{12})\ C_3$
$s_{13} = (4\ 3\ 1\ 2) = (1\ 4\ 2\ 3)$				$(s_{13})\ \boxed{S_4}$	s_{13}	s_{13}	$(s_{13})\ \boxed{S_4}$
$s_{14} = (3\ 4\ 2\ 1) = (1\ 3\ 2\ 4)$				$(s_{14})\ \boxed{S_4^3}$	s_{14}	s_{14}	$(s_{14})\ S_4$
$s_{15} = (2\ 3\ 4\ 1) = (1\ 2\ 3\ 4)$			4	s_{15}	$(s_{15})\ \boxed{C_4^1}$	$(s_{15})\ C_4\ \boxed{S_4}$	$(s_{15})\ S_4$	⎫ $6S_4$
$s_{16} = (4\ 1\ 2\ 3) = (1\ 4\ 3\ 2)$				s_{16}	$(s_{16})\ \boxed{C_4^3}$	$(s_{16})\ C_4^3\ \boxed{S_4^3}$	$(s_{16})\ S_4$
$s_{17} = (3\ 1\ 4\ 2) = (1\ 3\ 4\ 2)$				s_{17}	s_{17}	s_{17}	$(s_{17})\ S_4$
$s_{18} = (2\ 4\ 1\ 3) = (1\ 2\ 4\ 3)$				s_{18}	s_{18}	s_{18}	$(s_{18})\ S_4$
$s_{19} = (1\ 2\ 4\ 3) = (3\ 4)$				$(s_{19})\ \boxed{\sigma}$	s_{19}	s_{19}	$(s_{19})\ \sigma$
$s_{10} = (2\ 1\ 3\ 4) = (1\ 2)$				$(s_{20})\ \boxed{\sigma}$	s_{20}	s_{20}	$(s_{20})\ \sigma$
$s_{21} = (1\ 4\ 3\ 2) = (2\ 4)$			$1^2\cdot2$	s_{21}	$(s_{21})\ \boxed{\sigma_v}$	$(s_{21})\ C_2'\ \sigma_v$	$(s_{21})\ \sigma$	⎫ 6σ
$s_{22} = (3\ 2\ 1\ 4) = (1\ 3)$				s_{22}	$(s_{22})\ \boxed{\sigma_v}$	$(s_{22})\ C_2'\ \sigma_v$	$(s_{22})\ \sigma$
$s_{23} = (1\ 3\ 2\ 4) = (2\ 3)$				s_{23}	s_{23}	s_{23}	$(s_{23})\ \sigma$
$s_{24} = (4\ 2\ 3\ 1) = (1\ 4)$				s_{24}	s_{24}	s_{24}	$(s_{24})\ \sigma$

$$\left.\begin{array}{l} \fbox{}\!\!\!{}_{- - -} \\ \square \to \mathscr{D} \end{array}\right\} \mathscr{G}$$

$$\left.\begin{array}{l} \left.\begin{array}{l} (s_i) \to \mathscr{N} \\ (s_j) \to \mathscr{N}' \end{array}\right\} \mathscr{S}' \\ s_k \end{array}\right\} \mathscr{S}$$

† The set of numbers indicates ligand numbers distributed on the skeletal numbers (1, 2, 3, 4)

TABLE 2.4

Assignment of a D_{2d} molecular skeleton to cosets

$\mathcal{N}$	$\mathcal{N}'$	$\mathcal{N}_1$	$\mathcal{N}'_1$	$\mathcal{N}_2$	$\mathcal{N}'_2$
s	s'	(ss_5)	(ss_{17})	(ss_9)	(ss_{15})
s_1	s_{13}	s_5	s_{17}	s_9	s_{15}
s_2	s_{14}	s_8	s_{23}	s_{11}	s_{22}
s_3	s_{19}	s_6	s_{24}	s_{12}	s_{16}
s_4	s_{20}	s_7	s_{18}	s_{10}	s_{21}

skeletal numbering is disregarded (Table 2.3). Next, we construct a right coset of $\mathcal{N}$ in $\mathcal{S}_4$; this consists of $\mathcal{N}$, $\mathcal{N}'$, $\mathcal{N}_1$, $\mathcal{N}'_1$, $\mathcal{N}_2$, $\mathcal{N}'_2$, each element of $\mathcal{S}_4$ being classified into six cosets consisting of four elements as shown in Table 2.4.

Table 2.5 shows the results of cross multiplication of elements accord-

TABLE 2.5

Product table of elements in $\mathcal{S}_4$

j \ i	1	2	3	4	5	6	7	8	9	10	11	12	13	14	15	16	17	18	19	20	21	22	23	24
1	1	2	3	4	5	6	7	8	9	10	11	12	13	14	15	16	17	18	19	20	21	22	23	24
2	2	1	4	3	7	8	5	6	12	11	10	9	14	13	21	22	24	23	20	19	15	16	18	17
3	3	4	1	2	8	7	6	5	10	9	12	11	20	19	16	15	23	24	14	13	22	21	17	18
4	4	3	2	1	6	5	8	7	11	12	9	10	19	20	22	21	18	17	13	14	16	15	24	23
5	5	8	6	7	9	12	10	11	1	4	2	3	18	24	14	20	16	22	23	17	19	13	21	15
6	6	7	5	8	11	10	12	9	4	1	3	2	17	23	20	14	21	15	24	18	13	19	16	22
7	7	6	8	5	12	9	11	10	2	3	1	4	23	17	13	19	22	16	18	24	20	14	15	21
8	8	5	7	6	10	11	9	12	3	2	4	1	24	18	19	13	15	21	17	23	14	20	22	16
9	9	11	12	10	1	3	4	2	5	7	8	6	22	15	24	17	20	13	21	16	23	18	19	14
10	10	12	11	9	3	1	2	4	8	6	5	7	21	16	18	23	13	20	22	15	17	24	14	19
11	11	9	10	12	4	2	1	3	6	8	7	5	15	22	23	18	14	19	16	21	24	17	13	20
12	12	10	9	11	2	4	3	1	7	5	6	8	16	21	17	24	19	14	15	22	18	23	20	13
13	13	14	19	20	16	15	22	21	24	23	18	17	2	1	8	7	9	10	4	3	6	5	11	12
14	14	13	20	19	22	21	16	15	17	18	23	24	1	2	6	5	12	11	3	4	8	7	10	9
15	15	22	16	21	18	23	17	24	20	13	19	14	9	11	3	1	6	8	12	10	2	4	7	5
16	16	21	15	22	24	17	23	18	13	20	14	19	10	12	1	3	7	5	11	9	4	2	6	8
17	17	23	24	18	14	20	19	13	22	16	15	21	7	6	9	12	4	1	8	5	10	11	3	2
18	18	24	23	17	20	14	13	19	15	21	22	16	8	5	11	10	1	4	7	6	12	9	2	3
19	19	20	13	14	21	22	15	16	23	24	17	18	3	4	7	8	11	12	1	2	5	6	9	10
20	20	19	14	13	15	16	21	22	18	17	24	23	4	3	5	6	10	9	2	1	7	8	12	11
21	21	16	22	15	23	18	24	17	19	14	20	13	12	10	4	2	8	6	9	11	1	3	5	7
22	22	15	21	16	17	24	18	23	14	19	13	20	11	9	2	4	5	7	10	12	3	1	8	6
23	23	17	18	24	19	13	14	20	21	15	16	22	6	7	10	11	2	3	5	8	9	12	1	4
24	24	18	17	23	13	19	20	14	16	22	21	15	5	8	12	9	3	2	6	7	11	10	4	1

The numbers in this table show the number of elements obtained from $s_i \times s_j = s_k$.

ing to $s_i s_j = s_k$, the number k being given at the intersection of row j with column i. In Table 2.4, the distributions of ligands obtained by the application of elements belonging to the same coset, shown in the same column, give molecules which are indistinguishable if the skeletal numbering is disregarded. Fig. 2.14 shows isomers obtained by the operation of a representative element of each coset on the reference molecule. These

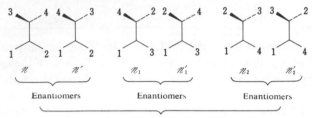

Fig. 2.14.   Isomers obtained by the operation of a representative element of each coset on the reference molecule.

molecules are not superposable in three-dimensional space and are stereoisomers. It can be seen that all the molecules $s'L$ given by the permutation s' belonging to the coset of $\mathcal{N}$ in $\mathcal{S}_n$ are stereoisomers of the molecules sL given by the permutation s belonging to $\mathcal{N}$. In general, when s_i and s_j do not belong to the same right coset of $\mathcal{N}$ in $\mathcal{S}_n$ the molecules $\{ss_iL\}$ and $\{ss_jL\}$ ($s \in \mathcal{N}$) are stereoisomers. Hence, the number of isomers in such a molecule having all different ligands is given by $Z_c = |\mathcal{S}_n|/|\mathcal{N}|$, where $|\mathcal{S}_n|$ and $|\mathcal{N}|$ represent the orders of $\mathcal{S}_n$ and $\mathcal{N}$, respectively.

As the allene skeleton used in the example satisfies $\mathcal{N} \subset \mathcal{S}$, the pairs of molecules $\mathcal{N}$, $\mathcal{N}'$ and $\mathcal{N}_1$, $\mathcal{N}'_1$ and $\mathcal{N}_2$, $\mathcal{N}'_2$ are enantiomers. Table 2.5 shows that when the elements $s' \in \mathcal{N}'$ (e.g., s_{13}) and $s_j \in \mathcal{N}'_1$ (e.g., s_{17}) are multiplied, the resulting element $s_i = s's_j$ (i.e., $s_{13}s_{17} = s_7$) belongs to $\mathcal{N}_1$. If the same calculation is applied to an element of $\mathcal{N}'_2$, the resulting element belongs to $\mathcal{N}_2$.

In general, when $\mathcal{N}$ is a normal subgroup with index 2 in $\mathcal{S}$, an element s' belongs to the right coset of $\mathcal{N}$ in $\mathcal{S}$, and $s_i = s's_j$ is satisfied, then the molecules $\{s_iL\}$ and $\{s_jL\}$ are enantiomers. On the other hand, when s_i and s_j belong to different right cosets of $\mathcal{N}$ in $\mathcal{S}_n$ but no s_i, s_j exist which satisfy the relation $s_i = s's_j$, then the molecules $\{s_iL\}$ and $\{s_jL\}$ are not enantiomers but are stereoisomers (in this case, diastereomers). Thus an allene skeleton having four different ligands yields six stereoisomers, all of which are chiral molecules (three pairs of diastereomers).

Generally, the number of enantiomeric pairs is given by $Z_p = |\mathcal{S}_n|/|\mathcal{S}|$, where $|\mathcal{S}|$ and $|\mathcal{S}_n|$ represent the orders of $\mathcal{S}$ and $\mathcal{S}_n$, respectively.

In the case of allene derivatives, $|\mathscr{S}_n| = 24$, $|\mathscr{N}| = 4$ and $|\mathscr{S}| = 8$, so that $Z_c = 24/4 = 6$ and $Z_p = 24/8 = 3$. When the skeletal point group is T_d, $\mathscr{N}$ contains twelve elements $\{s_1, s_2 \ldots s_{12}\}$ if the skeletal numbering is disregarded, and all the molecules given by these operations are the same. There is only one right coset $\mathscr{N}'$ of $\mathscr{N}$ in $\mathscr{S}_4$, containing twelve elements $\{s_{12}, s_{14}, \ldots s_{24}\}$; all the molecules $\{s'L\}$ given by these operations are the same. For a molecule with a T_d skeleton carrying four different ligands, two stereoisomers $\{sL\}$ and $\{ss'L\}$ exist and are enantiomers. The only coset of $\mathscr{N}$ is $\mathscr{N}'$, so the molecule has no diastereomers, i.e., $Z_c = 24/12 = 2$ and $Z_p = 24/24 = 1$.

In the case of molecules with D_{4h} skeletal symmetry, subject to $\mathscr{N} = \mathscr{S}$, there are no chiral structures, but achiral isomers can exist. In this case, $Z_a = |\mathscr{S}_n|/|\mathscr{S}| = 24/8 = 3$, so that the three achiral isomers shown in Fig. 2.15 exist.

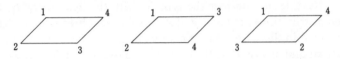

Fig. 2.15.   Three achiral isomers of a molecule with D_{4h} skeletal symmetry; no chiral isomers exist.

C.   Stereoisomers of molecules containing at least two identical ligands

The number and type of isomers of molecules in which two or more of the ligands are the same can be obtained by procedures similar to those described in the previous section. As the number of identical ligands increases, the number of isomers decreases, since the molecules obtained by interchanging similar ligands are indistinguishable. The number of isomers is given by Eq. 2.9 (known as the Polya equation), which can be derived as follows.

Consider a molecule with n skeletal positions and n ligands of m different kinds, $l_1, l_2 \ldots l_i \ldots l_m$. Here the number of a particular ligand l_i in the molecule is given by v_i ($\sum_i^m v_i = n$) and $h_{\mathscr{N}}(\lambda_1, \lambda_2, \ldots \lambda_n)$ represents the number of elements which belong to $\mathscr{N} \leftarrow \mathscr{D}$ grouped into classes expressed by $1^{\lambda_1}, 2^{\lambda_2}, \ldots n^{\lambda_n}$ (see Appendix B).

The expression $P_{v_1, v_2, \cdots v_m}(\lambda_1, \lambda_2, \ldots \lambda_n)$ gives the number of allotments which can be obtained using a set of ligands $(l_1, l_2, \ldots l_m)$ in such a way that the distribution of all ligands satisfies the type $1^{\lambda_1}, 2^{\lambda_2}, \ldots n^{\lambda_n}$ for each class.[†] For example, we will consider the members of $P_{222}(2, 2)$

---

† When a ligand is allotted to $\times$, other ligands of the same kind must be included in the same brackets until the number of $\times$ in brackets and the number of identical ligands become the same.

when the type is 1^2, 2^2 and the ligands comprise $(l_1, l_1, l_2, l_2, l_3, l_3)$. The distribution pattern is $[\times][\times][\times\ \times][\times\ \times]$. Keeping this distribution pattern, the ways in which the ligands can be allotted are as follows, and the number of distributions becomes six, i.e., $P_{222}(2, 2) = 6$ (see Table 8).

$[\times]$	$[\times]$	$[\times$	$\times]$	$[\times$	$\times]$
l_1	l_1	l_2	l_2	l_3	l_3
l_1	l_1	l_3	l_3	l_2	l_2
l_2	l_2	l_1	l_1	l_3	l_3
l_2	l_2	l_3	l_3	l_1	l_1
l_3	l_3	l_1	l_1	l_2	l_2
l_3	l_3	l_2	l_2	l_1	l_1

Next, let us consider the type 3^2 with the ligands $(l_1, l_1, l_2, l_2, l_3, l_3)$. The distributions are $[\times\ \times\ \times][\times\ \times\ \times]$. Since three similar ligands cannot be found in this case, ligands cannot be allotted to satisfy this distribution pattern and hence $P_{222}(0, 0, 2) = 0$ holds. When $h_{\mathcal{N}}(\lambda_1, \lambda_2, \ldots \lambda_n)$ and $P_{v_1, v_2, \cdots v_m}(\lambda_1, \lambda_2, \ldots \lambda_n)$ are decided by means of Polya's equation (see Appendix C), the total number of isomers, $\mathscr{C}_{\mathcal{N}}$, which cannot be superposed is given by Eq. 2.9.

$$\mathscr{C}_{\mathcal{N}} = \frac{1}{|\mathcal{N}|} \sum_{\lambda_1 + 2\lambda_2 + \cdots n\lambda_n = n} h_{\mathcal{N}}(\lambda_1, \lambda_2, \ldots, \lambda_n) P_{v_1, v_2, \cdots, v_m}(\lambda_1, \lambda_2, \ldots, \lambda_n)$$

(2.9)

Replacing $\mathcal{N}$ in Eq. 2.9 by $\mathscr{S}$, the number $\mathscr{C}_{\mathscr{S}}$ can be obtained as follows

$$\mathscr{C}_{\mathscr{S}} = \frac{1}{|\mathscr{S}|} \sum_{\lambda_1 + 2\lambda_2 + \cdots n\lambda_n = n} h_{\mathscr{S}}(\lambda_1, \lambda_2, \ldots, \lambda_n) P_{v_1, v_2, v_3, \cdots, v_m}(\lambda_1, \lambda_2, \ldots, \lambda_n)$$

(2.10)

This number equals the number of isomers, taking a pair of enantiomers as one isomer $(Z_a + Z_c/2)$. Thus, the number of chiral molecules Z_c and the number of achiral isomers Z_a are given by Eqs. 2.11 and 2.12.

$$Z_c = 2(\mathscr{C}_{\mathcal{N}} - \mathscr{C}_{\mathscr{S}})$$

(2.11)

$$Z_a = 2\mathscr{C}_{\mathscr{S}} - \mathscr{C}_{\mathcal{N}}$$

(2.12)

Polya's equation thus gives the total number of isomers, but not the

TABLE 2.6

Procedure for deciding h and P values for a D_{2d} allene skeleton having three kinds of ligands

Skeletal symmetry elements	Permutation	Class $(1^{\lambda_1}, 2^{\lambda_2}, \ldots, n^{\lambda_n})$	No. of classes $h(\lambda_1, \lambda_2, \ldots, \lambda_n)$	Distribution pattern	No. of distributions $P_{2,1,1}(\lambda_1, \lambda_2, \ldots, \lambda_n)$	$h \times P$
E	s_1 (1)(2)(3)(4)	1^1	1	[×][×][×][×] [1][1][2][3] [1][2][1][3]	$\dfrac{4!}{2!\,1!} = 12$	12
C_2	s_2 (1 2) (3 4)	2^2		$[\times \times][\times \times]$		
C_2'	s_3 (1 3) (2 4)	2^2	3	$\vdots$	0	0
C_2'	s_4 (1 4) (2 3)	2^2		none		
	$|\mathscr{N}| = 4$					
	$|\mathscr{S}| = 8$					
S_4	s_{13}(1 2 3 4)	4		$[\times \times \times \times]$		
S_4^3	s_{14}(1 3 2 4)	4	2	none	0	0
σ	s_{15}(1)(2)(3 4)	$1^2 \cdot 2$		[×][×][× ×] [3][2][1 1]		
σ	s_{20}(3)(4)(1 2)	$1^2 \cdot 2$	2	[2][3][1 1]	2	4

$$\mathscr{C}_\mathscr{N} = \frac{1}{|\mathscr{N}|}\left(h(4,0,0,\ldots) \cdot P_{2,1,1}(4,0,0,\ldots) + h(0,2,0,\ldots) \cdot P_{2,1,1}(0,2,0\ldots)\right) = \frac{1}{4}(1 \times 12 + 3 \times 0) = 3$$

$$\mathscr{C}_\mathscr{S} = \frac{1}{|\mathscr{S}|}\left(h(4,0,0,\ldots) \cdot P_{2,1,1}(4,0,0,\ldots) + h(0,2,0,\ldots) \cdot P_{2,1,1}(0,2,0,\ldots) + h(0,0,0,1,0,\ldots) \cdot P_{2,1,1}(0,0,0,1,0,\ldots)\right.$$
$$\left. + h(2,1,0,0,\ldots) \cdot P_{2,1,1}(2,1,0,0,\ldots)\right) = \frac{1}{8}(1 \times 12 + 3 \times 0 + 2 \times 0 + 2 \times 2) = 2$$

Then $\quad Z_c = 2(\mathscr{C}_\mathscr{N} - \mathscr{C}_\mathscr{S}) = 2$
$\qquad\quad Z_a = 2\mathscr{C}_\mathscr{S} - \mathscr{C}_\mathscr{N} = 1$

TABLE 2.7

Procedure for deciding h and P values for a D_{2d} allene skeleton having two kinds of ligands

Skeletal symmetry elements	Permutation	Class $(1^{\lambda_1}, 2^{\lambda_2}, \ldots, n^{\lambda_n})$	No. of classes $h(\lambda_1, \lambda_2, \ldots, \lambda_m)$	Distribution pattern	No. of distributions $P_{2,2}(\lambda_1, \lambda_2, \ldots, \lambda_m)$	$h \times P$
E	s_1 (1)(2)(3)(4)	1^4	1	[×][×][×][×] [1][1][2][2] [2][1][2][1] $\vdots$	$\dfrac{4!}{2!2!} = 6$	6
C_2 $\quad |\mathcal{N}| = 4$	s_2 (1 2) (3 4)	2^2		[× ×][× ×] [1 1][2 2] [2 2][1 1]	2	6
C_2'	s_3 (1 3) (2 4)	2^2	3			
C_2''	s_4 (1 4) (2 3)	2^2				
$\quad\quad |\mathcal{S}| = 8$						
S_4	s_{13}(1 2 3 4)	4		[× × × ×] none	0	0
S_4^3	s_{14}(1 3 2 4)	4	2			
σ	s_{19}(1)(2)(3 4)	$1^2 \cdot 2$	2	[×][×][× ×] [1][1][2 2] [2][2][1 1]	2	4
σ	s_{20}(3)(4)(1 2)	$1^2 \cdot 2$	2			

$C_{\mathcal{N}} = \dfrac{1}{4}(6 + 6) = 3$

$C_{\mathcal{S}} = \dfrac{1}{8}(6 + 6 + 4) = 2$

$Z_c = 2 \times (3 - 2) = 2$

$Z_a = 2 \times 2 - 3 = 1$

constitutional formula of each isomer. The constitutional formulae of all possible isomers can be obtained simply by enumerating the ways in which the ligands can be distributed among the skeletal positions.

We will consider an example with four ligands having D_{2d} symmetry, then a further example with six ligands having O_h symmetry. As the first example we will take an allene derivative (D_{2d} symmetry) having three kinds of ligands (l_1, l_1, l_2, l_3); then $v_1 = 2$, $v_2 = 1$, $v_3 = 1$. Table 2.6 shows the procedure for deciding h and P, and Fig. 2.16 shows the one

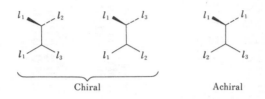

Chiral   Achiral

Fig. 2.16. Two chiral and one achiral isomers obtained from an allene having D_{2h} symmetry with three different kinds of ligands.

achiral isomer and two chiral isomers which are obtained. Another allene having D_{2d} skeletal symmetry but with only two kinds of ligands also has one achiral and two chiral isomers, as determined by the same procedure (Table 2.7 and Fig. 2.17).

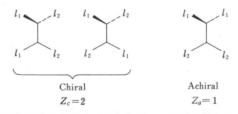

Chiral   Achiral
$Z_c = 2$ $Z_a = 1$

Fig. 2.17. Two chiral and one achiral isomers obtained from an allene having D_{2h} symmetry with two different kinds of ligands.

Next we will take a molecule having three kinds of ligands (l_1, l_1, l_2, l_2, l_3, l_3) arranged octahedrally on six vertices and having O_h skeletal symmetry; here $v_1 = 2$, $v_2 = 2$, $v_3 = 2$. As before, Table 2.8 shows how the values of h and P are decided and Fig. 2.18 shows the skeletal numbering. The table shows that there are two chiral isomers ($Z_c = 2$) and four achiral isomers. The distributions of ligands determined from an assignment table are shown in Table 2.9, where l_1, l_2 and l_3 are denoted by X, Y and Z, respectively.

To decide whether isomers with the same kinds of ligands are chiral, we can make use of the properties of a set of permutations giving equivalent

TABLE 2.8

Procedure for deciding h and P values for an O_h skeleton having three kinds of ligands arranged octahedrally

Skeletal symmetry elements	Permutation	Class $(1^{\lambda_1}, 2^{\lambda_2}, \ldots, n^{\lambda_n})$	No. of classes $h(\lambda_1, \lambda_2, \ldots, \lambda_n)$	Distribution pattern	P	$h \times P$
E	(1)(2)(3)(4)(5)(6)	1^6	1	$\dfrac{[\times][\times][\times][\times][\times][\times]}{(1)\ (1)\ (2)\ (2)\ (3)\ (3)}$ $(1)\ (2)\ (3)\ (1)\ (2)\ (3)$ $\ldots$	$\dfrac{6!}{2!2!2!}$ $=90$	90
$6C_1$	(1 5 4 2)(3)(6) (1 2 4 5)(3)(6) (1 3 4 6)(2)(5) (1 6 4 3)(2)(5) (2 3 5 6)(1)(4) (2 6 5 3)(1)(4)	$1^2 \cdot 4$	6	$[\times\times\times][\times][\times]$ none	0	0
$3C_2$	(1 4)(5 2)(3)(6) (1 4)(3 6)(2)(5) (2 5)(3 6)(1)(4)	$1^2 \cdot 2^2$	3	$[\times\times][\times\times][\times][\times]$ $(1\ 1)\ (2\ 2)\ (3)\ (3)$ $(1\ 1)\ (3\ 3)\ (2)\ (2)$ $\ldots$	$3! = 6$	18
$6C_2$	(1 3)(4 6)(2 5) (1 6)(3 4)(2 5) (1 5)(2 4)(3 6) (1 2)(4 5)(3 6) (2 6)(3 5)(1 4) (2 3)(5 6)(1 4)	2^3	6	$[\times\times][\times\times][\times\times]$ $(1\ 1)\ (2\ 2)\ (3\ 3)$ $(2\ 2)\ (1\ 1)\ (3\ 3)$ $\ldots$	$3! = 6$	36
$8C_3$	(1 2 3) (4 5 6) (1 3 2) (4 6 5) (1 2 6) (3 4 5) (1 6 2) (3 5 4) (1 5 6) (2 3 4) (1 6 5) (2 4 3) (1 5 3) (2 6 4)	3^2	8	$[\times\times\times][\times\times\times]$ none	0	0

(1 3 5) (2 4 6) | $|\mathcal{N}| = 24$

i							
i	(1 4)(2 5)(3 6)		2^3	1	$\dfrac{[× ×][× ×][× ×]}{(1\ 1)\ (2\ 2)\ (3\ 3)}\bigg\}\ \cdots$	$3! = 6$	6
$6S_4$	2.4	(1 3 4 6) (2 5) (1 6 4 3) (2 5) (1 2 4 5) (3 6) (1 5 4 2) (3 6) (2 3 5 6) (1 4) (2 6 5 3) (1 4)		6	$\dfrac{[× × × ×][× ×]}{\text{none}}$	0	0
$3\sigma_h$	(1 4) (2)(3)(5)(6) (2 5) (1)(3)(4)(6) (3 6) (1)(2)(4)(5)		$1^4 \cdot 2$	3	$\dfrac{[× ×][× ×][× ×][× × ×]}{\begin{matrix}(1\ 1)\ (2)\ (2)\ (3)\ (3)\\(1\ 1)\ (2)\ (3)\ (2)\ (3)\end{matrix}}\ \cdots$	$3 × 3! = 18$	54
$6\sigma_v$	(1 2)(4 5)(3)(6) (1 5)(2 4)(3)(6) (1 6)(3 4)(2)(5) (1 3)(4 6)(2)(5) (2 6)(3 5)(1)(4) (2 3)(5 6)(1)(4)		$1^2 \cdot 2^2$	6	$\dfrac{[× ×][× ×][× ×][×][×]}{\begin{matrix}(1\ 1)\ (2\ 2)\ (3)\ (3)\\(1\ 1)\ (3\ 3)\ (2)\ (2)\end{matrix}\bigg\}}\ \cdots$	$3! = 6$	36
$8S_3$	(1 3 2 4 6 5) (1 2 6 4 5 3) (1 6 5 4 3 2) (1 5 3 4 2 6) (1 5 6 4 2 3) (1 3 5 4 6 2) (1 5 6 4 2 3) (1 6 2 4 3 5)		6	8	$\dfrac{[× × × × × ×]}{\text{none}}$	0	0

$|\mathcal{S}| = 48$

$C_{\mathcal{N}} = \dfrac{1}{24}(90 + 18 + 36) = 6 \qquad C_{\mathcal{S}} = \dfrac{1}{48}(90 + 18 + 36 + 6 + 54 + 36) = 5 \qquad Z_c = 2 × 6 - 2 × 5 = 2 \qquad Z_a = 2 × 5 - 6 = 4$

Fig. 2.18. Skeletal numbering for a molecule arranged octahedrally on six vertices and having O_h symmetry.

TABLE 2.9

Distributions of ligands for an O_h skeleton having three kinds of ligands arranged octahedrally

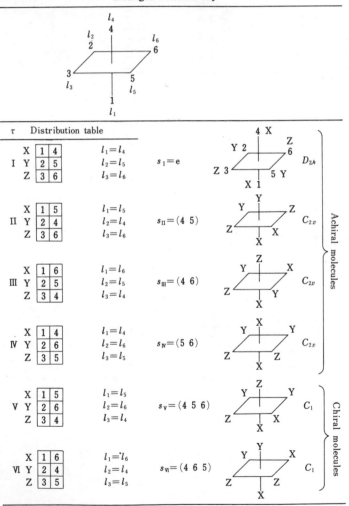

molecules, as described previously for a case where all the ligands were different. When all the ligands are different, the set of ligand permutations giving equivalent molecules is a function of skeletal symmetry and is given by the permutational group $\mathcal{N}$ mapped onto the skeletal rotational subgroup. For molecules with more than one of a given ligand, permutations of the same kinds of ligands in addition to the permutation $s \in \mathcal{N}$ give equivalent molecules. To judge the chirality of an isomer L_τ it is necessary to find a set Q_τ of permutations q_τ giving equivalent molecules when particular ligands are of the same type. Now $Q^{(r)}$ is a subgroup of $\mathcal{S}_n$ and is determined by the number and type of ligands. To determine this subgroup, an assignment table is used as follows. First a number of boxes equal to the skeletal number is prepared and arranged setting ligands of the same type in rows, so that the number of columns is equal to the number of different ligands. Let the number of boxes in the i-th row be v_i and the number in the j-th column μ_j; these are then arranged so that Eq. 2.13 is satisfied.

$$v_1 \geqq v_2 \geqq \ldots v_i \ldots \geqq v_n \geqq 0$$

$$\mu_1 \geqq \mu_2 \geqq \ldots \mu_j \ldots \geqq \mu_m \geqq 0 \qquad (2.13)$$

$$\sum_{i=1}^{n} v_i = \sum_{j=1}^{m} \mu_j = n$$

Next the boxes are assigned numbers under the conditions that the numbers increase from top to bottom and from left to right (this is not an absolute requirement, but in order to determine $Q^{(r)}$ it is necessary to use a standard numbering system). The assignment table obtained by putting a numbered ligand into the correspondingly numbered box is denoted by $\Gamma^{(r)}$. When the box numbers are also positioned to correspond to the skeletal numbers, the resulting molecule based on the assignment table $\Gamma^{(r)}$ is known as the reference molecule. Thus, $\Gamma^{(r)}$ is determined by the numbers and kinds of ligands, and from this a group $Q^{(r)}$ can be obtained which includes all the permutations q of ligand numbers in a given row of $\Gamma^{(r)}$.

Fig. 2.19 shows the assignment tables for molecules with a skeletal number of four for various kinds of ligands ranging from all different to all the same. Table 2.10 shows $Q^{(r)}$ corresponding to each type of assignment table. Since elements of $Q^{(1)}$ through $Q^{(5)}$ satisfy the conditions for a group, each of them is a subgroup of $\mathcal{S}_n$ ($Q^{(r)} \subseteqq \mathcal{S}_n$). In isomers (L) in which the ligand numbers of duplicated ligands correspond to box numbers in the same row in $\Gamma^{(r)}$, $Q^{(r)}$ becomes a group such that an element q operates to give an indistinguishable molecule due to the presence of identical

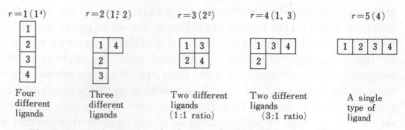

Fig. 2.19. Assignment tables for a skeletal number of 4 with various combinations of ligands.

ligands. For other isomers in which the ligand numbers of duplicated ligands do not coincide with box numbers in the same row in $\Gamma^{(r)}$, we can obtain a permutation s_τ which transforms $\Gamma^{(r)}$ into $\Gamma_\tau^{(r)}$, in which the ligand numbers of duplicated ligands do coincide with box numbers in the same row. Isomers of this type are denoted by L_τ. $Q_\tau^{(r)}$ is derived from $Q^{(r)}$ by means of the following equation.

$$Q_\tau^{(r)} = s_\tau^{-1} Q^{(r)} s_\tau \qquad (2.14)$$

The molecule L_τ gives an indistinguishable molecule under the operation $s \in \mathcal{N}$ and also $q_\tau \in Q_\tau^{(r)}$; this can be written as follows.

$$s L_\tau = q_\tau L_\tau = s q_\tau L_\tau = L_\tau \qquad (2.15)$$

This can be summarized by saying that the operation of an element sq_τ belonging to the set $\mathcal{T}_\tau = \{sq_\tau;\ s \in \mathcal{N},\ q_\tau \in Q_\tau^{(r)}\}$ on the molecule gives an indistinguishable molecule.

Next we will take a set $\mathcal{R}_\tau = \{sq_\tau,\ ss'q_\tau;\ s \in \mathcal{N},\ q_\tau \in Q_\tau\}$, where s' is an element of the coset of $\mathcal{N}$ in $\mathcal{S}$. Since $\mathcal{N}$ is a regular subgroup of index 2 the molecules $sq_\tau L$ and $ss'q_\tau L$ obtained by the operation of sq_τ and $ss'q_\tau$ in $\mathcal{R}_\tau$ are enantiomers. If $\mathcal{T}_\tau = \mathcal{R}_\tau$ holds for a certain isomer L_τ then the molecule and its mirror image are indistinguishable, and therefore if

TABLE 2.10
$Q^{(r)}$ for various ligand patterns

r	Kinds of ligands	$Q^{(r)}$
$r = 1$	W X Y Z	$Q^{(1)} = \{e\}$
$r = 2$	X X Y Z	$Q^{(2)} = \{e, (1\ 4)\}$
$r = 3$	X X Y Y	$Q^{(3)} = \{e, (1\ 3), (2\ 4), (1\ 3), (2\ 4)\}$
$r = 4$	X X X Y	$Q^{(4)} = \{e, (1\ 3), (3\ 4), (1\ 4), (1\ 3\ 4), (1\ 4\ 3)\}$
$r = 5$	X X X X	$Q^{(5)} = S_4$

the distribution of ligands on the skeleton satisfies this relationship, the molecule is achiral. However, if the distribution of ligands satisfies $\mathscr{T}_\tau \subset \mathscr{R}_\tau$ then the resulting molecule is chiral. These relationships are summarized in Eq. 2.16.

$$\mathscr{T}_\tau = \mathscr{R}_\tau \quad \text{(achiral molecule)}$$
$$\mathscr{T}_\tau \subset \mathscr{R}_\tau \quad \text{(chiral molecule)} \tag{2.16}$$

As an example, the isomers of allene derivatives will be examined in terms of Eq. 2.16 (it was previously shown that there are three isomers). The number of ligands is four and there are three types of ligands, so that $v_1 = 2$, $v_2 = 1$ and $v_3 = 1$. The assignment table is shown in Fig. 2.19 under $r = 2$. The isomer corresponding to the standard assignment table Γ is shown in Fig. 2.20. First we will examine the isomer I in Fig. 2.20

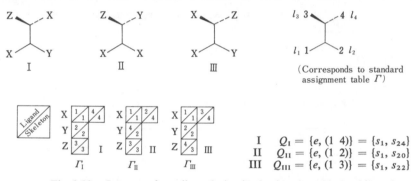

Fig. 2.20.   Isomers of an allene derivative having three kinds of ligands.

$(\tau = \text{I})$. This corresponds to the reference isomer if $X = l_1 = l_4$, $Y = l_2$ and $Z = l_3$. The assigment table Γ_I coincides with Γ and therefore $s_I = e$ and $Q_I = Q$. The isomer II becomes equivalent to the reference molecule if $X = l_1 = l_2$, $Y = l_4$ and $Z = l_3$, so s_{II} is found to be (2 4). In the same way, isomer III corresponds to the reference isomer when $X = l_1 = l_3$, $Y = l_2$ and $Z = l_4$, so that s_{III} is (3 4). The assignment tables Γ_I, Γ_{II} and Γ_{III} are shown in the same figure.

The skeletal symmetry of these isomers is D_{2d}. The elements of $\mathscr{N}$ and $\mathscr{S}$ are obtainable from Table 2.3 and are as follows.

$$\left. \begin{array}{l} \mathscr{N} = \{s_1, s_2, s_3, s_4\} \\ \mathscr{S} = \{s_1, s_2, s_3, s_4, s_{13}, s_{14}, s_{19}, s_{20}\} \end{array} \right\} \tag{2.17}$$

$\mathscr{R}_\tau$ and $\mathscr{T}_\tau$ can then be obtained as follows.

$$(s \times q)$$

q \ s	s_1	s_2	s_3	s_4	s_{13}	s_{14}	s_{19}	s_{20}
s_1	s_1	s_2	s_3	s_4	s_{13}	s_{14}	s_{19}	s_{20}
s_{24}	s_{24}	s_{18}	s_{17}	s_{23}	s_5	s_8	s_6	s_7

$L_{\rm I} \quad Q_{\rm I}$ 　　(2.18)

$$\mathscr{R}_{\rm I} = \{s_1, s_2, s_3, s_4, s_{13}, s_{14}, s_{19}, s_{20}, s_{24}, s_{18}, s_{17}, s_{23},$$
$$s_5, s_8, s_6, s_7\}$$

$$\mathscr{T}_{\rm I} = \{s_1, s_2, s_3, s_4, s_{24}, s_{18}, s_{17}, s_{23}\}$$

Since $\mathscr{T}_{\rm I} \subset \mathscr{R}_{\rm I}$, $L_{\rm I}$ yields chiral molecules. Similarly, for $L_{\rm II}$ and $L_{\rm III}$, we obtain

q \ s	s_1	s_2	s_3	s_4	s_{13}	s_{14}	s_{19}	s_{20}
s_1	s_1	s_2	s_3	s_4	s_{13}	s_{14}	s_{19}	s_{20}
s_{20}	s_{20}	s_{19}	s_{14}	s_{13}	s_4	s_3	s_2	s_1

$L_{\rm II} \quad Q_{\rm II}$ 　　(2.19)

$$\mathscr{T}_{\rm II} = \mathscr{R}_{\rm II} = \{s_1, s_2, s_3, s_4, s_{13}, s_{14}, s_{19}, s_{20}\}$$

q \ s	s_1	s_2	s_3	s_4	s_{13}	s_{14}	s_{19}	s_{20}
s_1	s_1	s_2	s_3	s_4	s_{13}	s_{14}	s_{19}	s_{20}
s_{22}	s_{22}	s_{15}	s_{21}	s_{16}	s_{11}	s_9	s_{10}	s_{12}

$L_{\rm III} \quad Q_{\rm III}$ 　　(2.20)

$$\mathscr{R}_{\rm III} = \{s_1, s_2, s_3, s_4, s_{13}, s_{14}, s_{19}, s_{20}, s_{22}, s_{15}, s_{21},$$
$$s_{16}, s_{11}, s_9, s_{10}, s_{12}\}$$

$$\mathscr{T}_{\rm III} = \{s_1, s_2, s_3, s_4, s_{22}, s_{15}, s_{21}, s_{16}\}$$

$$\mathscr{T}_{\rm III} \subset \mathscr{R}_{\rm III}$$

Thus tables $\Gamma_{\rm I}$ and $\Gamma_{\rm III}$ correspond to the chiral molecules $L_{\rm I}$ and $L_{\rm III}$ (I and III in Fig. 2.20), while $\Gamma_{\rm II}$ corresponds to the achiral molecule $L_{\rm II}$.

It is clear that point groups to which chiral molecules belong have only proper axes as symmetry elements, while the point groups of achiral molecules always contain improper axes. However, the larger n becomes, the more difficult it becomes to identify chiral molecules by this procedure.

In general, Z_a and Z_c can be obtained from Polya's equation (Appendix C), then s_r is determined from corresponding distribution tables. A molecule is then constructed and its point group checked. This procedure is shown in Table 2.9 for a molecule with a regular octahedral skeleton.

2.2.3 Description of isomers by permutational operations

As described in the previous section, different distributions of ligands in a molecule can be specified by considering the molecular skeleton and ligands separately and assigning numbers to each of the skeletal positions and ligands. The kinds of isomers are then described in terms of groups of symmetry operations. Ugi and co-workers[4] have proposed a method of describing molecules by using elements of symmetric permutational groups which permits isomers to be unequivocally specified. This method will be described briefly below.

Suppose that a molecule consists of skeletal positions X_s and ligands L_l. An isomer of this molecule is described by mapping the set of indices l of ligands L_l onto the set of indices s of skeletal positions X_s. This mapping is specified by a $(2 \times n)$ matrix notation using matrices $\binom{l}{s}$ of the indices l and s. When the permutational operations P_s of skeletal numbers s of a given molecule M give a matrix $\binom{l}{s}_M$ of M from the matrix $\binom{l}{s}_E$ of a standard molecule E, as shown in Eq. 2.21, P_s is taken as a descriptive element defining the molecule M.[†]

$$\binom{l}{s}_M = P_s \binom{l}{s}_E \tag{2.21}$$

The standard molecule is chosen such that the ligand numbers l coincide with the skeletal numbers s when they are arranged by means of the following rules, i.e.,

---

[†] Consider a molecule described by the following matrix:
$$\binom{l}{s}_M = \begin{pmatrix} 1 & 2 & 3 & 4 & 5 \\ 3 & 1 & 2 & 4 & 5 \end{pmatrix} = \begin{pmatrix} 3 & 1 & 2 & 4 & 5 \\ 1 & 2 & 3 & 4 & 5 \end{pmatrix}$$

Now $\binom{l}{s}_M$ can be converted to $\binom{l}{s}_E$ by substitution of skeletal numbers $(s, 1 \to 2 \to 3 \to 1)$ or of ligand numbers $(l, 1 \to 3 \to 2 \to 1)$. Therefore we can write

$$\binom{l}{s}_M = (1\ 2\ 3)_s \binom{l}{s}_E = (1\ 3\ 2)_l \binom{l}{s}_E$$

and the molecule can thus be described by P_s or P_l, where $P_l^{-1} = P_s$. In this section P_s is used. Since the operation s in the previous section is described in terms of P_l, it corresponds to s^{-1} in terms of P_s.

$$\binom{l}{s}_E = \begin{pmatrix} 1 & 2 & 3 & 4 & 5 \\ 1 & 2 & 3 & 4 & 5 \end{pmatrix} \qquad (2.22)$$

The rules are:

1) Ligand numbers, $l = 1, 2, \ldots, n$, are decided by using the priority sequence order of the R, S method for describing configuration. The ligands are thus placed in sequential priority (see section 3.1), $L_1, L_2, \ldots, L_n$. If there are two or more identical ligands, we define $L_i = L_{i+1}, \ldots, L_{i+n_i-1}$, where $l = i, \ldots, i + n_i - 1$.

2) An achiral skeleton is chosen and skeletal numbers $s = 1, 2, \ldots, n$, are assigned as follows.

    (a) The skeletal numbers are chosen so that the configuration of the standard molecule is R, as shown in Fig. 2.21.

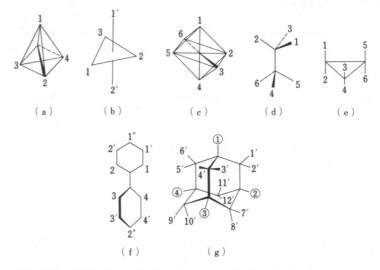

Fig. 2.21. Skeletal numbering of some standard molecules (see text).

    (b) If we look along a principal axis of the skeleton, the skeletal numbers are chosen so as to increase in a clockwise direction (Fig. 2.21(b)).

    (c) If there are more than two cyclic positions on the same principal axis C_n, the upper one is fully numbered first (Fig. 2.21(d)).

    (d) Equivalent pairs of skeletal positions are numbered consecutively, as in Fig. 2.21(e).

    (e) Skeletal numbers on a principal axis are denoted by s', s'', and ligands at similar points on the skeleton by l', l'' (Fig. 2.21(b) and (f)).

    (f) If there are several non-equivalent sets in symmetry-equivalent positions, the skeletal numbers are expressed in separate sets, i.e., $s = 1, 2, 3$; $s' = 1', 2', 3'$; $s'' = 1'', 2'', 3''$. Ligand numbers are divided into similar sets corresponding to the sets of skeletal numbers (Fig. 2.21(f) and (g)).

3) A matrix $\begin{pmatrix} l \\ s \end{pmatrix}_M$ is produced for an arbitrary molecule M from the ligand numbers and the skeletal numbers determined by means of the above rules, and similar matrices representing equivalent molecules are produced using different arrangements of l and s.[†] In order to determine a unique factor P_s relating $\begin{pmatrix} l \\ s \end{pmatrix}_M$ and $\begin{pmatrix} l \\ s \end{pmatrix}_E$ we must obtain a selection rule for $\begin{pmatrix} l \\ s \end{pmatrix}_M$. This is done by choosing the simplest matrix $\begin{pmatrix} l \\ s \end{pmatrix}_M$ in relation to $\begin{pmatrix} l \\ s \end{pmatrix}_E$. The resulting matrix is called the optimum matrix, $\begin{pmatrix} l \\ s \end{pmatrix}_{opt}$, and is obtained by applying the following selection rules.

(a) The optimum matrix is the matrix most similar to $\begin{pmatrix} l \\ s \end{pmatrix}_E$.

(b) It can be transformed into $\begin{pmatrix} l \\ s \end{pmatrix}_E$ by the simplest permutation of s (involving the smallest number of s and the fewest cycles).

(c) If several matrices still remain, the matrix in which the numbers of l and s coincide most times and in which the difference between l and s is smallest is taken.

(d) The optimum matrix of an isomer with non-equivalent sets in equivalent skeletal positions (Fig. 2.21(f) and (g)) is taken as the simplest permutation of unprimed s having the least difference in number (between l and s).

[†] Consider the molecules (a) and (b) shown below.

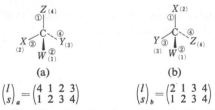

$$\begin{pmatrix} l \\ s \end{pmatrix}_a = \begin{pmatrix} 4 & 1 & 2 & 3 \\ 1 & 2 & 3 & 4 \end{pmatrix} \qquad \begin{pmatrix} l \\ s \end{pmatrix}_b = \begin{pmatrix} 2 & 1 & 3 & 4 \\ 1 & 2 & 3 & 4 \end{pmatrix}$$

The ligand priority is $W > X > Y > Z$, where $W = l_1$, $X = l_2$, $Y = l_3$ and $Z = L_4$. The corresponding matrices are also shown. When molecule (b) is rotated $120°$ on the W–C axis, it becomes identical to (a), and therefore the two matrices represent alternative descriptions of the same molecule. Here, the relation $\begin{pmatrix} l \\ s \end{pmatrix}_b = (1\ 2\ 3)\begin{pmatrix} l \\ s \end{pmatrix}_a$ holds.

The method for obtaining the optimum matrix in practice is as follows. First, we take four rows, α, β, γ and δ. The skeletal numbers are arranged along the α row. Next, the ligand numbers of a given molecule M are arranged along the β row (if M contains more than one of a given ligand, every possible ligand number is included). Row γ shows the arrangement of s produced by one of the proper rotations belonging to the symmetry elements of the skeleton on the skeletal numbers. This symmetry operation should be such that this row coincides with the previous row as far as possible. In this way we obtain optimum values, and ligand numbers are chosen to correspond to them in row δ. The resulting matrix $\begin{pmatrix} \delta \\ \gamma \end{pmatrix}$ becomes the optimum matrix. Finally, P_s can be obtained from the non-equivalent numbers in rows γ and δ.

Once the optimum matrix of a molecule has been determined, the chirality of the molecule can be determined directly from it, without using the method described previously. This is done as follows. We operate one of the elements of the permutation g' corresponding to the coset $\mathscr{C}$ of subgroup $\mathscr{D}$ belonging to the rotation of the point group $\mathscr{G}$ of the molecular skeleton on the optimum matrix, and if the matrix $\begin{pmatrix} l \\ s \end{pmatrix}$ thus produced can be transformed back to the optimum matrix by the operation of an element corresponding to $\mathscr{D}$, then the molecule is achiral. However, if the optimum matrix cannot be obtained by the operation of any element of this type, the molecule is chiral. This can be summarized as follows.

$$\left.\begin{array}{ll}
g'\begin{pmatrix} l \\ s \end{pmatrix}_{\text{opt}} = \begin{pmatrix} l \\ s \end{pmatrix} \qquad g\begin{pmatrix} l \\ s \end{pmatrix} = \begin{pmatrix} l \\ s \end{pmatrix}_{\text{opt}} \qquad \text{(achiral molecule)} \\[2mm]
\qquad g' \in \mathscr{C} \qquad\qquad\quad g \in \mathscr{D} \\[4mm]
g'\begin{pmatrix} l \\ s \end{pmatrix}_{\text{opt}} = \begin{pmatrix} l \\ s \end{pmatrix} \qquad g\begin{pmatrix} l \\ s \end{pmatrix} \neq \begin{pmatrix} l \\ s \end{pmatrix}_{\text{opt}} \qquad \text{(chiral molecule)}
\end{array}\right\} \quad (2.23)$$

Some examples of the application of these rules will be given. First, the chirality of the compound shown in Fig. 2.24 will be determined by this method.

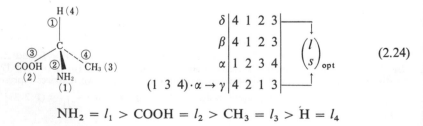

$$(2.24)$$

$$NH_2 = l_1 > COOH = l_2 > CH_3 = l_3 > H = l_4$$

The skeletal numbers are expressed as 1, 2, 3, 4, so that $\begin{pmatrix} l \\ s \end{pmatrix}_M$ becomes $\begin{pmatrix} 4 & 1 & 2 & 3 \\ 1 & 2 & 3 & 4 \end{pmatrix}$. Comparing the γ row, given by the operation (1 3 4) on the skeleton, expressed by the α row, with the β row, we obtain the δ row and hence $\begin{pmatrix} l \\ s \end{pmatrix}_{opt} = \begin{pmatrix} 4 & 1 & 2 & 3 \\ 4 & 2 & 1 & 3 \end{pmatrix}$. Since $\begin{pmatrix} l \\ s \end{pmatrix}_E = \begin{pmatrix} 1 & 2 & 3 & 4 \\ 1 & 2 & 3 & 4 \end{pmatrix}$, we have $\begin{pmatrix} l \\ s \end{pmatrix}_{opt} = (1\ 2)\begin{pmatrix} l \\ s \end{pmatrix}_E$, $P_s = (1\ 2)$ and this molecule is a (1 2) isomer. If we use $\begin{pmatrix} l \\ s \end{pmatrix}_M$ in the Eq. 2.24 directly, $\begin{pmatrix} l \\ s \end{pmatrix}_M = (1\ 2\ 3\ 4)\begin{pmatrix} l \\ s \end{pmatrix}_E$ and $P_s = (1\ 2\ 3\ 4)$ becomes not the simplest permutation, thus $\begin{pmatrix} l \\ s \end{pmatrix}_M$ is not equal to $\begin{pmatrix} l \\ s \end{pmatrix}_{opt}$. Since the standard molecule is defined with an R configuration for the skeletal structure, a (1 2) isomer will be in the S configuration.

As another example, the allenes shown in Eq. 2.25 will be considered. These were also discussed in section 2.2.2 (D_{2d} skeleton).

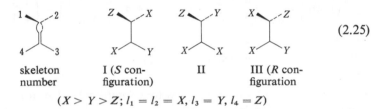

$$\text{(2.25)}$$

| skeleton number | I (S con-figuration) | II | III (R con-figuration) |

$$(X > Y > Z;\ l_1 = l_2 = X,\ l_3 = Y,\ l_4 = Z)$$

Now suppose that $X = C_6H_5$, $Y = CH_3$, $Z = H$. In the case where the molecule I is 2,3-methyldiphenylallene, we have

$$
\begin{array}{c|cccc}
\delta & 4 & 2 & 3 & 1 \\
\beta & 4 & 1,2 & 3 & 1,2 \\
\alpha & 1 & 2 & 3 & 4 \\
\hline
(1\ 4)(2\ 3)\cdot\alpha \to \gamma & 4 & 3 & 2 & 1
\end{array}
\qquad
\begin{pmatrix} l \\ s \end{pmatrix}_{opt} \quad P_s = (2\ 3),\ \begin{pmatrix} l \\ s \end{pmatrix}_{opt} = \begin{pmatrix} 4 & 2 & 3 & 1 \\ 4 & 3 & 2 & 1 \end{pmatrix}
$$

$$\text{(2.26)}$$

Thus, the compound I is the (2 3) isomer of methyldiphenylallene. We now choose one of the operations $g' = (3\ 4)$ of the coset in the rotational subgroup of D_{2d} and operate it on the optimum matrix.

$$(3\ 4)\begin{pmatrix} l \\ s \end{pmatrix}_{opt} = \begin{pmatrix} 4 & 2 & 3 & 1 \\ 3 & 2 & 4 & 1 \end{pmatrix} = \begin{pmatrix} l \\ s \end{pmatrix} \qquad \text{(2.27)}$$

We cannot find any elements which can transform a given molecule $\begin{pmatrix} l \\ s \end{pmatrix}$ into $\begin{pmatrix} l \\ s \end{pmatrix}_{opt}$ in the rotational subgroup of D_{2d} (see Table 2.4), and therefore the molecule is a chiral one.

Next, we will consider the molecule II of Eq. 2.25, and for this we have

$$
\begin{array}{c}
\delta \\
\beta \\
\alpha \\
(1\ 4)(2\ 3)\cdot\alpha \to \gamma
\end{array}
\begin{array}{|cccc|}
4 & 3 & 2 & 1 \\
4 & 3 & 1,2 & 1,2 \\
1 & 2 & 3 & 4 \\
4 & 3 & 2 & 1
\end{array}
\quad
\begin{pmatrix} l \\ s \end{pmatrix}_{opt} \quad P_s = e \quad \begin{pmatrix} l \\ s \end{pmatrix}_{opt} = \begin{pmatrix} 4 & 3 & 2 & 1 \\ 4 & 3 & 2 & 1 \end{pmatrix}
$$

$$\tag{2.28}$$

Thus, the compound II is e isomer. In the same way as Eq. 2.27,

$$
(3\ 4)\begin{pmatrix} l \\ s \end{pmatrix}_{opt} = \begin{pmatrix} 4 & 3 & 2 & 1 \\ 3 & 4 & 2 & 1 \end{pmatrix} = (1\ 2)(3\ 4)\begin{pmatrix} 4 & 3 & 2 & 1 \\ 3 & 4 & 2 & 1 \end{pmatrix} = \begin{pmatrix} 4 & 3 & 2 & 1 \\ 4 & 3 & 1 & 2 \end{pmatrix}
$$

$$
= \begin{pmatrix} 4 & 3 & 2 & 1 \\ 4 & 3 & 2 & 1 \end{pmatrix} = \begin{pmatrix} l \\ s' \end{pmatrix}_{opt}
$$

$$\tag{2.29}$$

In this case, $(1\ 2)(3\ 4)$ corresponds to a rotational subgroup of D_{2d}, and therefore this molecule is achiral.

For molecule III (Eq. 2.25), we have

$$
\begin{array}{c}
\delta \\
\beta \\
\alpha \\
(1\ 3)(2\ 4)\cdot\alpha \to \gamma
\end{array}
\begin{array}{|cccc|}
1 & 4 & 3 & 2 \\
1,2 & 3 & 4 & 1,2 \\
1 & 2 & 3 & 4 \\
3 & 4 & 1 & 2
\end{array}
\quad
\begin{pmatrix} l \\ s \end{pmatrix}_{opt} \quad P_s = (1\ 3) \quad \begin{pmatrix} l \\ s \end{pmatrix}_{opt} = \begin{pmatrix} 1 & 4 & 3 & 2 \\ 3 & 4 & 1 & 2 \end{pmatrix}
$$

$$\tag{2.30}$$

The compound III is $(1\ 3)$ isomer. In the same way as for molecule I, $\begin{pmatrix} l \\ s \end{pmatrix}$ given by the operation $(3\ 4)$ on the optimum matrix cannot be transformed into the optimum matrix again by operations in the rotational subgroup of D_{2d}, and the molecule is therefore chiral.

Finally, we will consider the isomers of octahedral structure (O_h skeletal symmetry) shown in Eq. 2.31 (see also Table 2.9).

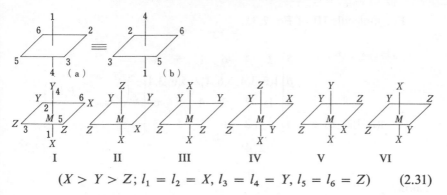

$$(X > Y > Z; l_1 = l_2 = X, l_3 = l_4 = Y, l_5 = l_6 = Z) \qquad (2.31)$$

The skeletal numbers are given by Eq. 2.31(a) or (b), and the following discussion will be based on (b). Now, for molecule I of Eq. 2.31 we have

$$
\begin{array}{c|cccccc}
\delta & 1 & 3 & 5 & 4 & 6 & 2 \\
\beta & 1,2 & 3,4 & 5,6 & 3,4 & 5,6 & 1,2 \\
\alpha & 1 & 2 & 3 & 4 & 5 & 6 \\
(2\ 6\ 5\ 3)\cdot\alpha \to \gamma & 1 & 3 & 5 & 4 & 6 & 2
\end{array}
\qquad \left(\dfrac{l}{s}\right)_{opt} = \left(\dfrac{l}{s}\right)_{E} \qquad (2.32)
$$

This corresponds to the standard molecule. Operating one of the elements of the coset of $\mathscr{G}$, $g' = (1\ 2)(3\ 4)$ on this, we have

$$(1\ 2)(4\ 5)\begin{pmatrix} 1 & 3 & 5 & 4 & 6 & 2 \\ 1 & 3 & 5 & 4 & 6 & 2 \end{pmatrix} = \begin{pmatrix} 1 & 3 & 5 & 4 & 6 & 2 \\ 2 & 3 & 4 & 5 & 6 & 1 \end{pmatrix} = \begin{pmatrix} 2 & 3 & 5 & 4 & 6 & 1 \\ 2 & 3 & 4 & 5 & 6 & 1 \end{pmatrix}$$

$$= (4\ 5)\left(\dfrac{l}{s}\right)_{E} \qquad (2.33)$$

This corresponds to the distribution of the molecule II (Eq. 2.31), and thus we find that the molecule I is chiral and that II is its enantiomer. The configuration of I is S, according to the rules for determining R,S. For the molecule II, we have

$$
\begin{array}{c|cccccc}
\delta & 1 & 4 & 6 & 5 & 2 & 3 \\
\beta & 1,2 & 3,4 & 5,6 & 5,6 & 1,2 & 3,4 \\
\alpha & 1 & 2 & 3 & 4 & 5 & 6 \\
(2\ 5)(3\ 6)\cdot\alpha \to \gamma & 1 & 5 & 6 & 4 & 2 & 3
\end{array}
\qquad \left(\dfrac{l}{s}\right)_{opt} = P_s\left(\dfrac{l}{s}\right)_{E} \quad P_s = (4\ 5)
$$

$$(2.34)$$

As already shown, it is the enantiomer of I. Its configuration is R.

For molecule III of Eq. 2.31,

$$
\begin{array}{r|cccccc}
\delta & 2 & 4 & 6 & 1 & 5 & 3 \\
\beta & 1,2 & 3,4 & 5,6 & 1,2 & 5,6 & 3,4 \\
\alpha & 1 & 2 & 3 & 4 & 5 & 6 \\
(1\ 5)(2\ 4)(3\ 6)\cdot\alpha \to \gamma & 4 & 2 & 6 & 1 & 5 & 3
\end{array}
\qquad
\left(\begin{array}{c} l \\ s \end{array}\right)_{opt} = P_s \left(\begin{array}{c} l \\ s \end{array}\right)_E
$$

$$ P_s = (2\ 4) \tag{2.35} $$

Taking $g' = (1\ 4)$,

$$
g' \times \left(\begin{array}{c} l \\ s \end{array}\right)_{opt} = (1\ 4)\left(\begin{array}{cccccc} 2 & 4 & 6 & 1 & 5 & 3 \\ 4 & 2 & 6 & 1 & 5 & 3 \end{array}\right) = \left(\begin{array}{cccccc} 2 & 4 & 6 & 1 & 5 & 3 \\ 1 & 2 & 6 & 4 & 5 & 3 \end{array}\right)
$$

$$
= e\left(\begin{array}{cccccc} 1 & 4 & 6 & 2 & 5 & 3 \\ 1 & 2 & 6 & 4 & 5 & 3 \end{array}\right) = \left(\begin{array}{c} l \\ s \end{array}\right)_{opt} \tag{2.36}
$$

Now e is an element of the rotational subgroup $\mathscr{D}$, so this molecule is an achiral one (see Table 2.9).

For the molecule IV of Eq. 2.31,

$$
\begin{array}{r|cccccc}
\delta & 2 & 3 & 6 & 5 & 4 & 1 \\
\beta & 1,2 & 3,4 & 5,6 & 5,6 & 3,4 & 1,2 \\
\alpha & 1 & 2 & 3 & 4 & 5 & 6 \\
(1\ 2)(4\ 5)(3\ 6)\cdot\alpha \to \gamma & 2 & 1 & 6 & 5 & 4 & 3
\end{array}
\qquad
\left(\begin{array}{c} l \\ s \end{array}\right)_{opt} = P_s \left(\begin{array}{c} l \\ s \end{array}\right)_E
$$

$$ P_s = (1\ 3) \tag{2.37} $$

and taking $g' = (1\ 4)$,

$$
g'\left(\begin{array}{c} l \\ s \end{array}\right)_{opt} = (1\ 4)\left(\begin{array}{cccccc} 2 & 3 & 6 & 5 & 4 & 1 \\ 2 & 1 & 6 & 5 & 4 & 3 \end{array}\right) = \left(\begin{array}{cccccc} 2 & 3 & 6 & 5 & 4 & 1 \\ 2 & 4 & 6 & 5 & 1 & 3 \end{array}\right)
$$

$$
= e\left(\begin{array}{cccccc} 2 & 4 & 6 & 5 & 3 & 1 \\ 2 & 4 & 6 & 5 & 1 & 3 \end{array}\right) = \left(\begin{array}{c} l \\ s \end{array}\right)_{opt} \tag{2.38}
$$

Thus, the molecule IV is also achiral.

Similarly, for molecule V of Eq. 2.31,

$$\begin{array}{c|cccccc|}
\delta & 2 & 4 & 5 & 3 & 1 & 6 \\
\beta & 1,2 & 3,4 & 5,6 & 3,4 & 1,2 & 5,6 \\
\alpha & 1 & 2 & 3 & 4 & 5 & 6 \\
\hline
(1\ 5\ 4\ 2)\cdot\alpha \to \gamma & 2 & 4 & 3 & 5 & 1 & 6
\end{array} \quad \begin{pmatrix} l \\ s \end{pmatrix}_{opt} \quad P_s = (3\ 5) \qquad (2.39)$$

and taking $g' = (3\ 6)$, we obtain

$$g'\begin{pmatrix} l \\ s \end{pmatrix}_{opt} = (3\ 6)\begin{pmatrix} 2 & 4 & 5 & 3 & 1 & 6 \\ 2 & 4 & 3 & 5 & 1 & 6 \end{pmatrix} = \begin{pmatrix} 2 & 4 & 5 & 3 & 1 & 6 \\ 2 & 4 & 6 & 5 & 1 & 3 \end{pmatrix}$$

$$= e\begin{pmatrix} 2 & 4 & 6 & 3 & 1 & 5 \\ 2 & 4 & 6 & 5 & 1 & 3 \end{pmatrix} = \begin{pmatrix} l \\ s \end{pmatrix}_{opt} \qquad (2.40)$$

and therefore this molecule is achiral.

Finally, for molecule VI of Eq. 2.31,

$$\begin{array}{c|cccccc|}
\delta & 1 & 4 & 6 & 2 & 3 & 5 \\
\beta & 1,2 & 3,4 & 5,6 & 1,2 & 3,4 & 5,6 \\
\alpha & 1 & 2 & 3 & 4 & 5 & 6 \\
\hline
(1\ 5)(2\ 4)(3\ 6)\cdot\alpha \to \gamma & 5 & 4 & 6 & 2 & 1 & 3
\end{array} \quad \begin{pmatrix} l \\ s \end{pmatrix}_{opt} = P_s\begin{pmatrix} l \\ s \end{pmatrix}_E$$

$$P_s = (1\ 3\ 5) \qquad (2.41)$$

Taking $g' = (1\ 4)(2\ 5)(3\ 6)$, we obtain

$$g'\begin{pmatrix} l \\ s \end{pmatrix}_{opt} = (1\ 4)(2\ 5)(3\ 6)\begin{pmatrix} 1 & 4 & 6 & 2 & 3 & 5 \\ 5 & 4 & 6 & 2 & 1 & 3 \end{pmatrix} = \begin{pmatrix} 1 & 4 & 6 & 2 & 3 & 5 \\ 2 & 1 & 3 & 5 & 4 & 6 \end{pmatrix},$$

$$(2\ 5)(3\ 6)\begin{pmatrix} 1 & 4 & 6 & 2 & 3 & 5 \\ 2 & 1 & 3 & 5 & 4 & 6 \end{pmatrix} = \begin{pmatrix} 1 & 3 & 6 & 2 & 4 & 5 \\ 5 & 1 & 6 & 2 & 4 & 3 \end{pmatrix} = \begin{pmatrix} l \\ s \end{pmatrix}_{opt} \qquad (2.42)$$

and $(2\ 5)(3\ 6) \in \mathscr{D}$ therefore this molecule is also achiral.

If we now assume that $X = Cl$, $Y = H_2O$, $Z = NH_2$ and $M = Co$ (III), the compounds I through VI in Eq. 2.31 become dichlorodiamino-cobaltates. The molecules I and II are S (*cis, cis, cis*) and R (*cis, cis, cis*), respectively. The molecules III and V are *trans, cis, cis* and *cis, cis, trans*, respectively.

The description of isomers by means of permutational operations is very useful for enantiomers, diastereomers and substitutional isomers, but it is a relatively new method, and various problems still remain in connection with deciding on a skeleton and obtaining the optimum matrix.

Furthermore, the expressions become very complex for large molecules and are intuitively unappealing. Thus, there is still scope for the development of improved methods in the future.

REFERENCES

1. G. E. McCasland and S. Proskow, *J. Am. Chem. Soc.*, **87**, 4688 (1955).
2. R. S. Cahn, C. K. Ingold and V. Prelog, *Angew. Chem. Intern. Ed. Engl.*, **5**, 385 (1966).
3. L. Kelvin, *Baltimore Lectures*, p. 436, 619, C. J. Clay, 1904.
4. I. Ugi, D. Marguarding, H. Klusacek, G. Gokel and P. Gillespie, *Angew. Chem. Intern. Ed. Engl.*, **9**, 703 (1970).

Appendix A

Set and group

A.1   Sets, Mappings and Equivalence

A collection of objects selected under well defined conditions is called a set. Each object is an element (or member) of that set. If x is an element of the set $\mathscr{A}$, we say that x belongs to $\mathscr{A}$ and designate this by $x \in \mathscr{A}$ or $\mathscr{A} \ni x$. If x has a certain property P, it is denoted by $P(x)$, and the set of all such x is denoted by $\{x; P(x)\}$. A collection which contains no elements is called an empty set, ϕ.

If for two sets $\mathscr{A}$ and $\mathscr{B}$, $x \in \mathscr{A}$ always implies $x \in \mathscr{B}$, then $\mathscr{A}$ is a subset of $\mathscr{B}$ and this is denoted by $\mathscr{A} \subset \mathscr{B}$ or $\mathscr{B} \supset \mathscr{A}$. The set composed of the sum of the elements of two sets $\mathscr{A}$ and $\mathscr{B}$ is called a union, denoted by $\mathscr{A} \cup \mathscr{B}$, i.e., we can write $\mathscr{A} \cup \mathscr{B} = \{x; x \in \mathscr{A} \text{ or } x \in \mathscr{B}\}$. The set of elements common to two sets $\mathscr{A}$ and $\mathscr{B}$ is called an intersection, denoted by $\mathscr{A} \cap \mathscr{B}$, i.e., $\mathscr{A} \cap \mathscr{B} = \{x; x \in \mathscr{A} \text{ and } x \in \mathscr{B}\}$. The set of elements which belong to $\mathscr{A}$ but not to $\mathscr{B}$ is called the difference, $\mathscr{A} - \mathscr{B}$, i.e., $\mathscr{A} - \mathscr{B} = \{x; x \in \mathscr{A}, x \notin \mathscr{B}\}$. The set of pairs (x, y) where x is an element of set $\mathscr{A}$ and y is an element of set $\mathscr{B}$ is called a product set, denoted by $\mathscr{A}\mathscr{B}$, i.e., $\mathscr{A}\mathscr{B} = \{(x, y); x \in \mathscr{A}, y \in \mathscr{B}\}$.

If a rule f exists such that the application of f to each element of $\mathscr{A}$ uniquely determines an element of $\mathscr{B}$, f is called a mapping (or function) of $\mathscr{A}$ onto $\mathscr{B}$ and is denoted by $f : \mathscr{A} \to \mathscr{B}$. If this mapping transforms an element a of $\mathscr{A}$ into an element b of $\mathscr{B}$, it is written as $f(a) = b$. When the mapping $f : \mathscr{A} \to \mathscr{B}$ satisfies $f(\mathscr{A}) = \mathscr{B}$, f is called the surjection of $\mathscr{A}$ into $\mathscr{B}$. When the mapping $f : \mathscr{A} \to \mathscr{B}$ transforms two different elements of $\mathscr{A}$ into two different elements of $\mathscr{B}$, i.e., if $a, a' \in \mathscr{A}$ $(a \neq a')$, then $f(a) \neq f(a')$, we call f the injection of $\mathscr{A}$ into $\mathscr{B}$. A mapping which is both a surjection and an injection is called a bijection.

When relations $\sim$ among arbitrary elements a, b, c of a set $\mathscr{M}$ satisfy the following conditions, the relations $\sim$ are called the equivalence relation.
1) $a \sim a$ (reflexive law)
2)   If $a \sim b$, then $b \sim a$ (symmetric law)
3)   If $a \sim b$ and $b \sim c$, then $a \sim c$ (transitive law)
Now, suppose that the equivalence relation is given for the set $\mathscr{M}$, and that a is an element of $\mathscr{M}$. The set of elements of $\mathscr{M}$ which are equivalent to a, $\{m; m \in \mathscr{M}, a \sim m\}$, is called the equivalence class decided by a, denoted by $[a]$. $\mathscr{M}$ can be divided into equivalence classes of each element m, and this is called a classification of $\mathscr{M}$ (by the equivalence relation $\sim$).

A.2   GROUPS

We shall next define a group, as follows. In a set $\mathscr{G}$, if a rule is decided which transforms two arbitrary elements g and h in to a product gh which is also a member of $\mathscr{G}$, then the set $\mathscr{G}$ is a group if the rule satisfies the following conditions.
1)   Every product of two elements belongs to the primary set.
2)   There exists at least one unit element E such that $Eg = gE = g$ for every element g of the set.
3)   The associative law is obeyed, i.e., $g(f \cdot h) = (g \cdot f)h$.
4)   For an arbitrary element g, there exists an inverse element $X = g^{-1}$, such that $gX = Xg = E$.
A group having elements consisting of finite numbers is called a finite group; the number of elements is called the order of the group, and is denoted by $|\mathscr{G}|$. If the number of elements is not finite, the group is called an infinite group.

We will now examine the point group C_{3v} introduced in section 2.2.1 to see whether it conforms to the above rules. The order is six, since there are six elements. A multiplication table of the elements is shown overleaf. It is clear that the table contains only elements of C_{3v}. There is also a unit element E such that $C_3^1 E = C_3^1$, $\sigma_1 E = \sigma_1$. The associative law holds,

i j	E	C_3^1	C_3^2	$\sigma 1$	$\sigma 2$	$\sigma 3$
E	E	C_3^1	C_3^2	$\sigma 1$	$\sigma 2$	$\sigma 3$
C_3^1	C_3^1	C_3^2	E	$\sigma 2$	$\sigma 3$	$\sigma 1$
C_3^2	C_3^2	E	C_3^1	$\sigma 3$	$\sigma 1$	$\sigma 2$
$\sigma 1$	$\sigma 1$	$\sigma 3$	$\sigma 2$	E	C_3^2	C_3^1
$\sigma 2$	$\sigma 2$	$\sigma 1$	$\sigma 3$	C_3^1	E	C_3^2
$\sigma 3$	$\sigma 3$	$\sigma 2$	$\sigma 1$	C_3^2	C_3^1	E

since for instance $C_3^1(C_3^2\sigma_1) = C_3^1\sigma_2 = (C_3^1 C_3^2)\sigma_1$. Finally, it can be seen that each element has an inverse element. For instance, C_3^2 is the inverse of C_3^1, since their product is E. Thus, all four requirements are met, and C_{3v} is indeed a group.

A.3   Subgroups, Cosets and Their Classification

When a subset $\mathscr{K}$ of a group $\mathscr{G}$ fulfils the requirements for a group given in the previous section, with respect to the same operations on its elements as in the case of $\mathscr{G}$, $\mathscr{K}$ is a subgroup of $\mathscr{G}$. A set of elements of a subgroup $\mathscr{K}$ of $\mathscr{G}$ obtained by connecting every element of $\mathscr{K}$ on the right or left side of one element of $\mathscr{G}$, i.e., $\{k \cdot g; \ k \in \mathscr{K}\}$ or $\{g \cdot k; \ k \in \mathscr{K}\}$, is called the left or right coset of $\mathscr{K}$ with g as the representative element ($g \cdot \mathscr{K}$ and $\mathscr{K} \cdot g$, respectively). If the group $\mathscr{G}$ is divided into several different right or left cosets of $\mathscr{K}$, this is called a classification of $\mathscr{G}$ based on $\mathscr{K}$.

Limiting the number of right cosets to a finite number, $\mathscr{K}g_1$, $\mathscr{K}g_2$, ..., $\mathscr{K}g_z$, the number z is called the index of $\mathscr{K}$ in $\mathscr{G}$. When the order is finite, the number of right cosets is equal to the number of left cosets.

When $g \cdot \mathscr{K} = \mathscr{K} \cdot g$ always holds for an arbitrary element g of $\mathscr{G}$, $\mathscr{K}$ is called a normal subgroup of $\mathscr{G}$. The necessary and sufficient condition for a subgroup $\mathscr{K}$ of $\mathscr{G}$ to be a normal subgroup is that for arbitrary elements g and k of $\mathscr{G}$ and $\mathscr{K}$, the relation $g^{-1}kg \ni \mathscr{K}$ holds.

We will again consider C_{3v} as an example. Since the set $\mathscr{K} = \{E, C_3^1, C_3^2\}$ satisfies the conditions for a group, as described above, this set is a subgroup of C_{3v}. Using an element σ_1 in C_{3v} which does not belong to $\mathscr{K}$, we can form a right coset of $\mathscr{K}$, $\mathscr{K}\sigma_1 = \{\sigma_1, \sigma_2, \sigma_3\}$. So C_{3v} is classified into $\mathscr{K}$ and $\mathscr{K}\sigma_1$ and in this case the index of $\mathscr{K}$ becomes 2. Next, with $\mathscr{G} = C_{3v}$, $\mathscr{K} = \{E, C_3^1, C_3^2\}$ we can calculate $g^{-1}\mathscr{K}g$ as follows.

Suppose $\qquad\qquad g = C_3^1, \qquad$ then $\qquad g^{-1} = C_3^2$

$$g^{-1}\mathscr{K}g = C_3^2\{E, C_3^1, C_3^2\}C_3^1 = \{E, C_3^1, C_3^2\} = \mathscr{K}$$

Suppose $\qquad\qquad g = \sigma_1, \qquad$ then $\qquad g^{-1} = \sigma_1$

$$g^{-1}\mathscr{K}g = \sigma_1\{E, C_3^1, C_3^2\}\sigma_1 = \{E, C_3^1, C_3^2\} = \mathscr{K}$$

The relations hold no matter which element we use, and therefore $\mathscr{K}$ is a normal subgroup of index 2 in C_{3v}. Since the elements of $\mathscr{K}$ consist only of rotational operations, such a subgroup is called a rotational subgroup. When two elements, g, h, of $\mathscr{G}$ are in a normal subgroup $\mathscr{K}$ of $\mathscr{G}$ and are in the relation $gh^{-1} \in \mathscr{K}$, then we define $g \sim h$, where $\sim$ is an equivalence relation. Classifying $\mathscr{G}$ by this equivalence relation, a set of this equivalence class $\mathscr{G}/\mathscr{K}$ is a group. The group $\mathscr{G}/\mathscr{K}$ is called a quotient group of $\mathscr{G}$ mod $\mathscr{K}$.

A.4   CONJUGATE CLASSES OF GROUPS

When for two elements g, $h \in \mathscr{G}$, an element f of $\mathscr{G}$ exists such that $h = f^{-1}gf$ exists, the element h is said to be conjugate to the element g. This conjugate relationship satisfies the following conditions and is an equivalence relation.
1) $g = e^{-1}ge = g$ (reflexive law)
2)   If h is conjugate to g, then from $h = f^{-1}gf$, $(f^{-1})^{-1}hf^{-1} = (f^{-1})f^{-1}gff^{-1} = g$, and thus g and h are conjugate to each other (symmetric law).
3)   If h is conjugate to g and k to h, then because elements k and p exist such that $h = f^{-1}gf$, $k = p^{-1}hp$ and $k = p^{-1}f^{-1}gfp = (fp)^{-1}g(fp)$, k is conjugate to g (transitive law). Therefore a group $\mathscr{G}$ can be classified based on a conjugate relation. A set of elements conjugate to g is called a conjugate class representing the element g.

Again we will consider C_{3v} as an example. E forms a class by itself $\mathscr{C}_1 = \{E\}$ and a class of C_3^1 is $\mathscr{C}_2 = \{C_3^1, C_3^2\}$ since $\mathscr{C}_2$ has the properties $EC_3^1E = C_3^1$, $C_3^1C_3^1C_3^2 = C_3^1$, $C_3^2C_3^1C_3^1 = C_3^1$, $\sigma_1C_3^1\sigma_1 = C_3^2$, $\sigma_2C_3^1\sigma_2 = C_3^2$ and $\sigma_3C_3^1\sigma_3 = C_3^2$. Similarly, a class $\mathscr{C}_3$ of σ_1 becomes $\{\sigma_1, \sigma_2, \sigma_3\}$. Therefore the elements of C_{3v} can be divided into three classes, $\mathscr{C}_1 = \{E\}$, $\mathscr{C}_2 = \{C_3^1, C_3^2\}$ and $\mathscr{C}_3 = \{\sigma_1, \sigma_2, \sigma_3\}$.

APPENDIX B

Symmetry permutational groups

The one-to-one mapping from a set of numbers $(1, 2, \ldots, n)$ onto itself is called a permutation s. There are $n!$ such permutations. For the numbers 1, 2, 3, the permutations s_1 through s_6 $(3! = 6)$ are as follows.

$$s_1 = \begin{pmatrix} 1 & 2 & 3 \\ 1 & 2 & 3 \end{pmatrix} \quad s_2 = \begin{pmatrix} 1 & 2 & 3 \\ 2 & 3 & 1 \end{pmatrix} \quad s_3 = \begin{pmatrix} 1 & 2 & 3 \\ 3 & 1 & 2 \end{pmatrix}$$

$$s_4 = \begin{pmatrix} 1 & 2 & 3 \\ 1 & 3 & 2 \end{pmatrix} \quad s_5 = \begin{pmatrix} 1 & 2 & 3 \\ 3 & 2 & 1 \end{pmatrix} \quad s_6 = \begin{pmatrix} 1 & 2 & 3 \\ 2 & 1 & 3 \end{pmatrix}$$

(B.1)

Since s_1 is identical with the original, i.e., $1 \to 1$, $2 \to 2$, $3 \to 3$, this is written as $s_1 = (1)(2)(3) = e$. Since s_2 involves $1 \to 2$, $2 \to 3$, $3 \to 1$, it is denoted by (1 2 3) to express the change of the primary order 1, 2, 3 to 2, 3, 1. In this way, s_1 through s_6 can be described as follows.

$$\left. \begin{aligned} s_1 &= (1)(2)(3) = e \\ s_2 &= (1\ 2\ 3) \\ s_3 &= (1\ 3\ 2) \\ s_4 &= (1)(2\ 3) = (2\ 3) \\ s_5 &= (2)(1\ 3) = (1\ 3) \\ s_6 &= (3)(1\ 2) = (1\ 2) \end{aligned} \right\}$$

(B.2)

Now, the product $s_2 s_6 = (1\ 2\ 3)(1\ 2)$ clearly becomes (1 3), since 1 is permuted to 2 and then 2 is permuted to 3. By forming products in this way, it can be shown that s_1 through s_6 conform to the definition of a group. Such a group is known as a permutational group, $\mathscr{S}_n$. The group containing 1, 2, 3 is called a cubic permutational group, since $n = 3$.

Next we shall obtain the inverse permutation s_i^{-1} of s_i. Since $s_i^{-1} s_i = e$, we have

$$s_i = \begin{pmatrix} 1 & 2 & \ldots & n \\ a_1 & a_2 & \ldots & a_n \end{pmatrix},$$

(B.3)

Now, if we assume $s_i^{-1} = \begin{pmatrix} a_1 & a_2 & \ldots & a_n \\ 1 & 2 & \ldots & n \end{pmatrix}$, we find $s_i^{-1} s_i = \begin{pmatrix} 1 & 2 & 3 & \ldots & n \\ 1 & 2 & 3 & \ldots & n \end{pmatrix}$

$= e$. In the case of $n = 3$, s_i^{-1} becomes as follows, and each element gives a member of the permutational group $\mathscr{S}_3$.

$$s_1^{-1} = s_1 = \begin{pmatrix} 1 & 2 & 3 \\ 1 & 2 & 3 \end{pmatrix} \qquad s_2^{-1} = \begin{pmatrix} 2 & 3 & 1 \\ 1 & 2 & 3 \end{pmatrix} = (1\ 3\ 2)$$

$$s_3^{-1} = \begin{pmatrix} 3 & 1 & 2 \\ 1 & 2 & 3 \end{pmatrix} = (1\ 2\ 3) \quad s_4^{-1} = \begin{pmatrix} 1 & 3 & 2 \\ 1 & 2 & 3 \end{pmatrix} = (2\ 3) \qquad \text{(B.4)}$$

$$s_5^{-1} = \begin{pmatrix} 3 & 2 & 1 \\ 1 & 2 & 3 \end{pmatrix} = (1\ 3) \quad s_6^{-1} = \begin{pmatrix} 2 & 1 & 3 \\ 1 & 2 & 3 \end{pmatrix} = (1\ 2)$$

Next we will divide $\mathscr{S}_3$ into different conjugate classes as follows. $s_1 = (1)(2)(3)$, $\{s_2, s_3\} = \{(1\ 2\ 3), (1\ 3\ 2)\}$ and $\{s_4, s_5, s_6\} = \{(1\ 2), (1\ 3), (2\ 4)\}$. The number of numerals in $(\times \cdots \times)$ is known as the length of the ring. In the case of $\mathscr{S}_3$, an element s_1 of three rings with length 1 (denoted by 1^3), an element $\{s_4, s_5, s_6\}$ of one ring with length 1 and another with length 2 (denoted by $1^1 2^1$) and an element $\{s_2, s_2\}$ of one ring with length 3 (denoted by 3^1) form conjugate classes.

In the case of an n-dimensional permutational group $\mathscr{S}_n$, an element s is expressed in the following general form.

$$s = \overbrace{[\times][\times][\times]\ [\times]}^{1}\underbrace{\overbrace{[\times\ \ \times][\times\ \ \times]\ [\times\ \ \times]}^{2}}_{\lambda_2} \cdots\cdots \overbrace{[\times \times \times \cdots \times]}^{k} \quad \text{(B.5)}$$

with underbraces λ_1, λ_2, λ_k.

In this equation, $\times$ appears n times, and replaces the numbers 1 through n. Therefore, there exist combinations of $\lambda_1, \lambda_2, \ldots, \lambda_k$ which satisfy the relation $1\lambda_1 + 2\lambda_2 \ldots + n\lambda_n = n$. Once $\lambda_1, \lambda_2, \ldots, \lambda_k$ are decided, we can obtain one form of a permutation and denote it by $(1^{\lambda_1}, 2^{\lambda_2}, \ldots, k^{\lambda_k})$. Permutations expressed by the same form are all included in the same conjugate class, and we can classify an element of $\mathscr{S}_n$ into several conjugate classes by the formation of such permutations. The number of classes is equal to the number of arrangements of n, and is given by the following expression.

$$h(\lambda_1, \lambda_2, \ldots \lambda_n) = \frac{n!}{1^{\lambda_1}\lambda_1! \cdot 2^{\lambda_2}\lambda_2! \ldots k^{\lambda_k}\lambda_k!} \qquad \text{(B.6)}$$

In the case of $\mathscr{S}_3$, the combinations of λ which satisfy $1\lambda_1 + 2\lambda_2 + 3\lambda_3 = 3$ are

$$\lambda_1 = 3 \quad \lambda_2 = \lambda_3 = 0$$

$$\lambda_1 = 1 \quad \lambda_2 = 1 \quad \lambda_3 = 0 \qquad (B.7)$$

$$\lambda_3 = 1 \quad \lambda_1 = \lambda_2 = 0$$

In the case of $\lambda_1 = 3$, the form becomes $[\times][\times][\times] \to 1^3$ with $h = 3!/(1^3 \cdot 3!) = 1$, corresponding to s_1. The form of $\lambda_1 = 1$, $\lambda_2 = 2$ becomes $[\times \ \times][\times] \to 1^1 2^1$ with $h = 3!/(1^1 \cdot 1! \cdot 2^1 \cdot 1!) = 3$; the elements s_4, s_5 and s_6 belong to this form. The form of $\lambda_3 = 1$ becomes $[\times \ \times \ \times] \to 3^1$ with $h = 3!/(3^1 \cdot 1!) = 2$; s_2 and s_3 belong to this form.

Next we will consider the relation between a permutational group $\mathscr{S}_3$ and a point group. Suppose that the numbers 1, 2, 3 form the vertices of a triangle. They now correspond to one of the symmetry operations of C_{3v}, as shown below.

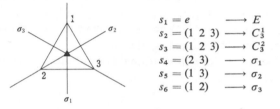

$$s_1 = e \qquad \longrightarrow \quad E$$
$$s_2 = (1\ 2\ 3) \longrightarrow \quad C_3^1$$
$$s_3 = (1\ 2\ 3) \longrightarrow \quad C_3^2$$
$$s_4 = (2\ 3) \qquad \longrightarrow \quad \sigma_1$$
$$s_5 = (1\ 3) \qquad \longrightarrow \quad \sigma_2$$
$$s_6 = (1\ 2) \qquad \longrightarrow \quad \sigma_3$$

In this case, $\mathscr{S}_3$ and C_{3v} are said to be isomorphous, and the relationship is denoted by $\mathscr{S}_3 \cong C_{3v}$.

In general, when a mapping of a group $\mathscr{G}$ onto a group $\mathscr{G}'$, $f : \mathscr{G} \to \mathscr{G}'$ always satisfies $f(g \cdot h) = f(g) \cdot f(h)$ for arbitrary elements $g, h \in \mathscr{G}$, f is called the homomorphism of $\mathscr{G}$ onto $\mathscr{G}'$. When the homomorphism $f : \mathscr{G} \to \mathscr{G}'$ is an injection or surjection, it is referred to as an injective or surjective homomorphism. When the homomorphism is a bijection, f is called an isomorphism. For example, if $\mathscr{K}$ is a subgroup of $\mathscr{G}$, $f' : \mathscr{K} \to \mathscr{G}$ is an injective homomorphism. If we assume that $\mathscr{K}$ is a normal subgroup, the mapping of g, an element of $\mathscr{G}$, onto an element g of $\mathscr{G}/\mathscr{K}$, $p : \mathscr{G} \to \mathscr{G}/\mathscr{K}$ is called a surjective homomorphism.

Appendix C

Polya's procedure

The number of isomers of a particular molecule can be determined by taking the set of ligands $\{1, 2, \ldots, n\}$ on the numbered skeleton as $\mathcal{U}$, the group of permutations in $\mathcal{U}$ as $\mathcal{G}$ and the set of ligand types $\{l_1, l_2, \ldots, l_m\}$ as $\mathcal{V}$. Assuming that the set of every mapping φ of $\mathcal{U}$ onto $\mathcal{V}$ is Φ and the equivalence class containing $\varphi_0 \in \Phi$ for $\mathcal{G}$ is $\bar{\varphi}_0$, then the set of φ such that $\bar{\varphi} = \varphi_0$ becomes $\{\varphi | \varphi \in \Phi$ for $g \in \mathcal{G}, \varphi g = \varphi_0\}$.

Taking a ligand l_i, it is weighted using the factor $w(l_i) > 0$ for an isomer containing v_i $(i = 1, 2, \ldots, m)$ ligands of kind l_i, and thus we have

$$W(\bar{\varphi}) = w(l_1)^{v_1} w(l_2)^{v_2} \ldots w(l_m)^{v_m} \tag{C.1}$$

Putting $w(l_i) = w_i$, the sum of the weights of an isomer $\bar{\varphi}$ for a group $\mathcal{G}$ which has permutations of ligands becomes

$$\sum W(\bar{\varphi}) = P\left(\mathcal{G} : \sum_{i=1}^{m} w_i, \sum_{i=1}^{m} w_i^2, \ldots\right) \tag{C.2}$$

The number of isomers $\bar{\varphi}$ with v_i ligands of type l_i is given by the coefficient of the appropriate term of the polynomial on the right-hand side of Eq. C.2. Taking this coefficient as $\mathcal{C}_{\mathcal{G}}$, then

$$|\mathcal{C}_{\mathcal{G}}| = \frac{1}{|\mathcal{G}|} \sum_{\lambda_1 + 2\lambda_2 + \cdots + n\lambda_n} h_{\mathcal{G}}(\lambda_1, \lambda_2, \ldots, \lambda_n) P_{v_1, v_2, \ldots, v_m}(\lambda_1, \lambda_2, \ldots, \lambda_n)$$

$|\mathcal{G}|$: order of $\mathcal{G}$ $\qquad$ (C.3)

where $h_{\mathcal{G}}(\lambda_1, \lambda_2, \ldots, \lambda_n)$ is the number of the permutation $g \in \mathcal{G}$ of the form $1^{\lambda_1}, 2^{\lambda_2}, \ldots, n^{\lambda_n}$ and $P_{v_1, v_2, \ldots, v_n}(\lambda_1, \lambda_2, \ldots, \lambda_n)$, using v_i ligands of type l_i, is the number of the allotment such that $P_{v_1, v_2, \ldots, v_n}(\lambda_1, \lambda_2, \ldots, \lambda_n)$ is constant in every class of a certain classification of $\mathcal{G}$ in the form $1^{\lambda_1}, 2^{\lambda_2}, \ldots, n^{\lambda_n}$. By using a subgroup $\mathcal{N}$ of rotational subgroup isomorphism with skeletal symmetry $\mathcal{S}_n$ as $\mathcal{G}$, we have

$$\mathcal{C}_{\mathcal{N}} = \frac{1}{|\mathcal{N}|} \sum_{\lambda_1 + 2\lambda_2 + \cdots + n\lambda_n} h_{\mathcal{N}}(\lambda_1, \lambda_2, \ldots, \lambda_n) P_{v_1, v_2, \ldots, v_m}(\lambda_1, \lambda_2, \ldots, \lambda_n)$$

$|\mathcal{N}|$; order of $\mathcal{N}$ $\qquad$ (C.4)

and the number of isomers can be calculated by regarding every molecule obtained by a permutational operation $s \in \mathcal{N}$ as the same; $\mathcal{C}_{\mathcal{N}}$ corresponds to isomers which cannot be spatially superimposed.

On the other hand, by using a subgroup $\mathcal{S}$ of $\mathcal{S}_n$ which is isomorphic with the group of every element of skeletal symmetry, we obtain

$$\mathscr{C}_{\mathscr{S}} = \frac{1}{|\mathscr{S}|} \sum_{\lambda_1 + 2\lambda_2 + \cdots + n\lambda_n} h_{\mathscr{S}}(\lambda_1, \lambda_2, \ldots, \lambda_n) P_{v_1 v_2 \cdots v_m}(\lambda_1, \lambda_2, \ldots, \lambda_n) \quad (C.5)$$

Thus, by regarding molecules obtained by every permutational operation $s \in \mathscr{S}$ as the same, $\mathscr{C}_{\mathscr{S}}$ corresponds to the number of isomers, counting a pair of enantiomers as one. Therefore, when the skeletal symmetry, the number of ligands and the types of ligands are given, the number of achiral isomers produced by arrangement of the ligands, Z_a, and the number of chiral isomers, Z_c, are given by

$$\mathscr{C}_{\mathscr{N}} = Z_a + Z_c$$
$$\mathscr{C}_{\mathscr{S}} = Z_a + \frac{1}{2} Z_c \quad (C.6)$$

These equations may be rearranged as follows.

$$Z_a = 2\mathscr{C}_{\mathscr{S}} - \mathscr{C}_{\mathscr{N}}$$
$$Z_c = 2(\mathscr{C}_{\mathscr{N}} - \mathscr{C}_{\mathscr{S}}) \quad (C.7)$$

CHAPTER 3

Nomenclature for Chirality and Prochirality

3.1 NOMENCLATURE FOR CHIRAL MOLECULES

Many attempts have been made to develop a notation suitable for the description of stereoisomers, and these have usually reflected the level of progress of stereochemistry at the time of their inception. Initially the direction of optical rotation by a compound was used to describe it, i.e., (+), (−) or *d*, *l*. However, the direction of optical rotation is often dependent on the conditions of measurement, and in addition, no consistent relationship could be found between the direction of optical rotation and stereo structure.

Rosanoff[1] attempted to overcome this difficulty by proposing that the configurations of hydroxy acids or amino acids which are chemically derived from an enantiomer of glyceraldehyde (A) arbitrarily designated as *d* should also be designated as *d*.

```
                                    CHO
                                     |
                                 H—C—OH
                                     |
                                HO—C—H
                                     |
            CHO                  H—C—OH
             |                       |
         H—C—OH                  H—C—OH
             |                       |
          CH₂OH                   CH₂OH

    d-glyceraldehyde (A)         d-glucose (B)
```

Later, Fischer arbitrarily chose the structure (B), the so-called Fischer projection for dextrorotatory glucose, and since that time, *d*-glucose has been used as a standard structure for the description of configuration.

By 1940, these descriptions were still in use, though the designations (+) and (−) were often added to *d* and *l* to distinguish them from the earlier usage of *d* and *l* referring simply to the direction of optical rotation. In the late 1940's, *d* and *l* used in the former, absolute sense were replaced by D and L, respectively. For a time, all three systems were in use for the description of enantiomers, causing great confusion.

The description of configuration by means of the D,L system is convenient for the homologs of amino acids and sugars, but presents problems with compounds having two asymmetric centers, one relating to sugars and the other to amino acids. For example, threonine has two such asymmetric centers and can be described in terms of D,L (amino acid) or L,D (sugar).

$$
\begin{array}{ccc}
\text{COOH} & & \\
| & & \\
\text{H}_2\text{N–C–H} & \text{CHO} & \text{COOH} \\
| & | & | \\
\text{H–C–OH} & \text{H–C–OH} & \text{H–C–NH}_2 \\
| & | & | \\
\text{CH}_3 & \text{CH}_2\text{OH} & \text{CH}_2\text{OH} \\
\text{L-threonine} & \text{D-glyceraldehyde} & \text{D-serine}
\end{array}
$$

To avoid confusion, the International Union of Chemistry proposed the nomenclature D_g, L_g and D_s, L_s based on D-glyceraldehyde (sugar) and D-serine (amino acid) as arbitrary standards, respectively.

In 1951, Bijvoet determined the absolute configuration of rubidium D-(+)-tartrate by X-ray crystallography and showed that the configurational standard arbitrarily chosen by Fischer was in fact coincident with the structure determined by X-ray crystallography.

However, the D,L nomenclature system still has difficulties. For instance, though D-malic acid can be derived from D-glyceraldehyde by conventional methods, it is also possible to obtain L-malic acid, so that there is ambiguity in determining the absolute configuration by correlation with the standard compound, and the problems are even greater with complex compounds.

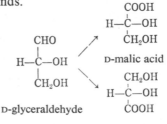

In 1956, dissymmetric compounds were discovered, and the need for a better system of nomenclature became pressing. In the same year, Cahn, Ingold and Prelog[2] proposed the *R,S* system, which not only permitted the stereo structure of an asymmetric compound to be determined directly from the designation, but was also applicable to dissymmetric compounds.

Even the *R,S* system has some disadvantages, though they are not as grave as those of the D,L system. For instance, the *R,S* system may change the designation of a reaction product in a reaction which shows retention of configuration, so that it is necessary to apply the system separately to all compounds which participate in the reaction when using this scheme.

$$CH_3COOH + HOCH_2 - \overset{\overset{\displaystyle CH_2OCH_3}{|}}{\underset{\underset{\displaystyle CH_3}{|}}{C}} - H \longrightarrow CH_3COOCH_2 - \overset{\overset{\displaystyle CH_2OCH_3}{|}}{\underset{\underset{\displaystyle CH_3}{|}}{C}} - H$$

(*R*)-3-methoxy-2-methylpropylalchohol   (*S*)-3-methoxy-s-methylpropyl acetate

Thus, in some respects the D,L system is more useful for describing asymmetric reactions, while the *R,S* system is better for describing the configurations of chiral compounds. Both systems are used in this book.

3.1.1 Nomenclature for using the *R,S* system

The *R,S* system was designed for use with tetrahedral chiral molecules, i.e., molecules with T_d skeletal symmetry and having four different ligands. However, the system can also be applied to chiral molecules with other skeletons.

Chiral molecules with skeletal symmetry such as T_d are said to have central chirality, and their configurations can be specified by the basic rules of the *R,S* system. Chiral molecules which do not have central chirality are first classified into two groups according to their structural appearance, i.e., axial chirality and planar chirality. Allenes and atrop isomers are included in the former group, while molecules such as that shown in the example below (section C) are in the latter. The *R,S* configurations of these groups can be specified by means of one additional rule for each group. The chirality of octahedral compounds can also be specified by the *R,S* system, but this will not be described here.

A. Specification of the central chirality of tetrahedral compounds

The sequence order of ligands is determined by means of the following rules (sequence rules).
1) Higher atomic number takes precedence.
2) Higher atomic mass number takes precedence.
3) Seq*cis* precedes seq*trans*.
4) Like pairs (*R,R* or *S,S*) precede unlike pairs (*R,S* or *S,R*).

5)   *R* and *M* precede *S* and *P*, respectively.

6)   The sequences of molecules containing doubly and triply bonded atoms are determined after adding one or two replica atoms, respectively, to the multiply bonded atoms. For example, $>C=O$ is considered as

$$>\underset{\substack{| \\ (O)}}{C}\!\!-\!\!\underset{\substack{| \\ (C)}}{O} .$$

7)   The sequence of a molecule having resonance structures is determined after the addition of replica atoms based on the classical formula, as shown below for furan.

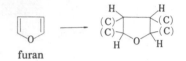

furan

In heterocyclic molecules, the nature of the replica atom may differ according to the direction in which the molecular structure is investigated, and in this case the sequence is determined using the average atomic number of the two different atoms. For example, the replica atoms at C-2 or C-6 of pyridine are different according to the direction of double bonding, so the sequence is determined using the average atomic number $(6 + 7)/2 = 6.5$ for C-2 and C-6.

8)   The configuration of a chiral center at an atom which is less than quadrivalent is determined after the addition of a phantom replica atom

of atomic number 0, e.g., $-\underset{\substack{| \\ O}}{\overset{\substack{(0) \\ |}}{S}}-$.

Using these sequence rules, atoms and groups can be ordered as shown in Table 3.1. When this is done, the designation *R* or *S* is decided by the following procedure. The tetrahedral molecule is viewed from the side remote from the ligand of lowest priority, as shown in Fig. 3.1, where the

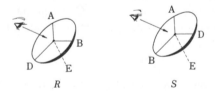

R   S

Fig. 3.1.   Determination of the configuration of a tetrahedral molecule with four different ligands.

order of priority is A > B > C > D, and the remaining ligands are counted from that of highest priority to that of lowest. If this counting is in a clockwise direction, the configuration is designated *R*, and if it is anticlockwise, the configuration is *S*.

B. Specification of axial chirality

The sequence rules for central chirality together with the following subrule are used to describe axial chirality. For example, the configuration of the compound shown below, together with its diagrammatic representation, is determined as follows. The representation is observed through the X–Y axis. The ligands nearest the observer precede those on the far side (see Table 3.1).

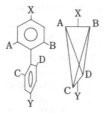

TABLE 3.1
Order of atoms and groups according to the sequence rules[3]

1	Hydrogen	26	3, 5-Xylyl	51	Dimethylammonio
2	Methyl	27	*m*-Nitrophenyl	52	Diethylamino
3	Ethyl	28	3, 5-Dinitrophenyl	53	Trimethylamino
4	*n*-Propyl	29	1-Propynyl	54	Phenylazo
5	*n*-Butyl	30	*o*-Tolyl	55	Nitroso
6	*n*-Pentyl	31	2, 6-Xylyl	56	Nitro
7	*n*-Hexyl	32	Trityl	57	Hydroxy
8	*iso*-Pentyl	33	*o*-Nitrophenyl	58	Methoxy
9	*iso*-Butyl	34	2, 4-Dinitrophenyl	59	Ethoxy
10	Allyl	35	Formyl	60	Benzyloxy
11	*neo*-Pentyl	36	Acetyl	61	Phenoxy
12	2-Propynyl	37	Benzoyl	62	Glycosyloxy
13	Benzyl	38	Carboxy	63	Formyloxy
14	*iso*-Propyl	39	Methoxycarbonyl	64	Acetoxy
15	Vinyl	40	Ethoxycarbonyl	65	Benzoyloxy
16	*sec*-Butyl	41	Benzyloxycarbonyl	66	Methylsulfynyloxy
17	Cyclohexyl	42	*tert*-Butoxycarbonyl	67	Methylsulfonyloxy
18	1-Propenyl	43	Amino	68	Fluoro
19	*tert*-Butyl	44	Ammonio	69	Mercapto
20	*iso*-Propenyl	45	Methylamino	70	Methylthio
21	Acetylenyl	46	Ethylamino	71	Methylsulfinyl
22	Phenyl	47	Phenylamino	72	Methylsulfonyl
23	*p*-Tolyl	48	Acetylamino	73	Sulfo
24	*p*-Nitrophenyl	49	Benzoylamino	74	Chloro
25	*m*-Tolyl	50	Benzyloxycarbonylamino	75	Bromo
				76	Iodo

Accordingly, if the molecule is observed from X, the order of ligand priorities is taken as A > B > C > D (if the molecule were to be viewed from Y, the order would be C > D > A > B). Next, the ligand of lowest priority is placed at the side remote from view and the remaining three ligands are counted from that of highest priority to that of lowest. As before, if this counting is in a clockwise direction, the configuration is designated as R, and if it is anticlockwise, the configuration is S.

C. Specification of planar chirality

Planar chirality is defined by means of the sequence rules for central chirality together with the following additional rules.
1)   Determine the chirality plane, which should be a natural plane of the molecule, e.g., the plane of the benzene ring in the example shown below.

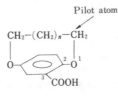

2)   Choose the atom of highest priority which is bound to the atom in the chiral plane as a pilot atom.
3)   Number the atoms in the chiral plane starting from the atom bound to the pilot atom and proceeding to the atom of next highest priority within the chiral plane.
4)   If the counting direction as observed from the pilot atom is clockwise, the molecule is designated as R, and if it is anticlockwise, as in the example shown above, then the configuration is S.

3.1.2   Nomenclature for conformational isomers

The historical nomenclature for conformational isomers, based on Newmann projections of the molecules, will be described first. Various terms have been used to express the spatial relationship of specific atoms or groups. These may be grouped as follows.
1)   *Anti, trans,* staggered: the conformation in which the two atoms or groups are on opposite sides in the Newmann projection.
2)   *Syn, cis,* skew: the conformation in which the two atoms or groups are on the same side in the Newmann projection.

However, the system presented by Cahn, Ingold and Prelog is more comprehensive. It is also based on the Newmann projections of molecules, and the designation is decided by means of the following rules.
1)   Representative atoms are chosen in the following way.

(a)   When the three atoms or groups in the same set are not identical, the atom or group with the highest priority is chosen.

(b)   When two atoms or groups in a set are identical, the remaining one is chosen.

(c)   When all three are identical, the atom or group at the minimal torsional angle with respect to another representative atom or group is chosen.

2)   As shown in Fig. 3.2(a), when two representative atoms or groups are located in the areas of $+30°$ to $-30°$ and $+150°$ to $-150°$ in the Newmann projection, the conformational isomer is specified as periplanar (p). When they are located in the areas of $+30°$ to $+150°$ and $-30°$ to $-150°$, the isomer is specified as clinal (c).

3)   As shown in Fig. 3.2(b), when the representatives are located in the same side, the isomer is specified as *syn* (s), and when they are on different sides, as *anti* (a).

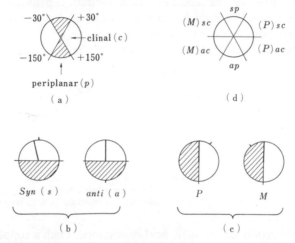

Fig. 3.2.   Nomenclature for conformational isomers according to Cahn, Ingold and Prelog.

4)   When the two representatives are observed from front to back according to their priority, if the remote representative atom or group appears in the clockwise sector, the isomer is designated as P or $(+)$, while if it appears in the anticlockwise sector, the isomer is designated as M or $(-)$, as shown in Fig. 3.2(c). These terms can be used in the specification of helicity.

Table 3.2 and Fig. 3.2(d) summarize the above rules for isomer nomenclature based on the torsional angles of substituents.

TABLE 3.2
Nomenclature for conformational isomers

Torsional angle	Term	Abbreviation	Historical abbreviation
$0° \pm 30°$	*syn*-periplanar	*sp*	*cis*
$+60° \pm 30°$	$+syn$-clinal	$(P)sc$	skew, *gauche*
$+120° \pm 30°$	$+anti$-clinal	$(P)ac$	—
$180° \pm 30°$	*anti*-periplanar	*ap*	*trans*
$-120° \pm 30°$	$-anti$-clinal	$(M)ac$	—
$-60° \pm 30°$	$-syn$-clinal	$(M)sc$	skew, *gauche*

3.2 PROCHIRALITY

There are many achiral molecules which can be converted to chiral molecules by a single operation such as replacement of a ligand with a different one, or the addition of a further substituent. For example, propionic acid is achiral because it has two hydrogen atoms, H^1 and H^2, which are positioned symmetrically with respect to the plane of symmetry, as shown in Fig. 3.3. However, if one of the hydrogens is replaced by a

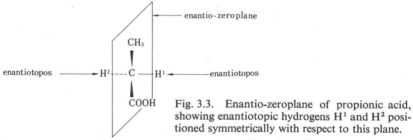

Fig. 3.3. Enantio-zeroplane of propionic acid, showing enantiotopic hydrogens H^1 and H^2 positioned symmetrically with respect to this plane.

hydroxyl group, chiral lactic acid is obtained. Such a topological property is termed prochirality, and molecules having such a property are termed prochiral molecules. The carbon atom involved is similarly known as a prochiral center (i.e., C-2 of propionic acid is a prochiral center).

Another example of a prochiral molecule is acetaminomalonic acid (again, C-2 is the prochiral center), since the two carboxyl groups are positioned symmetrically with respect to the plane of symmetry.

acetaminomalonic acid   acetaldehyde

In the case of acetaldehyde, the central carbon is sp^2 bonded and the molecule is planar, so that the molecule itself can be regarded as a plane of symmetry. Thus, if a fourth ligand is added, a chiral product can be obtained. Such a topological property is termed sp^2-prochirality. A carbon atom of this type is referred to here as an sp^2-prochiral center.

3.2.1 Enantiotopic relationships

If a molecule contains a prochiral center or an sp^2-prochiral center, the plane of symmetry which includes this center will be referred to as an enantio-zeroplane in this book. Objects which are positioned symmetrically with respect to this plane are said to be in an enantiomeric relationship, and the locations of such objects are said to be in an enantiotopic relationship. A location in an enantiotopic relationship is called an enantiotopos, and the object on each location is called an enantiotopic object. The relationship of the two sides of the enantio-zeroplane including the sp^2-prochiral center is enantiotopic and each of the two sides of this plane is called an enantioface in this book. These terms will be explained below by means of various examples.

A. Enantiotopos

If one of the two hydrogen atoms H^1 and H^2 attached to the prochiral center in propionic acid (Fig. 3.3) is replaced by an atom or group other than H, CH_3 or COOH, an optically active enantiomer is produced. Replacement of the other hydrogen atom instead yields the other enantiomer. These two hydrogens are therefore said to be in an enantiotopic relationship, and are enantiotopic hydrogens. The plane including C-1, C-2 and C-3 is the enantio-zeroplane. Similarly, a chiral sulfoxide is produced by oxidation of one of the sulfur lone pairs of an unsymmetrical thioether, and thus two lone pairs of electrons in such a situation are in an enantiotopic relationship.

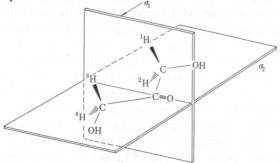

Fig. 3.4. A molecule divided into two identical prochiral regions by an unsaturated carbon. The molecule has two types of enantio-zeroplane, σ_1 and σ_2, and four pairs of enantiotopic hydrogens (see text).

A molecule which is divided into two identical prochiral regions by an unsaturated carbon has two types of enantio-zeroplane, as shown in Fig. 3.4. In this case there are four pairs of hydrogen atoms in enantiotopic relationships, i.e., $H^1 \parallel H^2$, $H^3 \parallel H^4$, $H^1 \parallel H^3$ and $H^2 \parallel H^4$.

B. Enantioface

The two sides of the molecular plane of acetaldehyde are symmetrical with respect to the molecular plane, as shown in Fig. 3.5. However, if the

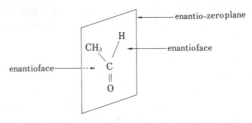

Fig. 3.5. Enantio-zeroplane and enantiofaces of a planar sp^2 bonded molecule.

acetaldehyde molecule is deuterated, two enantiomeric deuteroethanols are produced depending on which side of the molecular plane is attacked. The plane containing the sp^2 bond and the sp^2-prochiral center is the enantio-zeroplane, and the two sides of this plane are enantiotopic faces, and are referred to as an enantioface.

3.2.2   Diastereotopic relationships

In a molecule containing a chiral center and a prochiral or sp^2-prochiral center, if the enantio-zeroplane is placed under the influence of the chiral center, it is referred to as a diastereo-zeroplane in this book. As before, objects which are positioned symmetrically with respect to the diastereo-zeroplane are said to be in a diastereomeric relationship. The locations of these objects are in a diastereotopic relationship and such a location is known as a diastereotopos. If the diastereo-zeroplane contains an sp^2-prochiral center rather than a prochiral center, the two sides of the plane are known as diastereofaces. There are two different modes of diastereo-zeroplane. One arises in a molecule where both the prochiral center and the chiral center are in a considerably rigid structure, and is called a configurational plane; the other arises in a molecule which has no rigid structure between the prochiral and chiral centers, and is called a conformational plane (see Fig. 3.8, below). These various terms are explained in more detail next by means of examples.

A. Diastereotopos

The two hydrogens H^1 and H^2 attached to the prochiral center in the chiral molecule shown in Fig. 3.6 give the *trans* and *cis* methylated products, respectively, with respect to the chiral center if one of them is replaced by a methyl group, i.e., a diastereomer is produced. Hydrogen or other

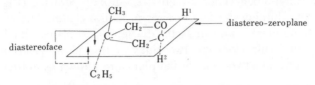

Fig. 3.6. Diastereo-zeroplane of a molecule containing a chiral and a prochiral center, showing the diastereotopic hydrogens H^1 and H^2.

atoms or groups which produce diastereomers when replaced in this way are known as diastereotopic. Their locations are in diastereomeric relationship and such a location is called a diastereotopos.

Diastereotopic ligands can exist in a molecule which has no rigid structure between the chiral center and the prochiral center, e.g., H^1 and H^2 in 3-methylvaleric acid

$$\underset{\underset{H}{|}}{\overset{\overset{CH_3}{|}}{C_2H_5-C}}-\underset{\underset{H^2}{|}}{\overset{\overset{H^1}{|}}{C}}-COOH \qquad \text{3-methylvaleric acid}$$

Fig. 3.7 shows a special example of a compound having two planes of symmetry. It consists of a prochiral center linked to two other prochiral centers, and yields a diastereomer on replacement of any one of the four

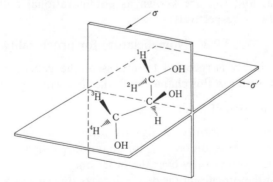

Fig. 3.7. A compound having a prochiral center linked to two other identical prochiral centers. There are two types of diastereo-zeroplane, σ and σ', and four pairs of diastereotopic hydrogens (see text).

hydrogens H^1 through H^4. Such a substitution forms two chiral centers simultaneously. One of the planes of symmetry (σ') includes C-1, C-2 and C-3, and the other, which is perpendicular to the first, contains H and OH (σ). Thus, there are four pairs of diastereotopic hydrogens, H^1/H^2, H^3/H^4, H^1/H^3 and H^2/H^4.

B. Diastereoface

3-Ethyl-6-benzyliden-3-methyl-2,5-dioxopiperazine (Fig. 3.8(a)) contains one chiral center and one sp^2-prochiral center. The molecular plane containing these is known as the diastereo-zeroplane, and hydrogenation of the molecule from the two sides of this plane yields a pair of diastereomers. The two sides of the plane are known as diastereofaces.

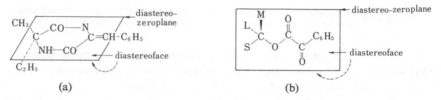

(a)   (b)

Fig. 3.8.   (a) Diastereo-zeroplane   of   3-ethyl-6-benzyliden-3-methyl-2,5-dioxopiperazine. This molecule contains configurational diastereofaces (see text). (b) Statistical diastereo-zeroplane of a non-rigid molecule. This molecule has conformational diastereofaces.

The molecule shown in Fig. 3.8(b) does not have a rigid molecular plane containing the chiral and sp^2-prochiral centers. However, as the molecule rotates internally such a plane is formed in a statistical sense, and the enantio-zeroplane therefore becomes a diastereo-zeroplane. Diastereofaces of rigid molecules and of the statistical type, shown in Fig. 3.8(a) and (b), are known as configurational and conformational diastereofaces, respectively.

3.2.3   Nomenclature for prochirality

Prochirality is specified by means of the general rules for chirality described earlier in this chapter with the following additional subrules.

A. Specification of Enantiotopos

1)   The relative priority of ligands follows the rules given in section 3.1.1(A) with the addition that for enantiotopic ligands, the ligand to be specified is given priority over the other one.
2)   When the prochirality is determined as described above, the notation pro-R or pro-S is used in place of R and S.
The expressions H_R and H_S are used in some cases

B. Specifications of Enantioface

The enantioface which is to be specified is taken as facing the observer. The priority of the three ligands of the sp^2-prochiral center is determined according to the general rules. The priority of ligands bonded to a center with a double bond is determined by means of rule 6 in section 3.1.1(A). When the order of the ligands is thus determined to be clockwise or anti-clockwise, the enantioface is designated as *re* or *si*, respectively, as shown below:

C. Specifications of Diastereotopos

A diastereotopos is similar to an enantiotopos, and is specified and described in a similar way.

D. Specifications of Diastereoface

A diastereoface is similar to an enantioface, and is specified in a similar way. For example, the upper sides of the compounds shown in Fig. 3.8 are both designated as *si*. In the case of a compound with two sp^2-prochiral centers on a double bond, as shown below, the two prochiral centers are expressed according to their relative priorities.

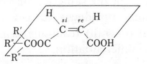

For example, the diastereoface on the upper or lower side of the compound shown above is *si-re* or *re-si*, respectively, because C-2 is prior to C-1 according to the sequence rule.

REFERENCES

1. *Cf.* M. Rosanoff, *J. Am. Chem. Soc.,* **28,** 114, (1906).
2. R. S. Cahn, C. K. Ingold and V. Prelog, *Angew. Chem. Intern. Ed. Engl.,* **5,** 385 (1966).
3. M. Nakazaki, *Kagaku Sosetsu* (Japanese) (ed. Chem. Soc. Japan), vol. 4, p. 13, Tokyo University Press, 1974.

CHAPTER 4

Nomenclature for Stereo-Differentiating Reactions

4.1 FUNDAMENTALS

The new concept of "differentiation" introduced in this book is indispensable for presenting a rational explanation of asymmetric reactions based on the well-established concepts of chemical reactions, i.e. in terms of the reagent (or catalyst), substrate and product. Essentially, it applies to the structural relationships involved in the reaction process, and emphasizes the roles of reagent and catalyst to carry out the reaction. For the purposes of the simple classification of observed, overall reactions, it is thus applied to the structural relationships between substrate and product. On the other hand, it also offers a new framework within which to describe the reaction mechanism, i.e., reaction steps, since these can be interpreted on the basis of the same terminology.

The present chapter introduces our new classification and nomenclature for the apparent (overall) reactions, while Chapter 7 considers the actual mechanisms of differentiation.

4.2 DEFINITION OF STEREO-DIFFERENTIATING REACTIONS

4.2.1 Reaction character and stereo-differentiation

Historically, asymmetric reactions have been associated with living systems (e.g., enzymatic reactions) and have been explained in terms of rather vague concepts. Indeed, the stereochemical characteristics of enzymatic reactions, particularly as regards the proper choice of substrate and precise mode of product formation, have been considered as something of a biological mystery. Thus, the character of the reactions has been explained in terms somewhat removed from those of general organic chemistry. We will begin by discussing the widely used terms "stereospecific reactions" and "stereoselective reactions", although this nomenclature has given rise to considerable confusion.

In the broadest sense, a sterospecific reaction is understood as one in which the stereo structure of the substrate influences that of the product to some extent. Examples are shown in Figs. 4.1 and 4.2. In the strict

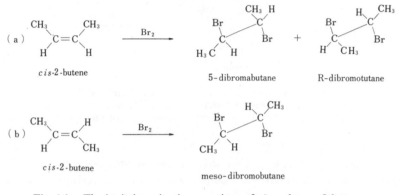

Fig. 4.1.   The ionic bromination reactions of *cis* and *trans* 2-butene.

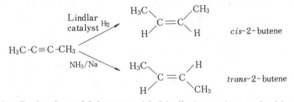

Fig. 4.2.   Reduction of 2-butene with Lindlar's catalyst and with sodium in liquid ammonia, yielding the *cis* and *trans* products, respectively.

sense, a stereospecific reaction is defined as a reaction in which only a specific isomer reacts, again in such a way that its stereo structure influences that of the product. The reaction in Fig. 4.3 gives an example of a

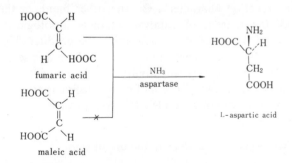

Fig. 4.3.   Conventional scheme for the reaction of aspartase on fumaric acid. Maleic acid does not react.

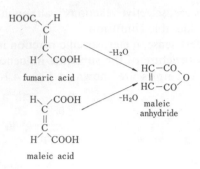

Fig. 4.4. Formation of maleic anhydride from maleic acid and fumaric acid. For steric reasons fumaric acid cannot form a *trans* anhydride, and under severe conditions it yields maleic anhydride via isomerization to maleic acid.

stereospecific reaction in this strict sense. A stereoselective reaction is defined as one in which one specific stereoisomer is produced to a greater extent than the other. An example is shown in Fig. 4.4.

These definitions are not derived from any strict principle for evaluating the reactions according to the nature of transformation of steric structure, and there is much divergence of opinion over how such reactions should be best understood and distinguished. In general, the substrate structure is transformed into the product structure by two different elemental processes. However, such a distinction is difficult to make without special information, such as that derived from the concepts presented in this book. In one reaction, the determining factor is not the particular steric structure of the substrate but resides in chemical structures such as Cl–$\overset{|}{\underset{|}{C}}$– or $\rangle C = C \langle$, and the reaction is governed by the nature of the reagent or catalyst, e.g. the substitution in steric inversion or retention and *cis* or *trans* addition. In this sense, the products are determined by the "reaction character". On the other hand, in the other elemental process, the reagent or catalyst interacts topologically with the steric structure of the substrate to determine the product. This latter process is designated by us as "stereo-differentiation", and is seen as a consequence of the "stereo-differentiating ability" of the reagent or catalyst.

This fundamental concept of stereo-differentiation is further explained below with the aid of examples. It also represents the central theme of the present book.

A. Stereospecific reactions in the broad sense

A special feature of the reactions shown in Figs. 4.1 and 4.2 is that the reactions do not depend on the steric structure of the substrates, but on the reaction character, i.e., the *trans* addition of bromine ions, *cis* hydrogenation with the Lindlar catalyst, *trans* reduction with sodium in liquid ammonia, etc. In these cases, the structure of the substrate is irrelevant as long as it contains a double bond. Thus, to term such reactions as "stereospecific" is not very meaningful.

B. Stereospecific reactions in the narrow sense

The reaction of aspartase shown in Fig. 4.3 can be divided into three elemental processes (Fig. 4.5), i.e., the differentiation of fumaric acid and other substances, the differentiation of the *si* and *re* faces of the fumaric

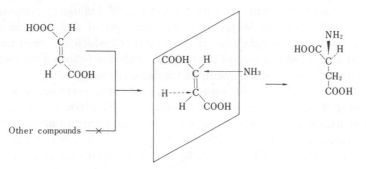

Fig. 4.5.   Elemental processes of the aspartase reaction in Fig. 4.3.

acid, and the *trans* addition of ammonia to the double bond. Regarding the first step, aspartase is not active on other substrates in addition to maleic acid, and so its differentiating ability is not simply a geometrical isomer-differentiating ability but a more general ability known as substrate specificity. The second step is, as mentioned, the stereo-differentiation of the *si* and *re* faces of fumaric acid. The third step, the *trans* addition reaction of ammonia, is essentially determined by the reaction character of ammonia.

C. Stereoselective reactions

The term "stereoselective" implies the more or less exclusive production of a certain product, as in Fig. 4.4. However, the maleic anhydride in this case is not obtained by any differentiating process: it is formed as a result of a thermodynamically controlled reaction. Thus, the factor governing this reaction is not meaningfully stereoselective, but it is in fact a case of thermodynamic control.

As discussed above, use of the terms "stereospecific reaction" and "stereoselective reaction" tends to confuse any discussion of the mechanisms involved in stereochemical reactions. Thus, we will therefore drop these terms.

D. Substrate-specific reactions

The term "substrate-specific reaction" is widely used in biochemistry for reactions which are catalyzed by enzymes having quite specific differentiating abilities. This term thus has a precise and objective meaning, and will be retained in this book.

4.2.2 Historical definitions of asymmetric reactions

Studies of asymmetric reactions developed from a desire to produce optically active compounds without the use of biological systems. The earliest definition, by Marckwald, was very broad and covered all the attempts to produce optically active compounds which had been made up to that time. Modifications of the original definition followed later with the development of analytical methods for diastereomers.

In 1904, Marckwald simply proposed that asymmetric synthesis included those reactions which produce optically active substances from symmetrically constituted compounds with the intermediate use of optically active materials but with the exclusion of all analytical processes. A typical example yielding optically active phenylalanine is shown below (Eq. 4.1).

$$\underset{CH_3}{\overset{C_2H_5}{\diagdown}}C\underset{NH-CO}{\overset{CO-NH}{\diagup}}C=CH-C_6H_5 \xrightarrow{H_2} \underset{CH_3}{\overset{C_2H_5}{\diagdown}}C\underset{NH-CO}{\overset{CO-NH}{\diagup}}CH-CH_2-C_6H_5 \qquad (4.1)$$

$$\xrightarrow{H_2O} C_6H_5-CH_2-\underset{NH_2}{\overset{|}{CH}}-COOH$$

phenylalanine

Marckwald excluded analytical operations such as optical resolution from his definition, but otherwise there was no restriction as to substrate, reagent or reaction process.

Later, Morrison and Mosher redefined an asymmetric reaction as a reaction in which an achiral unit in an ensemble of substrate molecules is converted by a reactant into a chiral unit in such a manner that the stereoisomeric products are produced in unequal amounts. Thus, in this definition, the requirements for an asymmetric reaction are the production of a chiral center from a prochiral center and the production of stereoisomers in unequal amounts. In a sense, this definition was based on the development of instrumental techniques which permitted the separation and analysis of different diastereomers. However, even the improved definition of Morrison and Mosher fails to provide a final and unambiguous rule for asymmetric reactions since it lacks a concept related to whether the reaction produces diastereomers or enantiomers, and it uses operational term "asymmetric synthesis" for the definition of transformation of stereostructure.

Prior to the definition of Morrison and Mosher, the present authors described a basic new system of classification to deal with the phenomena observed during hydrogenation reactions with asymmetrically modified Raney nickel catalysts.[1,2] This system was more in line with the classifications of general organic chemistry, and the classification scheme[3] developed on the basis of such studies will be described below.

4.2.3 New definition and classification for reactions of related asymmetic reactions

Our new classification is based on the newly introduced term "differentiation". It includes asymmetric reactions as defined by Morrison and Mosher, as well as kinetic resolutions of enantiomers and diastereomers. Special reactions which are closely related to differentiating reactions are dealt with in section 5.3. The classification is always decided simply on the basis of the apparent structural relationship between the substrate and product, and does not take into account the reaction mechanism (see Chapter 5).

Six categories of stereo-differentiating reactions[†] are distinguished by us, as follows.

1) When the chirality participating in the differentiation occurs in a reagent, the catalyst or the reaction medium, the reaction is classified as an enantio-differentiating reaction. Typically, it yields enantiomers as product.

[†] We exclude the racemic modification of a substrate with a stereochemically labile chiral center by reaction with an optically active reagent. In such a case, conversion to the diastereomer with the most favorable ground-state free energy occurs, so that the chiral center in the substrate ceases to be racemic. A reaction of this type is known as a first-order asymmetric transformation.

Similarly, we exclude cases where the solubility of one diastereomer in a solvent is less than that of the other, so that deposition of the less soluble isomer occurs until no more isomer deposits. Such a transformation is known as a second-order asymmetric transformation.

These asymmetric transformations apparently resemble stereo-differentiating reactions in that stereoisomers are produced in unequal amounts under the influence of chiral factors, but in fact these are thermodynamically controlled reactions, and there is no clear face-, topos- or isomer-differentiating process operating.

2) When the chirality related to the differentiation is present in the substrate, the reaction is classified as a diastereo-differentiating reaction. Typically, it yields diastereomers as product.

3) When the differentiation occurs at an sp^2-prochiral center, prochiral center or chiral center, the reaction is considered to be face-differentiating, topos-differentiating or isomer-differentiating, respectively.

The six reaction types are thus as follows:

   i) Enantio-differentiating reactions
        enantioface-differentiating reaction
        enantiotopos-differentiating reaction
        enantiomer-differentiating reaction
  ii) Diastereo-differentiating reactions
        diastereoface-differentiating reaction
        diastereotopos-differentiating reaction
        diastereomer-differentiating reaction

As shown in Fig. 4.6, face differentiation always introduces chirality

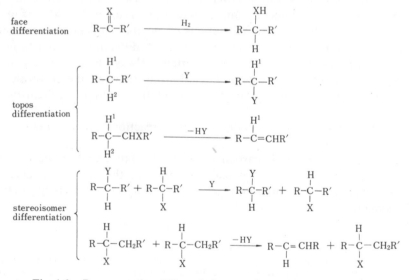

Fig. 4.6. Representative differentiating reactions. The moiety R must be chiral in the case of diastereo-differentiating reactions.

into the product. Topos-differentiating reactions can be divided into two types: one is substitution, introducing chirality into the product, and the other is elimination, preferentially removing either pro-R or pro-S substituents. Isomer-differentiating reactions can be similarly divided into those involving substitution and those involving elimination.

4.3 GENERAL DISCUSSION ON STEREO-DIFFERENTIATING REACTIONS

Stereo-differentiating reactions are not only valuable for the synthesis of isomers, but also for understanding the overall reaction mechanisms, since the products naturally reflect the steric interaction between the substrate and reagent.

4.3.1 Enantio-differentiating reactions

Enantio-differentiating reactions are those in which differentiation is produced by a chirality in the reagent or environment of the reaction system, and enantio-differentiating ability is defined as the ability of the reagent to attack one of the enantiofaces or enantiotopos in the substrate preferentially. Since almost complete enantio-differentiation is often found in enzyme reactions, such reactions have been described as biochemical asymmetric synthesis by some authors. They have also been described as absolute asymmetric synthesis in the broad sense. (In the strict sense, as used in this book, an absolute asymmetric reaction is one in which differentiation arises solely as a result of chiral physical force.)

In the typical enantio-differentiating reaction, it is required that the reagent or catalyst be optically active: enantiomers form the reaction product, and the differentiating ability is assessed according to the ratio of enantiomers produced. There is generally no way to determine the differentiating ability unless the reaction is performed with an optically active reagent, catalyst or reaction medium. However, in the reaction with reagent, where the product is obtained as diastereomers, optically active reagent is not required insofar as the determination of the differentiating ability is concerned.

Even when a substrate contains a chiral center besides a prochiral or chiral center where differentiation takes place, the reaction is referred to as enantio-differentiation if it occurs under the control of a chirality in the reagent, catalyst or reaction medium. However, when the substrate contains a chiral center and an sp^2-prochiral or prochiral center reacts with the optically active reagent or catalyst or the achiral reagent in the presence of optically active catalyst, the resulting differentiation is usually a combination of enantio-differentiation arising from the reagent, catalyst or medium and diastereo-differentiation induced by the chirality of the substrate itself (see sect. 8.2.5).

Enantio-differentiating reactions can be classified into three types, depending on whether a chiral reagent, a chiral catalyst or a chiral reaction medium is the controlling factor. Reactions with catalysts can be further subdivided into those with heterogeneous and homogeneous catalysts. Details of particular types of reactions will be given in Chapter 5.

For producing optically active materials, particularly on a large scale, enantio-differentiating catalytic reactions are obviously preferable, and these are a major goal of synthetic organic chemistry for optically active compounds.

The disadvantage of enantio-differentiation by the reagent, from an industrial point of view, is that one mole of optically active product is the most that can be obtained from one mole of optically active reagent. However, the mechanisms of such reactions are better understood than are the mechanisms of reactions with optically active catalysts, and they are therefore more useful for studies of the steric interactions of substrates and reagents and for determining the configurations of substrates or reagents by means of experimentally established relationships.

Since kinetic resolution is performed by enantiomer-differentiating reactions, these are considerably different in character from face- and topos-differentiating reactions. The time course of a typical isomer-differentiating reaction is shown in Fig. 4.7. As can be seen, differentiation is not apparent in the final stages of the reaction, and the apparent degree of differentiation changes throughout the reaction. This means that the differentiating ability must be evaluated in terms of differences in reaction rates or activation energies, which yield a qualitative, not quantitative, index of the extent of differentiation in related but nonidentical systems.

Enantiomer-differentiating reactions can be utilized for optical resolution or determination of the substrate configuration. However, in the former case, an extensive differentiating ability on the part of the reagent or catalyst is essential.

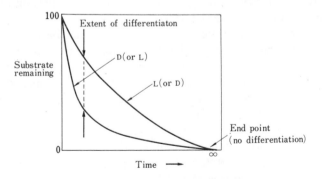

Fig. 4.7. Typical time courses for the enantiomers remaining in an enantiomer-differentiating reaction system. The apparent differentiation varies with time.

Quite apart from this, since the origin of optically active compounds can be said to have a close relation to the origin of life on the earth, knowledge of enantio-differentiating reactions will undoubtedly provide an important clue for understanding the true origin.

An enantio-differentiating reaction is the simplest system for investigating the recognition mechanism of molecular shape by a reagent, catalyst or enzyme, since the recognition of three sites on the substrate (ABC, ACD, BCD, or ABD in Fig. 4.8) by their correspondence with three sites on the reagent, catalyst or enzyme (abc, acd, bcd, or abd in Fig. 4.8) is sufficient to premit enantio-differentiation.

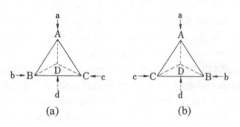

(a) (b)

Fig. 4.8. Substrate recognition by correspondence between recognition sites on the enzyme (a, b, c, d) and points on the substrate (A, B, C, D).

4.3.2 Diastereo-differentiating reactions

Reactions performed under the influence of chirality in the substrate molecule are referred to as diastereo-differentiating reactions. Typically, they yield diastereomeric products in unequal proportions. Since diastereomers have different physical and chemical properties, their proportions can be determined directly by current conventional methods. Accordingly, it is not always necessary to use optically pure substrates when studying diastereo-differentiating ability, and racemic substrates can be used. This is not the case with enantio-differentiating reactions, since the proportions of the products must usually be determined by measuring the optical rotation. However, for the purpose of preparing optically active compounds by this reaction, an optically active substrate is of course required.

In early work, it was always necessary to convert diastereomers to enantiomer derivatives in order to evaluate differentiating ability, and this has led to misunderstandings, e.g., that diastereo-differentiating reactions yield enantiomers rather than diastereomers as the final product. Marckwald in fact made no distinction, classifying all these reactions simply as asymmetric reactions, including of course the diastereoface-differentiating reactions and diastereotopos-differentiating synthesis.

Many investigators have paid attention only to the steric relation between the ligands linked to the chiral center of the substrate (R_1, R_2 and R_3 in Eq. 4.2) and the prochiral center.

$$
\begin{array}{c}
R_1 \quad\quad R_4 \\
R_2\text{--}\overset{A}{\underset{R_3}{C}}\text{--}\overset{\,}{C}{=}X
\end{array}
\longrightarrow
\begin{array}{c}
R_1 \quad\quad R_4 \\
R_2\text{--}\overset{A}{\underset{R_3}{C}}\text{--}\overset{B}{\underset{R_6}{C}}\text{--}R_5 + YZ
\end{array}
\longrightarrow
\begin{array}{c}
R_1 \quad\quad R_4 \\
R_2\text{--}\overset{A}{\underset{R_3}{C}}\text{--}Y + Z\text{--}\overset{B}{\underset{R_6}{C}}\text{--}R_5
\end{array}
\quad (4.2)
$$

However, the presence of chirality in the substrate molecule is not itself a sufficient condition for carrying out the differentiating reaction. Differentiation is achieved only when the substrate with a chiral center is made to react with a reagent that has sufficient diastereo-differentiating ability (see section 5.2.1B).

Two kinds of diastereo-differentiating reactions can be distinguished according to the mode of the diastereo-zeroplane on which the reagent reacts. One is the configurational type, found when the structure containing the chiral center is rigid (e.g., a ring structure), and the other is conformational, as in linear chiral molecules where the chiral and prochiral (or sp^2-prochiral) centers form a diastereo-zeroplane on a statistical basis. These two types of diastereo-zeroplane are illustrated in Figs. 4.9 and 4.10, respectively, which give typical models, and the reactions are known as configurational and conformational diastereo-differentiating reactions, respectively. The chiral effects involved are, likewise, termed configurational and conformational effects, respectively.

Diastereo-differentiating reactions are important not only for the synthesis of diastereomers but also in studies of interactions between substrate and reagent or catalyst in the transition state. The reactions are often useful for small-scale laboratory syntheses with high optical yield, since the mechanisms of the reactions have been extensively investigated

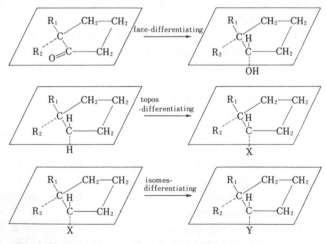

Fig. 4.9.   Examples of configurational differentiating reactions.

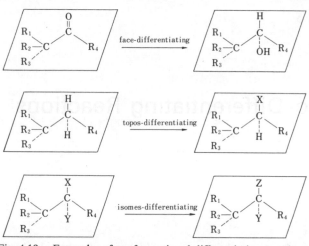

Fig. 4.10. Examples of conformational differentiating reactions.

and are generally fairly well understood. Diastereomers are in fact different compounds from the viewpoint of internal energy, and diastereomer-differentiating reactions are no different from competitive reactions between different compounds in a general sense.

REFERENCES

1. T. Tanabe, *Bull. Chem. Soc. Japan,* **46,** 1482 (1973).
2. Y. Izumi, *Angew. Chem. Intern. Ed. Engl.,* **10,** 871 (1971).
3. T. Izumi et al., *Kagaku Sosetsu* (Japanese), (ed. Chem. Soc. Japan), vol. 4, Tokyo University Press, 1974.

CHAPTER 5

Stereo-Differentiating Reactions

In this chapter, only reactions which are clearly valuable for understanding or for improving studies on stereo-differentiating reactions are considered as examples. Comprehensive reviews covering the majority of such reactions are available elsewhere.[1,2] The fundamental mechanisms of stereo-differentiating reactions are dealt with in Chapter 7.

In order to give the degree of stereo-differentiation, the optical yield (OY, %) as obtained by the polarimetric method is generally given. Stereochemical purity determined by analytical methods other than the polarimetric method is expressed as the enantiomer or diastereomer excess (e.e. or d.e., %) for enantiomers or diastereomers, respectively, instead of the optical purity (see sect. 8.2). In cases where the optical purity of the chiral factor[†] or the product cannot be determined since the specific rotation of the optically pure substance is unknown, or because of difficulties in enantiomer analysis, the optical purity or enantiomer excess, or the specific rotation, $[\alpha]$, or an observed value of the optical rotatory power, α, is used as a qualitative index of the degree of differentiation, respectively.

5.1 ENANTIO-DIFFERENTIATING REACTIONS

Since the chirality related to enantio-differentiating reactions is present in the reagent, catalyst or reaction medium, it is difficult to give a general treatment of the effects of chirality or substrate structure on the differentiation (or of the reaction mechanisms, as in the case of diastereo-differentiating reactions). Discussion of such effects (or of the reaction mechanisms) will thus be made according to groups of reactions that are closely related.

† A chiral moiety related to differentiation is called a chiral factor in this book.

5.1.1 Enantioface-differentiating reactions

Enantioface-differentiating reactions are reactions which produce enantiomers in different proportions from sp^2-prochiral compounds as a result of the action of chiral reagents, catalysts or reaction media. Each of these three groups will be discussed below.

A. Reactions with chiral reagents

a. Grignard reagents[3]

The use of optically active Grignard reagents to reduce sp^2-prochiral carbonyl compounds is well-known. Table 5.1 shows enantioface-differentiating reductions of alkylphenylketones by optically active Grignard

TABLE 5.1
Reduction of alkylphenylketones with Grignard reagents

Reagent / Substrate	(A)[4] C_2H_5 $HCCH_3$ CH_2MgCl		(B)[5] C_6H_5 HCC_2H_5 CH_2MgCl		(C)[6-9] $H_3C{-}CH_3$ $CH_3\,MgCl$		(D) $M(CH_2CH(CH_3)C_2H_5)_n$ $M{=}Be: n{=}2$ $M{=}Al: n{=}3$ Al[10,11]		Be[11]	
	OY	SY	OY	SY	OY	SY	OY	SY	OY	SY
$C_6H_5COCH_3$	4	37	47	19	36	55	8	83	—	—
$C_6H_5COC_2H_5$	6	59	52	89	19	50	13.2	75.6	14.8	88.5
$C_6H_5CO(n\text{-}C_3H_7)$	6	60	—	—	46	50	7	97	—	—
$C_6H_5CO(n\text{-}C_4H_9)$	6	61	—	—	52	44	—	—	—	—
$C_6H_5CO(iso\text{-}C_4H_9)$	10	89	53	63	—	—	—	—	—	—
$C_6H_5CO(iso\text{-}C_3H_7)$	24	78	82	69	55	80	30~45	61~93	46.2	97.5
$C_6H_5CO\text{-}\langle\;\rangle$	25	90	—	—	—	—	—	—	—	—
$C_6H_5CO(tert\text{-}C_4H_9)$	16	98	16	77	40~72	90	29.8	82.0	38.8	90.5

† OY: optical yield, SY; reaction yield

reagents, yielding alkylphenylcarbinols. The mechanism of these reactions has been explained in terms of a 6-membered cyclic transition complex, as shown in Fig. 5.1. As *1A* shows reduced steric hindrance compared to *1B*, the reaction proceeds more rapidly, so that more of product *2A* is produced than *2B*. This argument can be applied to many of the reactions, but the order of the bulk of the substituents is different in each case. Some of the reactions cannot be explained according to this order. For instance, in the reduction of *tert*-butylphenylketone or trifluoromethylketone, the relative bulk of the substituents is different in reactions with different Grignard reagents (see sect. 7.2.1).

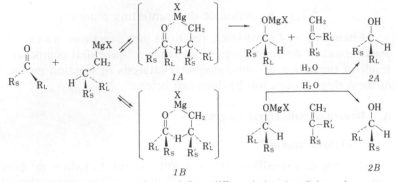

Fig. 5.1. Mechanism of enantioface differentiation by Grignard reagents as presented by Mosher *et al.* R_L, R'_L = larger groups; R_S, R'_S = smaller groups.

Table 5.1 shows that as the difference in the steric bulk of substituents on the chiral center increases and as the steric bulk of the alkyl substituent on the substrate increases, so the effectiveness of enantioface differentiation in the reaction increases. However, the size of the effects varies according to the reagents used and there does not seem to be any simple relationship between the nature of the substituents and the efficiency of differentiation.

Optically active Grignard reagents also differentiate between the enantiofaces of sp^2-prochiral alkenes. For example, (S)-alkanes are preferentially produced by the reduction of alkylidenecyanoacetic acid esters or alkylidenemalononitrile with (S)-2-methylbutylmagnesium chloride. According to Mosher's mechanism, the R isomer should be produced, as shown in Eq. 5.1. However, it was assumed that the reaction in fact proceeded by the mechanism shown in Eq. 5.2 due to electrostatic repulsion forces between the Grignard methylene group and the cyano group of the substrate. The factors affecting Grignard reactions are thus not fully understood at present.

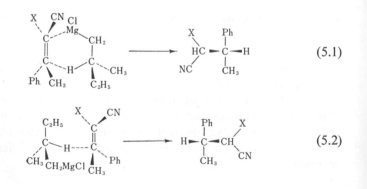

TABLE 5.2
Enantioface-differentiating reduction with optically active aluminium alkoxide

$$\begin{matrix} R^1 \\ R^2 \end{matrix}\!\!\diagdown\!\!CHOH + \begin{matrix} R^3 \\ R^4 \end{matrix}\!\!\diagdown\!\!CO \xrightarrow{\text{Al(OR)}_3} \begin{matrix} R^1 \\ R^2 \end{matrix}\!\!\diagdown\!\!CO + \begin{matrix} R^3 \\ R^4 \end{matrix}\!\!\diagdown\!\!CHOH$$

Reagent		Ketone		Optical yield (%)	Ref.
R^1	R^2	R^3	R^4		
CH_3	C_2H_3	CH_3	$iso\text{-}C_6H_{13}$	6	(12)
CH_3	$iso\text{-}C_3H_7$	CH_3	$cyclo\text{-}C_6H_{11}$	22	(12)
CH_3	$tert\text{-}C_4H_9$	CH_3	$n\text{-}C_6H_{13}$	6	(13)

TABLE 5.3
Enantioface-differentiating reduction with optically active alkoxymagnesium bromides[14]

$$\langle\!\bigcirc\!\rangle\text{-COCOOH} \xrightarrow{\text{R*-OMgBr}} \langle\!\bigcirc\!\rangle\text{-CHCOOH} \atop \underset{\text{OH}}{|}$$

	Product	
R*	Configuration	Optical yield (%)
(−)−methyl	S	33
(+)−neomenthyl	R	17
(−)−isobornyl	S	16

TABLE 5.4
Enantioface-differentiating reduction with optically active potassium alkoxides[15]

$$R^1COR^2 \xrightarrow{\text{R*OK}} R^1\!\!-\!\!\underset{\underset{CH_3}{\underset{|}{\overset{|}{OH}}}}{CH}\!\!-\!\!R^2$$

$$R^* \ldots (S)\text{—}C_6H_5\text{—}\overset{|}{CH}\text{—}$$

		Product	
R^1	R^2	Configuration	$[\alpha]_D$†
CH_3	$cyclo\text{-}C_6H_{11}$	S	+0.42
CH_3	α-naphthyl	S	−1.05
$tert\text{-}C_4H_9$	C_6H_5	S	−4.03
C_6H_5	o-tolyl	R	−1.38
C_6H_5	α-naphthyl	R	+3.18
$cyclo\text{-}C_6H_{11}$	C_6H_5	S	−4.55

† In ethanol

Reactions analagous to those shown above also occur with similar reagents containing beryllium or aluminium instead of magnesium (Table 5.1).

b. Meerwein-Ponndorf-Verley (MPV) reduction[3]

Many kinds of enantioface-differentiating MPV reductions of ketones using optically active aluminium, magnesium or potassium alcoholates have been reported, and several examples are listed in Tables 5.2, 5.3 and 5.4. The reaction mechanism of these MPV reductions is explained in terms of the intermediates A and B shown below. The reaction is considered to proceed faster through A than B, since B has a higher steric hindrance.

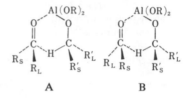

This mechanism accounts for the production of (S)-benzylalcohol-1-d from benzaldehyde-1-d by (−)-bornyloxymagnesium chloride (Eq. 5.3).

$$C_8H_5CDO \longrightarrow \underset{D}{C_6H_5-CH-OH} \qquad (5.3)$$

However, the reduction of benzaldehyde-1-d by (−)-bornyloxyaluminium bromide yielding (R)-benzylalcohol-1-d cannot be explained by the above reaction mechanism.

c. Reactions with alkylboranes[3]

Mono- or dialkylboranes prepared from optically active alkenes such as pinene act as enantioface-differentiating reagents during hydroboration.

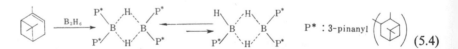

Such hydroboration proceeds with cis addition, and the orientation of the products follows the anti-Markovnikov rule in this case. Since the addition product can be converted to an alcohol, as shown in Eq. 5.5, by oxidation, the optical yield of the reaction can be determined from the optical purity of the resulting alcohol.

$$\underset{R'}{\overset{R}{\diagdown}} C = C \underset{R'''}{\overset{R''}{\diagup}} \xrightarrow{R_2^*BH} \underset{R'}{\overset{R}{\diagdown}} CH - \underset{\underset{BR_2^*}{|}}{\overset{R''}{C}} \underset{R'''}{\diagup} \xrightarrow{H_2O_2} \underset{R'}{\overset{R}{\diagdown}} CH - \underset{\underset{OH}{|}}{\overset{R''}{C}} \underset{R'''}{\diagup} \quad (5.5)$$

The hydroboration of *cis* alkenes usually proceeds in high optical yield (70–90%), but that of *trans* alkenes occurs in considerably lower yield. For example, (S)-2-butanol is obtained in 86% optical yield from *cis*-butene and di-(−)-pinanylborane, while (R)-2-butanol is obtained in only 13% optical yield from *trans*-2-butene.[16]

Since hydroboration follows the anti-Markovnikov rule and dialkyl-alkenes, $R(R')C^2 = C^1H_2$, have no sp^2-prochiral center at C^1 in the enantioface-differentiating addition of borane, addition of boron does not take place at C^2 but hydrogen addition occurs there. The optical yield of such reactions is thus usually very poor.[17]

Brown *et al.*[16] have speculated that hydroboration proceeds through the intermediates A and B shown below; as A is less sterically hindered, the reaction would proceed predominantly through this intermediate.

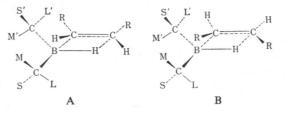

A   B

However, this suggestion is clearly not applicable in all cases, and the real mechanism of hydroboration has not yet been fully elucidated. Moore *et al.*[18] have recently presented a mechanism involving a nonplanar 4-center transition state.

Optically active alkylboranes act as enantioface-differentiating reagents toward carbonyl compounds as well as alkenes. For instance, the reduction of ethylpropylketone with dipinanylborane gives 3-hexanol in 62% optical yield.[19] However, the enantioface-differentiating abilities of alkylboranes are generally less with carbonyl compounds than with alkenes. It has also been found that the optical yields in a given reaction vary widely in experiments by different researchers, and in one case, different enantiomers were obtained predominantly in the same reaction carried out by different workers. It seems likely that the reaction species of pinanylborane depends sensitively upon the conditions of preparation, and several structures may exist in equilibrium, producing apparently contradictory results (for instance, see Eq. 5.4).

The enantioface-differentiating reduction of ketones has also been carried out with lithium alkyl(hydro)dipinan-3α-ylborate,[20] with the

TABLE 5.5

Enantioface-differentiating reduction of ketones with optically active borate complex[20]

$$\begin{array}{c}R_2 \\ C=O \\ R_3\end{array} + \begin{array}{c}R_1 \\ Li^+B^- \\ H \;\; P^* \; P^*\end{array} \longrightarrow \begin{array}{c}R_2 \\ CHOH \\ R_3\end{array}$$

Reagent (R_1)	Substrate		Product	
	R_2	R_3	Configuration	Optical yield (%)
CH_3	CH_3	C_2H_5	S	5–7
CH_3	CH_3	$iso\text{-}C_3H_7$	S	4
CH_3	CH_3	$tert\text{-}C_4H_9$	R	8–9
$n\text{-}C_4H_9$	CH_3	$tert\text{-}C_4H_9$	S	9
CH_3	C_2H_5	$iso\text{-}C_3H_7$	S	46–58

P* = (−)-α-pinanyl (83% optical purity)

$$Li^+ \quad \begin{array}{c}R \\ B^- \\ H \;\; P^* \; P^*\end{array}$$

results shown in Table 5.5. The optical yield of the products does not bear any simple relation to the bulk of the substituent on the reagent, or to the difference in the bulk of the substituents on the substrate.

Pinanylborane and lithium hydride-1-alkyl-di-3-pinanylborate have also been used in the enantioface-differentiating reduction of imines. The optical yields of reduction products of 2-alkyl-3,4,5,6-tetrahydropyridine obtained using these reagents are given in Table 5.6.[21] Again, no clear correlation between structure and optical yield could be found.

TABLE 5.6

Enantioface-differentiating reduction of imines with pinanylborane and borate complex[21]

R	Reagent[†1]	Product	
		Configuration	Optical yield[†2]
CH_3	P_2^*BH	S	2.0–3.3
CH_3	$P_3^*B_2H_3$	S	2.2–3.2
$n\text{-}C_3H_7$	$P_3^*B_2H_3$	S	2.9–10.7
CH_3	$P_2^*n\text{-}C_4H_7BH\text{-}Li^+$	S	19.5–24.0
$n\text{-}C_3H_7$	$P_2^*n\text{-}C_4H_7BH\text{-}Li^+$	R	4.0–4.3

[†1] P* = α-pinanyl (optical purity 74%)

[†2] Not corrected for the optical purity of the reagent

d. Reduction with amine-borane complexes

Optically active α-phenetylamine-borane complex can be used in the enantioface-differentiating reduction or reductive amination of ketones (Eq. 5.6).

$$
RCOR' + NH_3 \xrightarrow[\quad CH_3 \quad]{C_6H_5CHNH_2-BH_3} RCHR' \atop NH_2 \tag{5.6}
$$

The reduction of acetophenone or 2-heptanone with this reagent gives the corresponding alcohol in 2–3% optical yield.[22] The reductive amination of α-ketoglutaric acid with the same reagent gave glutamic acid in 3.1% optical yield.[22]

e. Reaction with amines or alcohols

Optically active amines or alcohols differentiate the enantiofaces of ketenes, yielding amide or ester diastereomers with an excess of one diastereomer. For example, the enantiofaces of phenylmethylketene are differentiated by optically active phenylethylamine, as shown in Eq. 5.7.[23]

$$
\begin{array}{c} C_6H_5 \\ CH_3 \end{array}\!\!\!>\!\!C\!=\!C\!=\!O + \begin{array}{c} C_6H_5 \\ CH_3 \end{array}\!\!\!>\!\!\overset{*}{C}H\!-\!NH_2 \longrightarrow \begin{array}{c} C_6H_5 \\ CH_3 \end{array}\!\!\!>\!\!\overset{*}{C}HNHCO\overset{*}{C}H\!\!<\!\!\begin{array}{c} C_6H_5 \\ CH_3 \end{array} \tag{5.7}
$$

The Strecker reaction of acetaldehyde, (S)-α-methylbenzylamine and hydrogen cyanide can be used to produce L-alanine in 68% optical yield, as shown in Eq. 5.8.[24]

$$
CH_3CHO + C_6H_5\overset{CH_3}{\underset{\;}{C}}HNH_2 + HCN \longrightarrow \left[\begin{array}{c} CH_3 \\ CH_3\!-\!CH\!-\!NH\!-\!CH\!-\!C_6H_5 \\ CN \end{array} \right]
$$

$$
\xrightarrow[Pd]{H_2O \quad H_2} CH_3CH\!-\!COOH \atop NH_2 \quad \text{alanine} \tag{5.8}
$$

As the reaction intermediate in this synthesis is very unstable optically, the optical yield of L-alanine may be controlled by the thermodynamic equilibrium of the diastereomeric intermediate, rather than by enantio-differentiation of the substrate.

Optically active amino acids can also be obtained by enantioface differentiation of α,β-unsaturated acids with optically active amines. For instance, (+)-β-aminobutyric acid can be obtained in 10% optical yield by the addition of (S)-methylbenzylamine to crotonic acid followed by hydrogenation of the addition product, as shown in Eq. 5.9.[25]

$$CH_3CH{=}CHCOOH + C_6H_5\overset{\underset{CH_3}{|}}{C}HNH_2 \longrightarrow CH_3\underset{\underset{NH_2CH(CH_3)C_6H_5}{|}}{C}HCH_2COOH$$

$$\overset{H_2}{\longrightarrow} CH_3\underset{\underset{NH_2}{|}}{C}HCH_2COOH \quad (5.9)$$

β-aminobutyric acid

The enantioface-differentiating addition is observed in the reaction of optically active sec-alcohols to phenyltrifluoroketene[26] (see sect. 7.2.2).

f. Reactions of haloesters

Michael-type condensation between haloesters and α,β-unsaturated esters occurs in the presence of bases to give cyclopropane derivatives, as shown in Eq. 5.10.

$$Cl\text{-}CH_2COOMen + CH_2{=}CHCOOC_2H_5 \longrightarrow$$

$$(5.10)$$

Although this reaction is classified as an enantioface-differentiating reaction, it actually consists of a conjugated enantioface-differentiating reaction at C-2 of the unsaturated ester and an enantiotopos-differentiating reaction at C-2 of the haloester. In fact, if the reaction proceeds through the intermediate shown in Eq. 5.10, as suggested by McCoy,[27] it may be regarded as an enantiotopos differentiation of the haloester and/or diastereo differentiation of the cyclopropane derivative. This differentiation reaction is strongly affected by the solvent. By Eq. 5.10 (−)-trans-product is obtained in toluene with 40% synthetic yield and 3.1% optical yield. While (+)-trans-product with 10.9% of optical purity is obtained with 51% synthetic yield in DMF.[28]

There are two cases in the enantioface-differentiating Reformatsky reaction; one uses the haloesters of optically active alcohols, and the other uses the esters of optically active haloacids. These are shown in Eqs. 5.11 and 5.12.

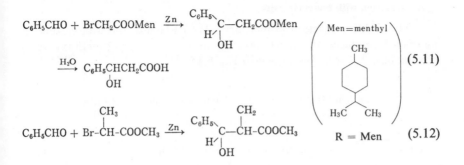

Eq. 5.11 proceeds in an optical yield of 15%,[29] while Eq. 5.12 yields a diastereomer ratio of 63:37.[30] While optical pure α-brom propionate gives 63 and 37% of (*RR, SS*) and (*RS, SR*) isomers respectively by Eq. 5.12 and the (*RR, SS*) isomer is produced with 2.2% optical yield (see p. 245).

g. **Aldol condensation and related active methylene reactions**

Enantioface differentiation of ketones or aldehydes with esters of optically active alcohols occurs with very high optical yield in reactions of this type, as illustrated by the condensation of acetophenone and acetic acid menthylester with diethylaminomagnesium bromide, yielding 3-hydroxy-3-phenylacetic acid in 93% optical yield (Eq. 5.13).[31]

$$C_6H_5COCH_3 + CH_3COOMen \xrightarrow{(C_2H_5)_2NMgBr} \underset{OH}{\overset{CH_3}{C_6H_5-\overset{|}{\underset{|}{C}}-CH_2COOMen}}$$

$$\xrightarrow{H_2O} \underset{OH}{\overset{CH_3}{C_6H_5-\overset{|}{\underset{|}{C}}-CH_2COOH}} \qquad \text{3-hydroxy-3-phenylacetic acid}$$

$$(5.13)$$

h. **Reactions with olefins**

Enantioface-differentiating addition of olefins to the carbon-carbon double bond of maleic anhydride occurs when the olefin is optically active, as shown in Eq. 5.14 for (*R*)-(−)-3-phenyl-1-butene[32] (product obtained with $[\alpha]_D^{20} = -14.5°$).

$$\underset{CH_3}{\overset{}{C_6H_5CHCH=CH_2}} + \left[\begin{array}{c} CO \\ \\ CO \end{array}\right]O \longrightarrow \underset{CH_3}{\overset{C_6H_5}{C}}=CH-CH_2-\left[\begin{array}{c} CO \\ \\ CO \end{array}\right]O \qquad (5.14)$$

i. Reaction with mercuric salts

The addition reaction of mercury (II) (+)-lactate and cyclohexene in methanol produces the diastereomers of lactomercury-2-methoxycyclo-hexane in a 2:1 ratio, as shown in Eq. 5.15.[33]

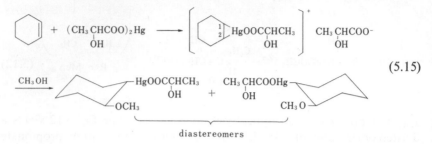

$$(5.15)$$

diastereomers

Although this reaction is formally an enantioface-differentiating reaction, the real process is presumably the diastereotopos (C-1 and C-2)-differentiating decomposition of the intermediate.

j. Reaction with peracids

Enantioface-differentiating oxidation of alkenes and imines with optically active peracids yields optically active epoxides and oxaziridines respectively. Eq. 5.16 shows the oxidation of styrene with peroxycamphoric acid yielding an epoxide in 4.6% optical yield.[34]

$$C_6H_5CH=CH_2 + \quad \overset{COOOH}{\underset{COOH}{\diagdown}} \quad \longrightarrow \quad C_6H_5 \diagdown_O \qquad (5.16)$$

mono perorychamphoric acid

Eq. 5.17 shows the oxidation of an imine, producing chiral centers on both carbon and nitrogen, as expected.

$$R-N=C\overset{R'}{\underset{R''}{\diagup}} \quad \longrightarrow \quad R-\ddot{N}—C\overset{R'}{\underset{O}{\diagup}}\overset{R'}{\underset{R''}{}} \qquad (5.17)$$

Disproportionate formation of enantiomeric configurations in the two centers can occur, as shown in Table 5.7.[35] Experiment no. 1 in the table shows that the nitrogen atom is a chiral center, since the product is optically active regardless of the presence or absence of a chiral carbon center. The formation of diastereomers in experiment no. 2 tends to confirm the presence of chiral centers on both carbon and nitrogen.

The effect of reaction temperature on the differentiation was investigated in the oxidation of 4-(p-nitrobenzylidene)-tert-butylamine (Eq. 5.17,

TABLE 5.7
Enantioface-differentiating oxidation with optically active peracids

Exp. no.	Substrate (Eq. 5.17)			Peracid	$[\alpha]_D$ of product (Eq. 5.17)
	R	R′	R″		
1	CH_3	C_6H_5	C_6H_5	$(1S)$-$(+)$-per-peroxycam-phoric acid	$-11.3(CHCl_3)$
2	iso-C_3H_7	p-$NO_2C_6H_4$	H		$\begin{cases} -12.0† \\ -1.3 \end{cases}$

† Optical activities of the diastereomers

R = *tert*-Bu, R′ = p-NO_2–C_6H_4–, R″ = H) with (+)-peroxycamphoric acid. The optical yields of oxidations at 29°C and −72°C were 34% and 66%, respectively.[36]

The enantioface differentiation of azo-type structures by oxidation with optically active peracids was confirmed by the reaction shown in Eq. 5.18, which gave a product with an optical activity of $[\alpha_D] = +1.2°$.[37]

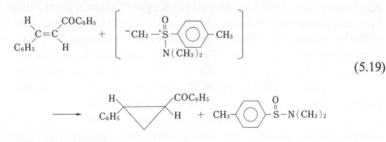

$$(5.18)$$

k. Reaction with S-ylides

The optically active ylide *N,N*-dimethylamino-*p*-tolyl oxosulfonium methylylide (Eq. 5.19) differentiates the enantiofaces of aldehydes, ketones and alkenes, yielding optically active epoxides from the former two, and cyclopropane derivatives from the latter. For example, the optically active cyclopropane derivative shown in Eq. 5.19 is obtained in 35% optical yield in the reaction shown.[38]

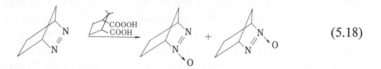

$$(5.19)$$

B. Reactions involving catalysts

The first discovery of a catalytic enantioface-differentiating reaction was made by Bredig in 1912,[39] only 18 years after the discovery of diastereoface-differentiating reactions by Fischer.[40] However, although reactions of this type are potentially the most valuable for the synthesis of optically active compounds, little progress was made until the discovery of the silk-palladium catalyst by the present author in 1956.[41] Subsequently, the modified Raney nickel catalyst (MRNi) was discovered,[42] and the relationship between the structure of the modifying reagent and the differentiating ability of MRNi was intensively investigated. In 1968, Knowles achieved nearly 100% optical yield in a reaction with a Wilkinson-type complex catalyst.[43] Such catalytic reactions now form the major field of investigation in research on stereo differentiating reactions.

a. Metal complex catalysts

Since the discovery of the powerful enantioface-differentiating Wilkinson-type catalysts by Knowles in 1968, a large number of catalytic enantioface-differentiating reactions have been reported using optically active organometallic catalysts. However, systematic studies on the activities and reaction mechanisms of these catalysts have not yet been performed, except in the case of the Wilkinson-type catalysts themselves.

Hydrogenation: Enantioface-differentiating hydrogenation of alkenes can be achieved by Wilkinson catalysts carrying an optically active phosphine ligand. Remarkably high differentiating ability can be obtained: for instance, an optical yield of 95% was obtained in the hydrogenation of 4-hydroxy-3-methoxy-α-benzoylaminocinnamic acid to 3,4-dihydroxy-phenylalanine (DOPA).[44]

Optically active Wilkinson catalysts of this type are classified in terms of the structures of their phosphine ligands. Two cases may be considered; one in which the phosphorus atom itself is chiral, and one where the phosphorus carries an optically active alkyl group. The properties of these catalysts are listed in Table 5.8. In general, the former type seems superior, but the last catalyst listed in Table 5.8, which has the chiral center in an alkyl group (the DIOP developed by Kagan[46]), gives good results. However more data are needed to clarify the relationship between the structure of the ligand and the differentiating ability of the catalyst.[43-45] The mechanism of hydrogenation by Wilkinson complex catalysts is believed to be as shown in Fig. 5.2. The results in Table 5.8 show that the differentiating ability depends strongly upon the precise structure of ligands in the complex, but the reasons for these effects are not clear.

A new type of hydrogenation catalyst which consists of cobalt complexes with dimethylglyoxime (DMG) and quinine (A)[47] or cyanide and optically active propylenediamine (B)[48] as ligands has also been developed. The reactions shown in Eqs. 5.20 and 5.21 gave optical yields of 61.5% and 7.1%, respectively.

TABLE 5.8
Enantioface-differentiating hydrogenation reactions

Catalyst (configuration)	R₁	R₂	R₃	Solvent	Temp (°C)	Pressure (atm)	Substrate	Product	Configuration	OY [†7]	SY [†8]	Ref.
RhL₄Cl L=P⟨R¹ R² R³⟩	CH₂	CH₃-CH₂-CH₂	Ph	B-E [†2]	60	26	CH₂ Ph-C-COOH CH₂ HOOC-CH₂-C-COOH	CH₂ Ph-CH-COOH CH₃ HOOC-CH₂-CH-COOH		22 [†1] 4.4 [†1]		(43)
	Ph	CH₃ C₂H₅-C-CH₃ H	CH₃ C₂H₅-C-CH₂ H				CH₂ Ph-C-COOH⁺	CH₃ Ph-CH-COOH		~15 [†1]		
(Rh(1,5-hexadiene)L₂)Cl L=P⟨R₁ R₂ R₃⟩	Ph	Ph	neomenthyl	B-E [†2]	60	20	CH₃ (E) Ph-C=CH-COOH	CH₃ Ph-CH-CH₂-COOH	S	61	80	(45)
							CH₃ (E) Ph-CH=C-COOH	CH₃ Ph-CH₂-CH-COOH	R	52		
							CH₂ (E) Ph-C-COOH	CH₃ Ph-CH-COOH	S	28		
	CH·	Ph	CH₃-CH₂-CH₂-	M	25	4	OCH₃ HO-⟨◯⟩-CH=C-COOH NHCOPh	OCH₃ HO-⟨◯⟩-CH₂-CH-COOH NHCOPh		3.1 [†1]		(44)
			CH₃-C CH₂ H	M						31 [†1]		
			⟨◯⟩ CH₂O							1.2 [†1]		
			⟨◯⟩ OCH₃ CH₂	M [†3]						61 [†1]		
			⟨◯⟩	M						43 [†1]		
		⟨◯⟩	⟨◯⟩ OCH₃	iso-Pro						92 [†1]		
				M						95 [†1]		
	Ph	CH₃ CH₃⟩CH	⟨◯⟩ OCH₃							1.2 [†1]		
		Ph	⟨◯⟩ OCH₃	M [†3]	50					58 [†1]		
		CH₃-CH₂-CH₂	⟨◯⟩ OCH₃	M		4	OCH₃ CH₃COO-⟨◯⟩-CH=C-COOH NHCOCH₃	OCH₃ CH₃COO-⟨◯⟩-CH₂-CH-COOH NHCOCH₃		21 [†1]		
				M [†2]	25					85 [†1]		
				95%E						90 [†1]		
				iso-Pro		0.7				93 [†1]		
	CH₂	⟨◯⟩	⟨◯⟩ OCH₃				Ph-CH=C-COOH NHCOCH₃	Ph-CH₂-CH-COOH NHCOCH₃		90 [†1]		
				95%E [†2]	25	0.7	Ph-CH=C-COOH NHCOPh	Ph-CH₂-CH-COOH NHCOPh		90 [†1]		
							CH₂=C-COOH NHCOCH₃	CH₂-CH-COOH NHCOCH₃		63 [†1]		
Ph(L-L)ClS (S=solvent) L-L=D-H-C⟨CH₂ CH₂ / O O⟩ ⟨CH₂ CH₂ / P P⟩ Ph Ph Ph Ph				B-E [†2]	Room temp	Atmospheric	CH₂ Ph-C-COOH	CH₃ Ph-CH-COOH	S	63	[†6]	(46)
							CH₂ Ph-C-COOCH₃	CH₃ Ph-CH-COOH	S	7	[†6]	
							Ph-CH=C-COOH NHCOCH₃	Ph-CH₂-CH-COOH NHCOCH₃	R	72	[†6]	
							PhCH=C-COOH NHCOCH₂-Ph	Ph-CH₂-CH-COOH NHCOCH₂-Ph		68 [†5]	[†6]	

[†1] Optical purity of the catalyst was not 100%; optical yield is adjusted accordingly
[†2] In the presence of triethylamine or other bases   [†3] In the presence of NaOH
[†4] B = benzene, E = ethanol, M = methanol, Pro = propanol
[†5] As phenylalanine   [†6] Semiquantitative
[†7] OY: Optical yield   [†8] SY: Synthetic yield

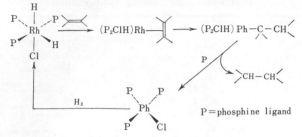

Fig. 5.2.   Mechanism of hydrogenation by Wilkinson-type complex catalysts.

$$C_6H_5COCOC_6H_5 \xrightarrow{A} \underset{\underset{\text{benzoin}}{OH}}{C_6H_5COCHC_6H_5} \tag{5.20}$$

$$\underset{C_6H_5\overset{\overset{CH_2}{\|}}{C}COOH}{} \xrightarrow{\cdot B} \underset{C_6H_5\overset{\overset{CH_3}{|}}{C}HCOOH}{} \tag{5.21}$$

Methyl acetoacetate has been hydrogenated to yield methyl 3-hydroxybutyrate using an aqueous solution of an optically active polyimine in the presence of ruthenium chloride, but the nature of the catalytic species is not known. An optical yield of 3.5% was obtained.[49]

Hydroformylation: Rhodium-phosphine complex catalyzes the hydroformylation of alkene. The use of optically active phosphine gives an optically active aldehyde (Eq. 5.22). Intensive studies of this reaction have been performed with various complexes of Co, Rh, Pd and Pt[50] by Pino *et al.* Representative results are summarized in Table 5.9.

$$PhCH=CH_2+CO+H_2 \longrightarrow \underset{PhCHCH_3}{\overset{\overset{CHO}{|}}{}} \tag{5.22}$$

Hydrosilylation: Metal complexes can catalyze the enantioface-differentiating silylation of alkenes and ketones. Eq. 5.23 shows an example in which the monochloro product is obtained in 17.6% optical yield using the nickel complex catalyst (A)[52] or in 5.3% using the platinum complex (B).[53]

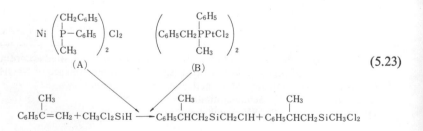

$$\tag{5.23}$$

TABLE 5.9

Enantioface-differentiating hydroformylation with metal complexes

Substrate	Catalyst	Product	Configuration	Optical yield	Ref.
Styrene	RhH(CO)(PPh$_3$)$_3$/DIOP = 1/4	Ph—CH—CH$_3$ with CHO	(R)	25.2	(51a)
	RhH(CO)$_{4-n}$(P—Ph with CH$_2$Ph, CH$_3$)$_n$ (Rh/P = 1/10)	"	(S)	25.0	(51b)
	[RhCO(P—Ph with Ph, Neomen)$_2$Cl]$_2$	"	(S)	1.2	(51c)
α-Methylstyrene	RhH(CO)(PPh$_3$)$_3$/DIOP = 1/4	Ph—CH—CH$_2$—CHO with CH$_3$	(R)	1.6	(51a)
Allyl benzene	"	Ph—CH$_2$—CH—CHO with CH$_3$	(R)	15.5	(51a)
1-Butene	RhH(CO)(PPh$_3$)$_3$/DIOP = 1/4	CH$_3$—CH$_2$—C—CHO with CH$_3$	(R)	18.8	(51d)
cis-Butene	"	"	(S)	27.0	(51d)
trans-Butene	"	"	(S)	3.2	(51d)

The same platinum complex (B) also catalyzes the enantioface-differentiating silylation of the ketone shown in Eq. 5.24, yielding the silylether in 18.6% optical yield.[53)]

$$tert\text{-}C_4H_9COC_6H_5 + CH_3Cl_2SiH \longrightarrow \overset{\overset{\displaystyle tert\text{-}C_4H_9}{|}}{C_6H_5\overset{}{C}HOSiCH_3Cl_2} \qquad (5.24)$$

Enantioface-differentiating hydrosilylation is performed efficiently by rhodium complexes. Eq. 5.25 shows an example which proceeds in 61.8% optical yield.[54)]

$$C_6H_5CO\text{-}tert\text{-}C_4H_9 + (CH_3)_3SiH$$

$$\xrightarrow[\quad]{[Rh[(C_6H_5CH_2)(CH_3)C_6H_5P]_2H_2S_2]^+ \text{ (S: solvent)}} \quad \overset{C_6H_5}{\underset{tert\text{-}C_4H_9}{\diagup}}CH\text{-}OSi(CH_3)_3 \;{}^-$$

$$(5.25)$$

Oligomerization: Diethylaluminium chloride - titanium menthoxide system catalyzes the enantioface-differentiating oligomerization of 1,3-pentadiene as shown in Eq. 5.26, yielding a mixture of trimers.[55)]

$$CH_2\text{=}CHCH\text{=}CHCH_3 \xrightarrow{(C_2H_5)_2AlCl\text{-}Ti(Men)_4} \qquad \qquad + \qquad \qquad (5.26)$$

Alkylation: Optically active 3-vinylcyclooctene is obtained by the addition of ethylene to 1,3-cyclooctadiene catalyzed by a nickel complex, as shown in Eq. 5.27. When dimenthyl-isopropylphosphine is used as a ligand on the catalyst, the product is obtained in 27% optical yield, and optically active 3-methyl-1-pentene is released simultaneously as a by-product.[56)]

$$\bigcirc + \overset{\diagdown}{\diagup}C\text{=}C\overset{\diagup}{\diagdown} \xrightarrow[\pi\text{-}C_3H_5\text{-}NiX\text{-}AlX_3]{iso\text{-}C_3H_7\text{-}P\text{=}(Men)_2} \qquad$$

$$(5.27)$$

3-vinylcyclooctene

Carbene reactions: Enantioface differentiation occurs in the addition of a carbene to an alkene under the influence of an optically active copper complex. This reaction is achieved first by the following catalyst.[57a)]

Recently, this reaction has been applied to the synthesis of optically active chrysanthemic acid (Eq. 5.28). The chrysanthemic acid (*trans*) was obtained in 68% optical yield.[57b)]

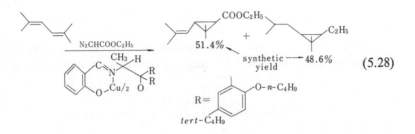

$$(5.28)$$

Cobalt complex also catalyzes enantioface-differentiating addition of carbene. In the presence of (+)-camphorquinone-α-oxime, *trans*-2-phenylcyclopropane carboxylate is obtained from styrene with neopentyl diazoacetate in an optical yield of 88%.[58)]

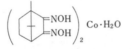

Other reactions: Palladium, cobalt and nickel complexes with the Schiff base of an optically active amine and salicylaldehyde catalyze *trans* hydrogenation between a prochiral alkene and tri-*tert*-butylaluminium, as shown in Eq. 5.29.

$$CH_2=C \underset{R''}{\overset{R'}{\diagup}} + Al[C(CH_3)_3]_3 \rightleftharpoons Al\left(CH_2-CH\underset{R''}{\overset{R'}{\diagup}}\right)_3 + CH_2=C\underset{CH_3}{\overset{CH_3}{\diagup}}$$

$$M = Pd, Co, Ni$$

$$(5.29)$$

The enantioface differentiation was detected in terms of the optical activity of the alkane obtained by acid hydrolysis of the alkylaluminium.[59)]

b. Homogeneous catalysts

Cyanhydrin synthesis: Enantioface-differentiating cyanhydrin synthesis was first carried out by Bredig in 1912.[39] This was the first time that a catalyst capable of enantio-differentiation was used. The reaction (Eq. 5.30) was carried out in chloroform solution with an optically active amine as a catalyst.

$$R \cdot CHO + HCN \longrightarrow R\text{-}\underset{\underset{\displaystyle OH}{|}}{\overset{\overset{\displaystyle H}{|}}{C}}\text{-}CN \qquad (5.30)$$

cyanhydrin

Table 5.10 presents some results on enantioface-differentiating benzcyanhydrin synthesis.[60,61] Since cyanhydrin is optically unstable,

TABLE 5.10
Enantioface-differentiating benzcyanhydrin synthesis

Catalyst	Example	Optical purity of product (%)	Ref.
Quinine derivative	{Cinchonine {Cinchonidine	6.9 8.7	(61)
Aminocellulose derivative	Diethylamino-cellulose		
	(R = Et$_2$N–)	6.8	(60)
Condensation product of glucosamine with a copolymer of styrene and divinylbenzene		1.1	
Polyaziridine derivative	*iso*-Bu $-(NH\overset{|}{C}HCH_2)_n-$	19.6	

the enantioface-differentiating abilities of the catalyst cannot be estimated from the apparent optical purity of the product. The optical yield for the quinine derivative in the table was obtained by extrapolation (see section 8.2.6(A)). Since precise determination of the optical yield is difficult, this reaction is not entirely suitable for research into differentiation processes, despite its historical importance.

Optically active benzcyanhydrin is obtained even with an achiral base as the catalyst if the optically active bis-iodomethylquinine derivative shown

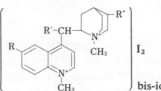 I_2

) bis-iodomethylquinine derivative

is present in the reaction system, even though it does not form a salt with hydrogen cyanide. The reaction mechanism is thought to be as shown in Eq. 5.31, one mole of the catalyst associating with one mole of the aldehyde, and another acting as the base catalyst.[61]

$$R-CHO \xrightarrow[\text{chiral factor}]{\boxed{B^*}} RCHO \xrightarrow[\boxed{B^*}]{HCN} \underset{OH}{R-CH-CN} \quad (5.31)$$

basic catalyst

Kinetic studies have shown that this reaction proceeds by different mechanisms in polar and nonpolar solvents. Highly polar solvents prevent the optically active base-aldehyde association, and no differentiation occurs at all.[62−64]

Bromination: Optically active bases catalyze the enantioface-differentiating bromination of alkenes. For example, in the bromination of (*RS*)-4-methylcyclohexene with dihydrocinchonine as a catalyst, *trans* and *cis* dibromides are obtained in 0.66% and 10.76% optical yield, respectively (Eq. 5.32).[65]

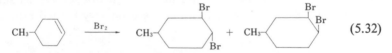

$$(5.32)$$

Enantioface differentiation certainly occurs in this reaction, since optically active products are obtained from (*RS*)-substrate. However, since the substrate contains a chiral center at C-4, diastereoface differentiation must also be affecting the results (see section 8.2.5).

Addition of mercaptan: Optically active amines catalyze the enantioface-differentiating addition of mercaptans to alkenes. Thus, an optically active thioester is obtained by the addition of laurylmercaptan to methyl crotonate (Eq. 5.33) using an optically active amine polymer as a catalyst.[66]

$$CH_3CH=CHCOOH_3+n\text{-}C_{12}H_{25}SH \xrightarrow{\underset{(-\overset{*}{C}H-CH_2-NH-)_n}{\overset{\underset{CH_2}{\overset{CH_3}{\diagdown}CH}}{}}} \underset{S-C_{12}H_{25}}{CH_3CHCH_2COOCH_3}$$

$$(5.33)$$

c. Heterogeneous catalysts

The first successful enantioface-differentiating reaction with a heterogeneous catalyst was carried out by the authors' research group in 1956, using a silk-palladium catalyst.[41] Subsequent investigations led to the discovery of asymmetrically modified Raney nickel (MRNi) catalysts,[42,67] which have better reproducibility than silk-palladium as regards differentiation.

There are two types of enantioface-differentiating heterogeneous catalysts; in one, the metal is deposited on a chiral support (silk-palladium type), and in the other the chiral modifier is adsorbed on the metal catalyst (MRNi type). Both types can hydrogenate $C=O$, $C=N$ and $C=C$ bonds.

Silk-palladium type catalysts: Silk-palladium catalyst is prepared by the hydrogenation of silk-palladium chloride obtained by boiling silk in aqueous palladium chloride.[41] Similar catalysts have been obtained by using optically active polyamino acids,[68] polymers which have optically active amino acid residues,[69] or silica gel acid-precipitated from sodium silicate solution containing an optically active quinine base.[70] The enantioface-differentiating capabilities of these catalysts with various substrates are listed in Tables 5.11 through 5.13. The silk-palladium catalysts were designed in the expectation that the chiral environment supplied by the supporting polymer would influence the catalytic reaction due to the metal. Similarly, Beamer et al.[71] supposed that the silica gel support provides sites where the substrate can bind in a special orientation, as shown in Fig. 5.3. However, this mechanism is based on very limited experimental

TABLE 5.11
Enantioface-differentiating hydrogenation with silk-Pd catalyst[41]

Substrate	Product		
	Structure	$[\alpha]_D$	Optical yield (%)
$C_2H_5OOCCH_2CH_2CCOOC_2H_5$ $\quad\quad\quad\quad\quad$ $\overset{\|}{N}OCOCH_3$	$HOOCCH_2CH_2CHCOOH$ $\quad\quad\quad\quad\quad$ $\overset{\|}{N}H_2$	+2.25	6.1
$C_6H_5CH_2CCOOC_2H_5$ $\quad\quad\quad$ $\overset{\|}{N}OCOCH_3$	$C_6H_5CH_2CHCOOH$ $\quad\quad\quad\quad$ $\overset{\|}{N}H_2$	+9.25	26.3
$C_6H_5CH=C-CO$ $\quad\quad\quad$ $\overset{\|}{N}\ \overset{\|}{O}$ $\quad\quad\quad\quad$ $\overset{\|}{C}$ $\quad\quad\quad\quad$ CH_3	$C_6H_5CH_2CHCOOH$ $\quad\quad\quad\quad$ $\overset{\|}{N}H_2$	+12.5	35.6
$C_6H_5C\text{——}CC_6H_5$ $\ \ \overset{\|}{HON}\quad \overset{\|}{N}OH$	$C_6H_5CHCHC_6H_5$ $\ \ \overset{\|}{H_2N}\ \ \overset{\|}{N}H_2$	+8.75	——

TABLE 5.12
Enantioface-differentiating hydrogenation with catalysts supported on silica gel
and polyamino acids

Catalyst	A[†1]		B[†1]		Ref.
	$[\alpha]_D$	OY(%)	$[\alpha]_D$	OY(%)	
PdQN/0.5 silica gel[†2]	+0.87	3.21	—	—	(70)
PdQDN/0.5 silica gel[†2]	+0.45	1.66	—	—	(70)
PdCN/0.5 silica gel[†2]	+0.47	1.74	—	—	(70)
PdCDN/0.5 silica gel[†2]	+0.88	3.25	—	—	(70)
Poly-L-leucine/Pd[†3]	−0.320	1.18	−2.50	5.16	(68a)
Poly-γ-benzyl-L-glutamic acid/Pd[†3]	−1.124	4.15	−2.9	6.0	(68b)
Poly-β-benzyl-L-aspartic acid/Pd[†3]	+0.388	1.43	+0.46	0.95	(68b)
Poly-L-valine/Pd[†3]	+0.245	0.90	+2.06	4.25	(68a)

[†1] A: Substrate = α-methylcinnamic acid, product = α-methyl-β-phenylpropionic acid; B: substrate = α-acetoaminocinnamic acid, product = phenylalanine

[†2] Palladium catalyst supported on silica gel (from 42 g of sodium silicate in the presence of 0.5 g of alkaloid sulfate). QN = quinine salt, QDN = quinidine salt, CN = cinchonine salt, CDN = cinchonidine salt

[†3] Palladium catalyst supported on polyamino acid

TABLE 5.13
Enantioface-differentiating hydrogenation[†1] with IRC-50/amino acid/Pd catalysts[69]

Catalyst	Substrate	Product		
		Structure	$[\alpha]_D$	OY(%)[†2]
IRC-50/L-alanine/ Pd	$C_6H_5COCOOH$	$C_6H_5CHCOOH$ OH	−0.04	0.03
	$CH_2{=}CCOOH$ NHCOCH$_3$	$CH_3CHCOOH$ NH$_2$	−0.27[†]	0.19
IRC-50/L-phenylalanine/Pd	$C_6H_5COCOOH$	$C_6H_5CHCOOH$ OH	−0.12	0.08
	$CH_2{=}CCOOH$ NHCOCH$_3$	$CH_3CHCOOH$ NH$_2$	−0.16[†2]	0.11

[†1] Reaction conditions, 60°C, 80 atm

[†2] The optical rotation was measured in 1 N NaOH as the DNP-amino acid (OY = optical yield)

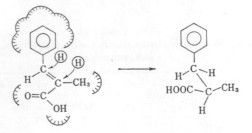

Fig. 5.3.   The mechanism of enantioface-differentiating hydrogenation with poly-L-leucine-Pd catalyst according to Beamer *et al.*[71]

results, and gives no indication of the steric relation between the palladium catalyst and the substrate or supporter.

MRNi-type catalysts: Enantioface-differentiating reactions by MRNi catalysts are the most extensively investigated catalytic reactions among those by MRNi-type catalyst. MRNi catalysts are generally easy to prepare and have good enantioface-differentiating ability. For example, erythro-2-methyltartaric acid-MRNi performs the enantioface-differentiating hydrogenation of methyl acetoacetate (MAA) into methyl 3-hydroxybutyrate (MHB) (Eq. 5.34, I) with 53% optical yield.[72a] In 1976, an optical yield of 87.2% was obtained in the liquid phase hydrogenation of MAA in THF with nickel-palladium (95:5) catalyst supported on silica gel with tartaric acid.[72b]

$$CH_3COCH_2COOR \xrightarrow{\text{H}_2/\text{MRNi}} CH_3\underset{OH}{CH}CH_2COOR \qquad (5.34)$$

I; R=CH$_3$
II; R=C$_2$H$_5$

The enantioface-differentiating ability of MRNi is greatly affected by the conditions of preparation of the Raney nickel (R,Ni) catalyst. This is illustrated in Fig. 5.4,[73] which shows optical yield of HMB against conversion of MAA in the hydrogenation over D$_S$-tartaric acid-MRNi prepared from various kinds of Raney nickel alloy. The MRNi discussed in this section except special cases, the RNi was prepared from 1.5 g alloy (40% Ni) by development with 20 ml of 20% sodium hydroxide at 80°C for 40–60 min. MRNi-type catalyst can be prepared from various metals, as shown in Table 5.14.

MRNi is prepared from RNi by treatment with aqueous solutions of optically active compounds (modifying solutions), such as α-amino acids or α-hydroxy acids, at specified temperature and pH (the modifying temperature and pH) for a given time, followed by washing with water and

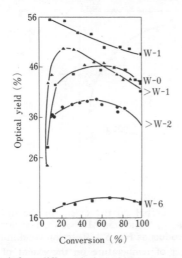

Fig. 5.4. Enantioface-differentiating abilities of D_S-tartaric acid-MRNi prepared from various kinds of Raney Ni. Substrate, methyl acetoacetate; product, methyl 3-hydroxybutyrate (MHB). modifying conditions, 100°C, pH 4.9; reaction conditions, 60°C, 90 atm.

methanol successively.[75] The differentiating ability of the MRNi is greatly affected by changes in the modifying pH and temperature.[67] The effect of the modifying pH on the differentiating ability of MRNi modified with L-malic acid and L-aspartic acid is shown in Fig. 5.5.[76] Fig. 5.6 shows the effect of the modifying temperature on the differentiating abilities (2*S*, 3*S*)-erythro-2-methyltartaric acid-MRNi,[72a] D_S-tartaric acid-MRNi,[77] (+)-2-methylglutamic acid-MRNi,[78] L-valine-MRNi[77] and L-glutamic acid-MRNi.[42] For the MRNi discussed in this section, catalysts modified at about pH 5 and 0°C are considered.

TABLE 5.14
Enantioface-differentiating hydrogenation of ethyl acetoacetate with various tartaric acid-modified metal catalysts[†1]

Catalyst	Modifying conditions[†2]		Optical yield (%)	Reaction temp. (°C)	Ref.
	pH	Temp. (°C)			
Raney-Co	3.6	5	8.1	80	(74a)
Raney-Cu	4	0	14	115	(74b)
Raney-Ru	5.5	—	3.0	80	(74c)
Ru/silica gel	5.5	—	4.2	75	(74d)
Raney-Ni	5.0	0	18	60	(74e)

[†1] Substrate, ethyl acetoacetate (Eq. 5.34, II); reaction conditions, 60°C, 80 atm
[†2] Conditions for preparing the catalyst (see the text)

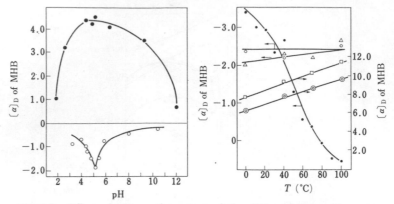

Fig. 5.5.   Effect of pH on the extent of the differentiation with various catalysts.[80] ●, L$_S$-Malic acid-MRNi; ○, L-aspartic acid-MRNi. Substrate and product as Fig. 5.4. Reaction conditions, 60°C, 80 atm.

Fig. 5.6.   Effect of temperature on the extent of the differentiation with various catalysts. ☐, (2S,3S-Erythro-2-methyltartaric acid-MRNi;[72] ◎, D$_S$-tartaric acid-MRNi;[77] △, (+)-2-methyl-glutamic acid-MRNi;[78] ○, L-valine-MRNi;[77] ●, L-glutamic acid-MRNi.[42] Substrate and product as Fig. 5.4. Reaction conditions, 60°C, 80 atm.

The optical yield of product is often strongly affected by the solvent[79a] or by the reaction temperature[79b] (Fig. 5.7 and 5.8, respectively). As shown, the directions of differentiation of MRNi may be reversed by the solvent or reaction temperature. The following discussion will be made with the results obtained by the hydrogenation of MAA without solvent at 60°C and 80 atom except in special cases.

The differentiating ability of MRNi is closely related to the structure of the modifying reagent. For example, the following structural factors are required in the modifying reagent to obtain high optical yield in the hydrogenation of methyl acetoacetate: (1) the modifying reagent should be optically active, (2) if the modifying reagent has two or more chiral centers, they should have similar directions of enantioface differentiation[67] (for instance, *meso*-tartaric acid has no differentiating ability since its two chiral centers produce equal differentiating abilities in opposite directions), and (3) α-amino acids or α-hydroxy acids are the preferred modifying reagents for high differentiating ability.[67] The presence of substituents on C-1 and N or O decreases the optical yield. (4) The molecule with simple structure is enough to be a superior modifying reagent. For example, the differentiating ability of the polypeptide-MRNi[72] does not seem to be much greater than that of amino acid-MRNi, even though there is a greater number of chiral centers, and in addition, the direction of differentiation is not simply controlled by the chirality of the polypeptide

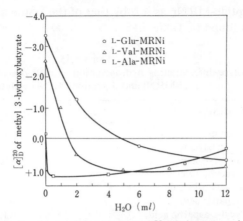

Fig. 5.7. Effect of water on the optical yield.[79a] Substrate and product as Fig. 5.4. Reaction conditions, 60°C, 80 atm.

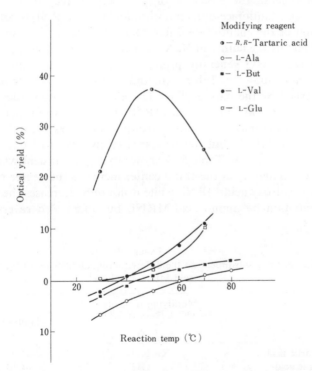

Fig. 5.8. Effect of hydrogenation temperature on the optical yield.[79b] Hydrogenation pressure, atmospheric; substrate and product as Fig. 5.4.

in polypeptide-MRNi, as it is by that of the amino acid in the amino acid-MRNi as found in Table 5.15.

TABLE 5.15

Enantioface-differentiating hydrogenation of methyl acetoacetate with amino acid-MRNi and dipeptide-MRNi catalysts[†][80]

Modifying reagent	$[\alpha]_D$ of product
L-Leucine	−1.10
L-Leucylglycine	+1.12
Glycyl-L-leucine	−0.41
L-Leucyl-L-leucine	−0.42
L-Leucyl-D-leucine	+1.20
L-Aspartic acid	−1.10
Glycyl-L-aspartic acid	−1.47

† Substrate, methyl acetoacetate (Eq. 5.34, I)

In general, the structure of a modifying reagent R–CHX–COOH affects the enantioface-differentiating ability of MRNi as follows: (1) the direction of the enantioface-differentiation is determined by the chirality at C-2 and by the nature of X. Amino and hydroxy groups (X = NH$_2$ and OH, respectively) generally produce differentiating abilities opposite in direction, as shown in Table 5.16, and (2) the extent of the differentiation is governed by the nature of R. The effects of a substituent R are very often opposite in amino acid-MRNi and in hydroxy acid-MRNi. An increase in the bulk of the alkyl group R increases the differentiation by amino acid-MRNi but decreases that by the hydroxy acid-MRNi, as shown in Fig. 5.9. Table 5.17 shows that a substituent which decreases the electron density at the chiral center increases the degree of differentiations by hydroxy acid-MRNi, while it not only decreases the extent of the differentiation by amino acid-MRNi, but often also reverses its direction.[85]

TABLE 5.16

Effect of the α-substituent on the differentiating abilities of α-substituted fatty acid-MRNi catalysts[†]

	Modifying reagents HOOC–CH$_2$–CH–COOH $\overset{\|}{X}$	$[\alpha]_D$ product	Ref.
L-aspartic acid	X=NH$_2$	−1.9	(76)
L$_s$-malic acid	OH	+4.15	(81)
L-methylsuccinic acid	CH$_3$	+0.20	(81)

† Substrate, methyl acetoacetate; reaction conditions, 60 °C, 80 atm

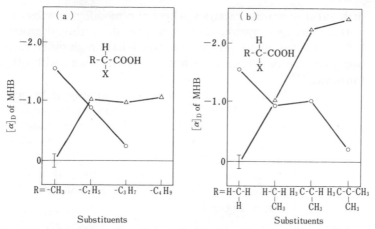

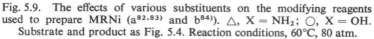

Fig. 5.9. The effects of various substituents on the modifying reagents used to prepare MRNi (a[82,83] and b[84]). △, X = NH₂; ○, X = OH. Substrate and product as Fig. 5.4. Reaction conditions, 60°C, 80 atm.

TABLE 5.17

Electronic effects of substituents on the modifying reagents on the differentiating abilities of MRNi[85]

Modifying reagents	Modifying conditions†		
	pH	Temp. (°C)	$[\alpha]_D$ of product
⟨○⟩-CHCO₂H NH₂	5.62	0	−0.17
Cl-⟨○⟩-CHCO₂H NH₂	6.05	0	+0.45
CH₃O-⟨○⟩-CHCO₂H NH₂	6.15	0	−0.28
⟨○⟩-CHCO₂H OH	2.9	0	+0.14
Cl-⟨○⟩-CHCO₂H OH	2.9	0	+2.48
CH₃O-⟨○⟩-CHCO₂H OH	2.9	0	+0.25

† Substrate, methyl acetoacetate; reaction conditions, 60°C, 80 atm

Analogous relationships between the enantioface-differentiating ability of MRNi and the structure of the modifying reagent can be found in the hydrogenation of analogous substrates such as acetoacetic esters and acetylacetone. However, completely different relationships hold for different types of substrates, such as those containing $\rangle C=O$, $\rangle C=N$ and $\rangle C=C\langle$. Table 5.19 shows some results for the hydrogenation of such substrates with catalysts where the modifier is substituted with a methoxy group at various distances from the carbonyl group.[86a)] Table 5.20 shows that modifying reagents which have high differentiating ability with one type of substrate do not necessarily do so with other types of substrate.[86b)]

Various data, including the amount of modifying reagent adsorbed on the nickel surface, IR[87a,b)] and UV[87c)] spectra, the epitaxy[87d)] of the modifying reagent on nickel films and the fact that one mole of modifying reagent is released when one mole of nickel complex M_2Ni is used as a further modifier,[88)] suggest that (1) the modifying reagent is adsorbed via a chelate-like linkage with the nickel, and (2) the modifying reagent rises above the nickel lattice to provide a chiral environment at the catalyst surface, as shown in Fig. 5.10.

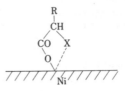

Fig. 5.10.   Adsorption state of the modifying reagent on the surface of the nickel catalyst.

Several reaction mechanisms have been proposed. Klabunovsky[89)] suggested that the MAA absorbed on the catalyst surface may chelate with nickel, as shown in Fig. 5.11, and may be given the effect of chilarity of the modyfying reagent through the nickel in the transition state. However, the chelating power of the substrate has no correlation with the differentiating ability, as shown by the results in Table 5.18. MRNi is much less cor-

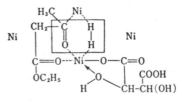

Fig. 5.11.   Mechanism of enantio-differentiating hydrogenation by MRNi catalysts according to Klabunovskii.[89)]

roded by acetylacetone during hydrogenation compared with unmodified Raney nickel,[90)] and the resistance to corrosion correlates with the chelating power of the modifying reagent.   All the above observations suggest

TABLE 5.18
Enantioface-differentiating abilities of tartaric acid–MRNi[1]

Substrate	Product	OY(%)	Ref.
$CH_3COCH_2COCH_3$	$CH_3CHCH_2COCH_3$ $|$ OH	33	(78)
$CH_3COCH_2COOCH_3$	$CH_3CHCH_2COOCH_3$ $|$ OH	33	(78)
$CH_3COC(CH_3)_2COOCH_3$	$CH_3CHC(CH_3)_2COOCH_3$ $|$ OH	23.6[2]	(87a)

[1]   Modified at pH 5 and 0 °C
[2]   This substrate is well-differentiated even though it cannot form an enolate
      Reaction conditions, 60 °C, 80 atm

that the chelation of substrate is not essential for the enantioface-differentiating reaction of MAA.

The interaction between substrate (MAA or acetyl aceton) with modifying reagent on the catalyst surface was demonstrated by the UV spectrometric study using a thin nickel film deposited on a quartz plate by us.[87c]

Recently, Sachtler *et al.*[91a] and Yasumori *et al.*[91b] have independently studied modified nickel catalysts at gas phase reaction by physicochemical methods such as IR spectroscopy or ESCA, and proposed that the substrate adsorbed with catalyst interacts with the modifying reagent by hydrogen bonding. However, there still remains the problem of differences between gas and liquid phase reactions.

Similarly, we can rule out the suggestion that pairs of sites in enantiomeric relation exist on the catalyst surface and that one of these is occupied by the modifying reagent, leaving the other to provide a chiral environment for hydrogenation of the substrate, or that the modifying reagent "corrodes" the surface of the nickel catalyst to provide a chiral cavity for the reaction, since a catalyst modified successively with two different modifying reagents acquires the enantio-differentiating effect of the second modifier, and since removal of the modifying reagent from the surface of the modified catalyst leaves a catalyst with low differentiating ability.[42]

In any case, the proposed reaction mechanisms using empirical models leave the difficulty of completely explaining the phenomena found in reactions with MRNi, such that the direction of differentiation reverses with the change of reaction temperature. Thus, although there are still unsolved problems, we have proposed a mechanism whereby differentiation must take place prior to the transition state in hydrogenation. Our proposal is discussed in detail in section 7.3 and 8.21.

TABLE 5.19

Relationship between the type of substrate and enantioface-differentiating ability of $CH_3-CH(OCH_3)-(CH_2)_nCOOH-MRNi$ catalyst[†] [91d)]

n	$CH_3COCH_2COOCH_3$	$C_6H_5-C(=CH_2)-COOCH_3$	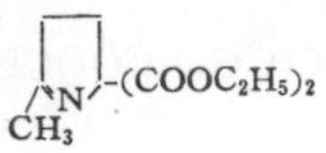 $\underset{CH_3}{N}-(COOC_2H_5)_2$
0	+0.51	+0.03	−0.74
1	+0.76	+0.02	+0.47
2	±	−0.09	+0.09

† In the table, + and − represent directions of optical rotation and the numbers give the optical yield (%)
Reation conditions, 60 °C, 80 atm

TABLE 5.20

Relationship between the type of substrate and enantioface-differentiating abilities of various catalysts[87a)]

	$CH_3COCH_2COOCH_3 \rightarrow CH_3CH(OH)CH_2COOCH_3$		
	Modifying conditions		
Modifying reagent	pH	Temp. (°C)	OY(%)[†]
L-Alanine	6.0	0	0.45
L-Glutamic acid	5.0	0	11.4
L-Tartaric acid	5.0	0	33.0

	$C_6H_5-C(=CH_2)-COOC_2H_5 \rightarrow C_6H_5-CH(CH_3)-COOC_2H_5$		
	Modifying cnoditions		
Modifying reagent	pH	Temp. (°C)	OY(%)[†]
L-Alanine	6.0	0	0.44
L-Glutamic acid	3.2	0	0.14
L-Tartaric acid	1.8	0	0.01

† OY: Optical yield

C. Reactions in optically active solvents

Several enantioface-differentiating reactions have been attempted in optically active solvents, but all the reactions which have been successful involved the use of organometallic reagents which could form complexes with the optically active solvent molecules. Thus, these reactions are no different from reactions involving optically active reagents, even though the optically active compound is initially introduced as the solvent. An example is provided by the formation of ethyl 2-hydroxy-2-phenylbutyrate in 5% optical yield in the Grignard reaction of C_2H_5MgCl with ethyl phenylglyoxylate in (+)2,3-dimethoxybutane as a solvent.[92,93] This reaction was actually carried out in benzene containing (+)2,3-dimethoxy-butane equimolar with respect to the Grignard reagent, and it seems clear that an optically active complex was formed which acted as a reagent.

Various experiments have been carried out using solvents which should readily form complexes with the reagent. These include 1,2,5,6-di-O-isopropylidene-α-D-glucofuranone and sparteine.

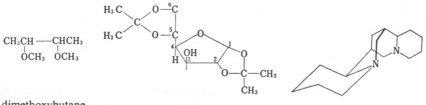

CH₃CH—CHCH₃
| |
OCH₃ OCH₃

dimethoxybutane

1,2,5,6-di-O-
isopropylidene-α-D-
glucofuranone (DIPG)

sparteine

When two moles of CH_3MgBr were subject to reaction with cyclohexyl-phenylketone (Eq. 5.35) in ether solution containing one mole of DIPG, one mole of CH_3MgBr reacts with the 3-OH group of DIPG to form DIPG–O–MgBr, and this coordinates with another mole of CH_3MgBr to form the optically active complex which actually reacts with cyclohexyl-phenylketone to give cyclohexylmethylphenylcarbinol in 70% optical yield.[94]

$$C_6H_5CO-\!\!\big<\!\!\big> + CH_3MgBr \longrightarrow C_6H_5-\overset{\overset{\displaystyle CH_3}{|}}{\underset{\underset{\displaystyle OH}{|}}{C}}-\!\!\big<\!\!\big> \quad (5.35)$$

Similar reactions occur with other organometal reagents. For example, (R)-(+)-1-phenyl-1-propanol is obtained in 2.5% optical yield by the reaction of diphenylcadmium with propionaldehyde in the presence of dimethyl O-dimethyltartrate, as shown in Eq. 5.36.[95]

$$C_2H_5CHO+(C_6H_5)_2Cd \longrightarrow \underset{\underset{OH}{|}}{C_2H_5CHC_6H_5} \qquad \text{1-phenyl-1-propanol} \qquad (5.36)$$

Similarly, the Reformatsky reaction of ethyl bromoacetate and benzaldehyde in the presence of sparteine yields ethyl 3-hydroxy-3-phenylpropionate in 98% optical yield, as shown in Eq. 5.37.[96a]

$$BrCH_2COOC_2H_5+C_6H_5CHO \longrightarrow \underset{\underset{OH}{|}}{C_6H_5CHCH_2COOC_2H_5} \qquad (5.37)$$

Enantioface differentiation can take place in the Simmons-Smith reaction in the presence of optically active compounds. For example, dimethoxycarbonylcyclopropane is produced in 3.4% optical yield by the reaction of the Simmons-Smith reagent obtained from CH_2I_2 and Cu–Zn with dimethyl fumarate in the presence of menthol, as shown in Eq. 5.38.[96b]

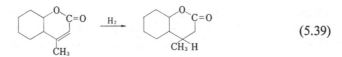

$$(5.38)$$

Enantioface differentiation has also been observed in electrochemical reduction performed in the presence of optically active compounds. The electrochemical reduction of 4-methylcoumarin in the presence of sparteine gives 3,4-dihydromethylcoumarin in 17% optical yield, as shown in Eq. 5.39.[97]

$$(5.39)$$

4-methylcoumarin

$$\underset{\underset{}{|}}{\overset{CH_3}{C_6H_5}} \underset{}{\overset{}{C}} = NCH_2C_6H_5 \xrightarrow[\underset{\underset{OH}{\overset{|}{N(CH_3)_3I}}}{\overset{H\ H}{|\ \ |}}{\underset{C_6H_5-C-C-CH_3}{\overset{H_2}{}}}] \underset{\overset{CH_3}{|}}{C_6H_5\overset{}{C}HNHCH_2C_6H_5} \qquad (5.40)$$

Eq. 5.40 shows the reduction of a Schiff base to N-benzylmethylbenzylamine in 8.1% optical yield in the presence of an optically active 1-phenyl-2-trimethylaminopropanol salt.[98] Differentiation is expected to proceed through an optically active radical produced in the electric double layer.[99]

5.1.2 Enantiotopos-differentiating reactions

Two kinds of enantiotopos-differentiating reactions may be considered: one is the differentiating substitution of enantiotopic ligands, and the other is differentiating addition to enantiotopic lone pair electrons. Some typical enantiotopos-differentiating reactions will be discussed below.

A. Reaction with optically active reagents

The complex of butyl lithium and sparteine shown in Eq. 5.41 differentiates the enantiotopic hydrogens at the 3,4-position on isopropyl ferrocene, producing an excess of optically active 3,1′-dilithio-1-isopropyl ferrocene.

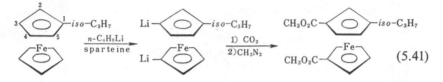

$$(5.41)$$

The optical yield was estimated to be 3% by determination of the optical purity of 3-isopropyl-1,1′-ferrocene dicarboxylic acid dimethyl ester obtained by successive treatments with CO_2 and diazomethane (Eq. 5.41).[100]

When the acid anhydride of a dicarboxylic acid having enantiotopic carboxyl groups, such as 3-phenylglutaric acid anhydride, is treated with an optically active alcohol or amine, as in Eq. 5.42, differentiation of the enantiotopic carboxyl moieties occurs: in the case of Eq. 5.42, a 60:40 diastereomer ratio of the product amide is obtained.[101] Reaction of the same anhydride with (−)-menthol (Eq. 5.43) similarly gives a 54:46 diastereomer ratio of the ester product.[102]

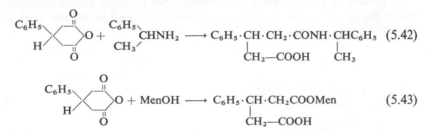

The differentiation of enantiotopic carbonyl groups can be performed by chiral hydrazides. Eq. 5.44 shows the reaction of an 8,14-secosteroid with L-tartaric acid amide hydrazide, producing an optically active steroid hydrazide. Since this product is easily epimerized, the optical yield was estimated from the optical purity of the stable product obtained by *N*-methylation immediately after formation of the steroid hydrazide (Eq. 5.44).[103]

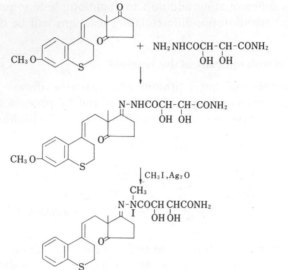

$$+ \quad NH_2NHCOCH\text{-}CH\text{-}CONH_2$$
$$\qquad\qquad\qquad OH\ OH$$

N-NHCOCH-CH-CONH₂
$$\qquad\qquad OH\ OH$$

(5.44)

$$\big\downarrow CH_3I, Ag_2O$$

CH₃
N-NCOCH CHCONH₂
$$\quad\ \overset{\parallel}{O}\ \overset{|}{I}\ \ \overset{|}{OH}\overset{|}{OH}$$

Eq. 5.45 shows the differentiation of the enantiotopic carbon atoms of 4-methylcyclohexanone by the optically active reagent 2-octyl nitrite yielding an optically active oxime; $[\alpha]_D = +17°$.[104]

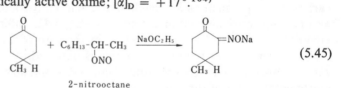

$$+ \ C_6H_{13}\text{-}CH\text{-}CH_3 \quad \xrightarrow{\ NaOC_2H_5\ }$$
$$\qquad\qquad \overset{|}{ONO}$$

=NONa

(5.45)

2-nitrooctane

Ephedrine differentiates the enantiotopic chlorine atoms of thionyl chloride, as shown in Eq. 5.46, yielding oxathiazoline-2-oxide with an 80:20 diastereomer ratio.[105]

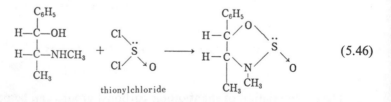

thionylchloride

(5.46)

Thio-ethers having enantiotopic lone pair electrons can be oxidized by optically active peracids to yield optically active sulfoxides. For example, the oxidation of methyloxycarbonylbenzyl-*tert*-butyl sulfide with mono-peroxy camphoric acid (Eq. 5.47) yields an optically active sulfoxide in 5.4% optical yield.[106]

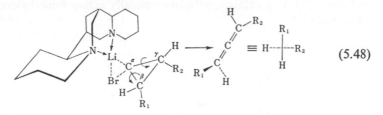

$$\text{(5.47)}$$

The formation of an optically active allene by reaction between *gem*-dibromocyclopropane derivatives and an optically active complex consisting of sparteine and butyl lithium[107] is considered to be a special case of enantiotopos differentiation. The reaction proceeds through the intermediate shown in Eq. 5.48.

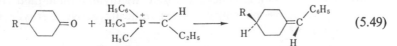

$$\text{(5.48)}$$

Since the formation of the allene is known to proceed through the fission of the C_β–C_γ σ bond with the subsequent formation of π bonds along the C_α–C_β and C_α–C_γ bonds, the configuration of the resulting allene depends on the direction of conrotatory rotation about these bonds. In this case, the directions of rotation with respect to the α carbon are enantiotopic, so the formation of an optically active allene can be attributed to enantiotopos differentiation by the optically active reagent.

Eq. 5.49 shows the formation of an optically active cycloalkylidene compound by the Witting reaction of an optically active phosphoylide derived from (S)-$(+)$-benzylphenylpropylphosphonium bromide with 4-alkylcyclohexanone.

$$\text{(5.49)}$$

Since the ylide is considered to differentiate unoccupied enantiotopic spaces of the cyclohexanone, the reaction is classified as an enantiotopos-differentiating reaction.[108]

B. Reactions with catalysts

Enantiotopos-differentiating oxidation of cycloalkenes with achiral peroxides can be carried out in the presence of an optically active catalyst. For example, the oxidation of cyclohexene with *tert*-butyl hydroperoxide in the presence of the cupric salt of (+)-α-camphoric acid monoethyl ester yields 2-cyclohexene-1-ol in 6.6% optical yield (Eq. 5.50).[109]

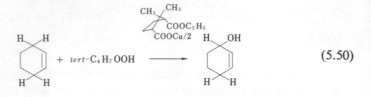

$$\text{(5.50)}$$

Enantiotopos-differentiating dehydration of cyclic alcohols with an optically active catalyst gives optically active cycloalkenes. An example is provided by the dehydration of 4-methylcyclohexanol with camphor sulfonic acid as a catalyst, yielding optically active 4-methylcyclohexene (Eq. 5.51).[110]

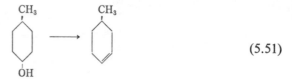

$$\text{(5.51)}$$

Differentiation of the enantiotopic carbonyl groups of 2-*n*-butyl-2-methylpentadione occurs during cyclization with L-proline as a catalyst, giving 1-methylbicyclo[4,3,0]nona-5-en-2-one in 84% optical yield (Eq. 5.52).[111a]

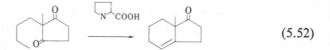

$$\text{(5.52)}$$

The codimerization of norborane with ethylene in the presence of a catalyst consisting of π-allylnickel chloride, triethyldialuminium trichloride and optically active phosphane, gives optically active 2-ethylidenenorbornane (Eq. 5.53).[111b]

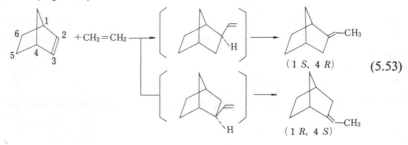

$$\text{(5.53)}$$

When the reaction is carried out under the control of $(-)$-isopropyl-dimenthyl phosphane,

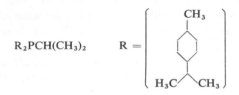

$$R_2PCH(CH_3)_2 \qquad R =$$

the $(+)$ isomer $(1S, 4R)$ is obtained in excess, with an optical yield of 45.4% (reaction temp. $-25°C$) to 86% (reaction temp. $-65°C$). Since the addition of ethylene at the first step produces *exo*-isomer exclusively and the resulting chirality at the C_2 carbon disappears at the final stage, optical active product is obtained by differentiation of two carbon atoms (C-1 and C-2, respectively) in enantiotopic relation. Thus, the apparent differentiation of this reaction is classified as an enantiotopos-differentiating reaction.

As a special case of enantiotopos-differentiating reactions, the reaction in Eq. 5.54 is given as an example. Here, the chiral reagent differentiates the enantiotopic spaces spread on both sides of the enantio zero plane of the substrate. However, since there are very few examples of this type of reaction, it is included among enantiotopos-differentiating reactions.

Eq. 5.54 consists of the Wittig reaction of 4-alkylcyclohexanone with carboethoxymethylene triphenylphosphine in the presence of an optically active catalyst (mandelic acid, thiazolidene-4-carboxylic acid or *N*-naphthoylamino acid) to give an optically active 4-alkylcyclohexylidene acetic acid ester.[112)]

$$R-\left\langle\begin{array}{c}\end{array}\right\rangle=O + (C_6H_5)_3P=CH-COOC_2H_5 \longrightarrow \quad \overset{H_{\diagdown}}{\underset{R}{\diagup}}\left\langle\begin{array}{c}\end{array}\right\rangle=CH-COOC_2H_5 \quad (5.54)$$

In this reaction, as mentioned, enantiotopic spaces adjacent to the alkyl-cyclohexanone are differentiated, and the product possesses axial chirality.

5.1.3 Enantiomer-differentiating reactions

As discussed in Chapter 4, kinetic resolution is performed by enantiomer-differentiating reactions. Thus, most of the results presented in this section show only the trend of differentiation.

A. Reactions with chiral reagents

Enantiomer differentiation of chiral halogen compounds occurs when an optically active amine is used as a reagent. For example, partial reaction of brucine with the α isomer of 1,2,3,4,5,6-hexachlorocyclohexane leaves the unchanged (−)-enantiomer.[113]

The enantiomers of chiral *sec*-alcohols are differentiated during partial acylation with acid anhydrides in the presence of optically active amines, leaving unchanged optically active alcohol. It has been proposed that the actual differentiating reagents in these reactions are acylated optically active amines formed from the acid anhydride and the optically active amine during the reaction, as shown in Eq. 5.55.[114]

$$(RCO)_2O + B^* \longrightarrow RCOB^* \quad (B^* = \text{optically active amine})$$

$$RCOB^* + \begin{array}{c} R' \\ \diagdown \\ CHOH \\ \diagup \\ R'' \end{array} \longrightarrow \begin{array}{c} R' \\ \diagdown \\ CHOCOR \\ \diagup \\ R'' \end{array} + \begin{array}{c} R' \\ \diagdown \\ CHOH \\ \diagup \\ R'' \end{array} \qquad (5.55)$$

Optically active amines also differentiate the enantiomers of reactive derivatives of chiral fatty acids (e.g., that shown in Eq. 5.56). The intermediate in Eq. 5.56 was treated with 2/3 molar equivalent of optically active α-methylbenzylamine and optically active starting material was regenerated from the remaining unchanged intermediate after its recovery.[115]

Optically active acid chlorides differentiate the enantiomers of chiral amines. For example, when (+)-camphor-10-sulfonyl chloride was treated with an excess of (RS)-α-methylbenzylamine, unchanged α-methylbenzylamine containing excess S-isomer was recovered (Eq. 5.57).[116]

$$C_6H_5\underset{\underset{NH_2}{|}}{CHCH_3} + \text{[structure: } CH_2SO_2Cl \text{ camphor]} \longrightarrow C_6H_5\underset{\underset{CH_3}{|}}{CHNHSO_2}CH_2\text{-[structure]} \qquad (5.57)$$

Enantiomer-differentiation of chiral acyl halides occurs when they are treated with less than equivalent amounts of optically active alcohols, as shown in Eq. 5.58.

$$Men\text{-}OCH_2COCl + Men\text{-}OH \longrightarrow Men\text{-}OCH_2COO\text{-}Men \qquad (5.58)$$

TABLE 5.21

Effects of various solvents on the reaction shown in Eq. 5.57

Solvent	k_{dl}/k_{ll}
Chloroform	10.8
Benzene	5.8
n-Hexane	4.9
Acetonitrile	4.3
Methyl ethyl ketone	1.04
Liquid SO_2	0.46

In Eq. 5.58, $(-)$-menthol was treated with the $(\pm)$-acid chloride and unchanged acid chloride was recovered. Table 5.21 shows the effect of the solvent conditions on the enantiomer-differentiating ability of the reagent. Change of the solvent not only affects the differentiating ability of the reagent but can also reverse the direction of differentiation.[117]

Optically active ketones undergo enantiomer-differentiating reactions with chiral Grignard reagents. Eq. 5.59 shows the partial reaction of (RS)-phenylbutylmagnesium chloride with $(+)$-camphor. Recovery of unchanged Grignard reagent (containing the S isomer in excess) and treatment with carbon dioxide gave optically active 3-phenylvaleric acid.[118]

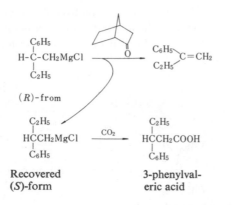

$$(5.59)$$

Enantiomer differentiation occurs in the Meerwein-Ponndorf-Verley reaction of a chiral ketone with an optically active alkoxide. The ratio of the reduction rates of (S)-$(+)$- and (R)-$(-)$-4′,1″-dinitro-1,2,3,4-dibenz-1,3-cycloheptadiene-6-one was found to be 0.34 (Eq. 5.60). Thus reduction of 80% of the (RS)-form with (S)-$(+)$-methyl-*tert*-butylcarbinol-aluminium alkoxide left the (S)-$(+)$-form of the substrate in excess. The enantiomers of this compound are also atrop isomers.[119]

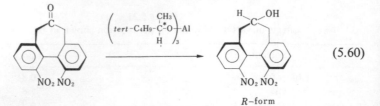

$$(5.60)$$

R–form

Pinanylborane, which can be prepared from optically active pinene and diborane, as shown in Eq. 5.61, performs the enantiomer-differentiating reduction of chiral alkenes. The reaction in Eq. 5.62 leaves (S)-3-methylcyclopentene in 65% optical yield by the partial reduction of (±)-3-methylcyclopentene with (+)-tetra-3-pinanylborane (see Eq. 5.61).[120]

$$(5.61)$$

$$(5.62)$$

The results of enantiomer-differentiating reduction of (±)-*trans*-cyclo-octene (Eq. 5.63)[121] or (±)-1,3-dimethylallene (Eq. 5.64)[122] with (+)-tetra-3-pinanylborane show that this reducing agent possesses enantiomer-differentiating ability for compounds with axial chirality.

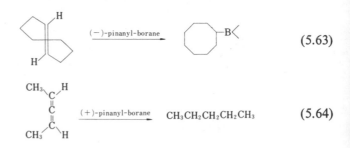

$$(5.63)$$

$$(5.64)$$

Partial reductions as shown leave excesses of the unchanged (R)-forms of the substrates (21% optical yield in Eq. 5.63).[121]

Optically active thiols or O-alkyl-alkanthiophosphonic acids preferentially reduce one enantiomer of chiral sulfoxides, so that if excess sulfoxide is used, the recovered sulfoxide will be optically active. Eq. 5.65 shows the reaction of excess (RS)-methylphenylsulfoxide with L-cystine, leaving excess unreacted (R)-methylphenylsulfoxide.[123] Optically active (S)-ethylmethylsulfoxide can similarly be recovered from the partial reduction of (±)-ethylmethylsulfoxide with (S)-O-ethylethanthiophosphonic acid (Eq. 5.66).[124]

$$C_6H_5SOCH_3 \xrightarrow{\overset{\displaystyle HSCH_2CHCOOH}{\underset{\displaystyle NH_2}{|}}} C_6H_5\text{-}S\text{-}CH_3 \qquad (5.65)$$

$$C_2H_5SOCH_3 \xrightarrow{\overset{\displaystyle OC_2H_5}{\underset{\displaystyle OH}{\overset{|}{C_2H_5\text{-}P=S}}}} C_2H_5SCH_3 \qquad (5.66)$$

Optically active peracids carry out the enantiomer-differentiating oxidation of chiral sulfoxides to achiral sulfones. For instance, optically active peroxy-2-phenylpropionic acid partially oxidized methylphenylsulfoxide, as shown in Eq. 5.67, leaving (R)-methylphenylsulfoxide.[125]

$$C_6H_5\text{-}SO\text{-}CH_3 \xrightarrow{\overset{\displaystyle CH_3}{\overset{|}{C_6H_5CHCO_3H}}} C_6H_5SO_2CH_3 \qquad (5.67)$$

Fehling's solution prepared from optically active tartaric acid and copper sulfate performs the enantio-differentiating oxidation of sugars. For example, D-altrose is oxidized 1.6 times faster by Fehling's solution prepared from sodium (S,S)-tartrate than that from sodium (R,R)-tartrate.[126,127]

$$
\begin{array}{c}
CHO \\
| \\
HO-C-H \\
| \\
H-C-OH \\
| \\
H-C-OH \\
| \\
H-C-OH \\
| \\
CH_2OH
\end{array}
$$

D-altrose

B. Reactions with catalysts

Optically active catalysts can also differentiate the enantiomers of chiral substrates. For instance, (+)-camphorsulfonic acid catalyzes the dehydration of (±)-phenylethanol, leaving an excess of (−)-phenylethanol, as shown in Eq. 5.68.[128a]

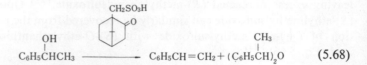

$$\begin{matrix} & & CH_2SO_3H \\ OH & & \\ | & & \\ C_6H_5CHCH_3 & \longrightarrow & C_6H_5CH=CH_2 + (C_6H_5CH)_2O \end{matrix} \qquad (5.68)$$

Optically active metal complexes can also perform enantiomer-differentiating catalytic oxidation. For instance, $(-)$-$[Co(en)_2NH_3Cl]Br_2$ (en = $NH_2CH_2CH_2NH_2$)[128b] and L-lysine-Cu^{2+} [129] perform the enantiomer-differentiating catalytic oxidation of 3,4-dioxyphenylalanine (DO PA) as shown in Eq. 5.69.

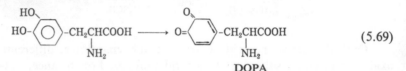

$$HO-\langle O \rangle-CH_2CHCOOH \longrightarrow O=\langle O \rangle-CH_2CHCOOH \qquad (5.69)$$

DOPA

Enantiomer-differentiating polymerization by chiral metal complexes is well known (see section 6.1), but similar complexes prepared from tetra-$(-)$-menthoxytitanium and tri-*iso*-butylaluminium also carry out the enantiomer-differentiating isomerization of double bonds of chiral olefins, as shown in Eq. 5.70 for 3,4-dimethyl-1-pentene.[130]

$$CH_2=CHCHCH(CH_3)_2 \xrightarrow{(iso-C_4H_9)_3Al} CH_3CH=CCH(CH_3)_2 \qquad (5.70)$$
$$\qquad | \qquad\qquad\qquad\qquad\qquad\qquad |$$
$$\quad CH_3 \qquad\qquad\qquad\qquad\qquad\qquad CH_3$$

C.   Reactions in chiral solvents

Enantiomer-differentiating reactions of chiral substrates can also occur in optically active solvents. For example, the partial oxidation of $(\pm)$-2,2'-dichlorobenzoin with oxygen in the presence of α-cyclodextrin leaves $(+)$-2,2'-dichlorobenzoin in excess in the unchanged material (Eq. 5.71).[131]

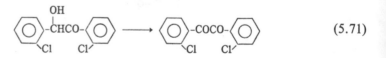

$$\langle O \rangle-CHCO-\langle O \rangle \longrightarrow \langle O \rangle-COCO-\langle O \rangle \qquad (5.71)$$

5.2   DIASTEREO-DIFFERENTIATING REACTIONS

In this section, a general discussion of diastereo-differentiating reaction is given first, followed by typical examples of such reactions.

The addition reaction of HCN to L-arabinose followed by hydrolysis of the addition product yields more L-mannoic acid than L-gluconic acid, as shown in Eq. 5.72.

$$
\begin{array}{c}
\text{CHO} \\
| \\
\text{H-C-OH} \\
| \\
\text{HO-C-H} \\
| \\
\text{HO-C-H} \\
| \\
\text{CH}_2\text{OH}
\end{array}
\xrightarrow{\text{HCN}}
\left[
\begin{array}{cc}
\text{CN} & \text{CN} \\
| & | \\
\text{H-C-OH} & \text{HO-C-H} \\
| & | \\
\text{H-C-OH} & \text{H-C-OH} \\
| & | \\
\text{HO-C-H} & \text{HO-C-H} \\
| & | \\
\text{HO-C-H} & \text{HO-C-H} \\
| & | \\
\text{CH}_2\text{OH} & \text{CH}_2\text{OH}
\end{array}
\right]
\xrightarrow[\text{H}^+]{\text{H}_2\text{O}}
\begin{array}{cc}
\text{COOH} & \text{COOH} \\
| & | \\
\text{H-C-OH} & \text{HO-C-H} \\
| & | \\
\text{H-C-OH} & \text{H-C-OH} \\
| & | \\
\text{HO-C-H} & \text{HO-C-H} \\
| & | \\
\text{HO-C-H} & \text{HO-C-H} \\
| & | \\
\text{CH}_2\text{OH} & \text{CH}_2\text{OH}
\end{array}
$$

L-arabinose   L-mannoic acid   L-gluconic acid

(5.72)

This was first recognized by Fischer in 1894.[40] It was the first discovery of a diastereo-differentiating reaction, and marked the beginning of research in the field of asymmetric synthesis.

5.2.1 Factors affecting the efficiency of diastereo-differentiation

In diastereo-differentiating reactions, the chirality of the substrate (chirality factor) contributes to the stereo-differentiation, but the role of the reagent or catalyst is also important. Thus, we may say that a substrate chirality factor is a necessary but not sufficient condition for diastereo-differentiation. The importance of the reagent or catalyst should be emphasized, since the nature of the reagent or catalyst not only changes the degree of the diastereo-differentiation, but may also change the direction of differentiation in reactions with a given substrate. The reaction conditions can also greatly affect the results.

A. Effect of substrate structure

The efficiency of diastereoface differentiation depends on the probability of formation of a diastereo-zeroplane in the substrate. Thus, a substrate which provides a configurational diastereo-zeroplane will be differentiated by a reagent or catalyst with higher efficiency than one which provides a conformational diastereo-zeroplane.

Examples of conformational diastereoface-differentiation are given in Eqs. 7.73 and 5.74. Eq. 5.73 shows the formation of L-alanine and L-phenylalanine in 8% and 5% optical yield, respectively, from the corresponding α-ketoacid hydrazones of (*R*)-2-methyl-3-phenylpropionyl hydrazide.[132]

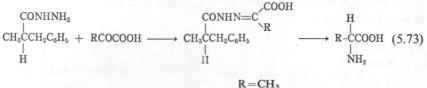

Eq. 5.74 shows a similar type of reaction using (S)-N-aminoanabasine. This yields alanine with an optical yield of 40%, an excellent result for a conformational diastereoface-differentiating reaction.[133]

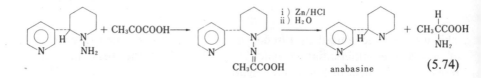

Eq. 5.75 shows the configurational diastereoface-differentiating hydrogenation of a cyclic compound obtained from $(+)$-1,2-diphenyl-ethanolamine and dimethylacetylenedicarboxylate, yielding L-aspartic acid in 98% optical yield. This reaction was carried out by Horeau et al.[134]

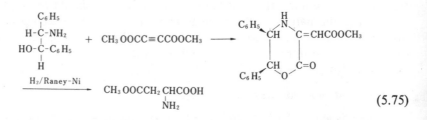

Corey et al.[135] have obtained good results in the configurational diastereoface-differentiating hydrogenation of the optically active cyclic hydrazone shown in Eq. 5.76; the hydrazone was prepared from an N-amino-2-hydroxymethylindoline derivative and an α-ketoester.

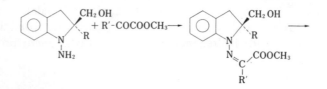

TABLE 5.22

Optical yields of amino acids prepared by the procedure of Corey et al. (Eq. 5.74)

α-Ketoester		Product	Optical yield (%)
(S)-(−), R = H	Methyl pyruvate	(R)-alanine	78–82
(S)-(−), R = CH₃	Methyl pyruvate	(R)-alanine	75
(S)-(−), R = H	Methyl α-ketobutyrate	(R)-α-aminobutyric acid	89–90

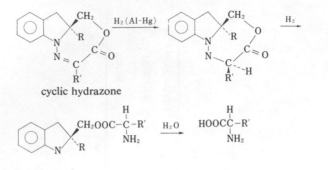

cyclic hydrazone

$$(5.76)$$

After reduction of the cyclic hydrazone with Al–Hg, the product was further hydrogenated and then hydrolyzed. Very high optical yields were obtained, as shown in Table 5.22. Even better results were obtained with the cyclic hydrazone shown in Table 5.23. This also contains a seven-membered lactone ring, but in this case, two chiral centers are present in the ring rather than one.

TABLE 5.23
Optical yields of amino acids prepared by the procedure of Corey *et al.*

	Product	Optical yield (%)
R=CH$_3$	Alanine	96
C$_2$H$_5$	α-Aminobutyric acid	97
iso-C$_3$H$_7$	Valine	97
sec-C$_4$H$_9$	Isoleucine	99

Historically, conformational diastereoface-differentiating reactions have been well investigated and various empirical rules, such as those of Cram or Prelog, have been established. One of the most extensively investigated reactions is the diastereoface-differentiating reaction of the α-ketoacid ester of an optically active alcohol with a Grignard reagent. Differentiation is performed with high efficiency for substrates which can form a suitable diastereoface with high probability. Table 5.24 shows the optical yields obtained with ketoesters having various substituents; R_L should be bulky and R_L and R_M should differ greatly in bulk for good differentiation (Eq. 5.77).

TABLE 5.24
Effects of various substituents R_L in the reaction shown in Eq. 5.77

R	R_L	R_M	Optical yield (%)	Ref.
C_6H_5	C_6H_5	CH_3	6.8	(136)
C_6H_5	α-Naphthyl	CH_3	12	(137)
C_6H_5	n-C_6H_{13}	CH_3	18	(137)
C_6H_5	$C(CH_3)_3$	CH_3	24	(138)
C_6H_5	Mesityl	CH_3	30	(137)
C_6H_5	$C(C_6H_5)_3$	CH_3	49	(137)

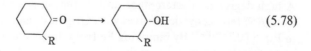

$$\tag{5.77}$$

Table 5.25 shows that the group R has relatively little effect on differentiation in this reaction.

TABLE 5.25
Effects of various substituents R in the reaction shown in Eq. 5.77

R	$R_L R_M CH-$	R'	Product Configuration	Product OY[†1] (%)	Ref.
C_6H_5	(−)-Menthyl	CH_3	*R*	30.5	(139)
p-$CH_3C_6H_4$	(−)-Menthyl	CH_3	*R*	25	(140)
p-$CH_3OC_6H_4$	(−)-Menthyl	CH_3	*R*	26	(141)
α-Naphthyl	(−)-Menthyl	CH_3	*R*	29[†2]	(142)

[†1] Optical yield
[†2] Estimate

In the reduction of 2-substituted cyclohexanone with $LiAlH_4$ (Eq. 5.78),[143] the bulk of the group R, which acts as a differentiating factor, markedly affects the results, as shown in Table 5.26.

$$\tag{5.78}$$

TABLE 5.26
Effect of various substituents R on the configurational diastereoface-differentiating reduction of 2-substituted cyclohexanone (Eq. 5.78)

R	trans	cis
CH_3	3	1
$\dfrac{CH_3}{CH_3}$>CH	2	1
$(CH_3)_3C$	1	1

Table 5.27 shows the results of the diastereoface-differentiating reaction of a Grignard reagent with various ketoacid menthylesters.[144] It can be seen that the degree of differentiation decreases as the distance from the chiral center to the carbonyl group increases, reducing the probability of diastereoface formation.

TABLE 5.27
Effect of the distance between the chiral center and the reaction center on differentiation

Substrate $CH_3CO(CH_2)_n COOMen$	Optical yield (%)
$n = 0$	18
1	—[†1]
2	5–13[†2]
3	1.6–16[†2]
4	$[\alpha]_{5760}^{25} = +0.09°$
6	(±)

[†1] In this case the addition compound cannot be produced through the enol form
[†2] The product is obtained as the lactone and the optical yield of the lactone is shown

Axial and planar chiralities can also be utilized in diastereoface-differentiating reactions, as well as chiral centers. For instance, D-alanine is obtained in 6.4% optical yield by the hydrogenation of $(-)$-N-(4-[2-2]-para-cyclophanecarbonyl)-dehydroalanine with a platinum catalyst, as shown in Eq. 5.79.[145]

$$\text{(structure)} \quad \begin{array}{c} i\,)\ H_2 \\ ii\,)\ HCl \end{array} \quad CH_3CHCOOH \quad NH_2 \qquad (5.79)$$

A high degree of diastereoface-differentiation is also observed in the reaction (RS)-ferrocenyl derivatives with Grignard reagents or KBH_4, as shown in Eq. 5.80.[146,147] By removing Fe from the product, a tert- or sec-alcohol of high optical purity is obtained.

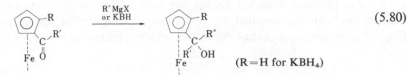

$$(5.80)$$

$$(R=H \text{ for } KBH_4)$$

Diastereoface-differentiation can also occur in unsaturated compounds with a chiral hetero atom rather than a chiral carbon atom. For example, (R)-$(+)$-benzoyl-1-naphthylphenylmethylsilane gives an alcohol containing the (R,S)-isomer in excess on reaction with CH_3MgBr (Eq. 5.81).[148]

$$(5.81)$$

Diastereoface-differentiation also occurs in the reaction of (S)-$(+)$-p-tolylvinylsulfoxide with Br_2, giving the $(+)$ product in 32% diastereomer excess[149] (Eq. 5.82).

$$p\text{-}CH_3C_6H_4\overset{\uparrow}{\underset{}{S}}CH=CH_2 \xrightarrow{\ Br_2\ } p\text{-}CH_3C_6H_4\overset{\uparrow}{\underset{}{S}}\underset{\underset{Br}{|}}{C}H\underset{\underset{Br}{|}}{C}H_2 \qquad (5.82)$$

B. Effects of reaction conditions and differentiating ability of the reagent or catalyst

Although diastereoface-differentiation is affected by the bulk of ligands at the chiral center, as described in the previous section, the degree of differentiation depends more on the differentiating ability of the reagent or catalyst than on the structure of the substrate. For instance, Table 5.28

TABLE 5.28

Diastereoface-differentiating ability of various reagents and catalysts in Eq. 5.83

Reducing reagent	Product cis/trans ratio	Ref.
$LiAlH_4$	21/79	(150)
$NaBH_4$	26/74	(151)
B_2H_6	31/69	(152)
$Al(O\text{-}iso\text{-}C_3H_7)$	76/24	(153)
$H_2(PtO_2)$	28/72	(151)
$H_2(Pt, HCl)$	65/35	(153)

shows that different reducing agents not only affect the proportions of diastereomers in the product of reduction of 2-methylcyclopentanone (Eq. 5.83) but also may change the isomer which is preferentially produced.

$$\text{(cyclopentanone with CH}_3\text{)}=O \longrightarrow \text{(cyclopentanol with CH}_3\text{)}-OH \qquad (5.83)$$

Table 5.29 shows that the degree of differentiation depends greatly on the reagent or catalyst in the reduction of the optically active imine shown in Eq. 5.84, and also that the features of the relationship between structure and degree of differentiation depend on the reagent or catalyst.[154,155]

$$\begin{array}{c}R\\ \diagdown\\ \quad \ \ C=O+NH_2CHCH_3 \longrightarrow\end{array} \begin{array}{c}R\\ \diagdown\\ \quad \ \ C=N\\ CH_3 \diagup \qquad \diagdown CHCH_3\end{array} \xrightarrow[\substack{\text{or}\\ \text{reagent}}]{\substack{H_2\\ \text{catalyst}}}$$

$$\begin{array}{c}R \quad \overset{H}{\underset{|}{}}\\ \diagdown\\ \quad \ \ CHNHCC_6H_5\\ CH_3 \diagup \quad \underset{|}{} \\ \qquad \quad CH_3\end{array} \xrightarrow{H_2 Pd-C} \begin{array}{c}R\\ \diagdown\\ \quad \ \ CHNH_2\\ CH_3 \diagup\end{array} \qquad (5.84)$$

TABLE 5.29

Effect of catalyst (or reagent)-substrate interaction on diastereoface-differentiation (Eq. 5.82)

Substrate		Product[†] configuration and optical yield (%)		
R	Configuration	Pd-C	LiAlH$_4$	B$_2$H$_6$
C$_2$H$_5$	S	S-28	S-10	S-20
iso-C$_3$H$_7$	R	R-55	R-10	R-50
tert-C$_4$H$_9$	R	R-56	R-5	R-50

† After reduction with H$_2$/Pd-C, as shown in Eq. 5.82

Diastereo-differentiation may also be influenced by the solvent. For example, Eq. 5.85 shows the diastereoface-differentiating reaction of (S)-1-tert-butyl-n-valerylester phenylglyoxalate with CH$_3$MgI. (S)-Atrolactic acid is obtained in 7.5% excess in ether solution, according to Prelog's rule, but no differentiation occurs in THF.[156]

$$C_6H_5COCOOCH\begin{array}{c}(CH_2)_3CH_3\\ \diagup\\ \diagdown\\ C(CH_3)_3\end{array} \xrightarrow{CH_3MgI} C_6H_5\overset{\overset{OH}{|}}{\underset{\underset{CH_3}{|}}{C}}COOCH\begin{array}{c}(CH_2)_3CH_3\\ \diagup\\ \diagdown\\ C(CH_3)_3\end{array}$$

$$\longrightarrow C_6H_5\overset{\overset{OH}{|}}{\underset{\underset{CH_3}{|}}{C}}COOH \qquad (5.85)$$

atrolactic acid

Sometimes the direction of differentiation is reversed by the solvent effect. For example, in the case of Michael's reaction of menthyl monochloroacetate with acrylic acid ester, *trans*-cyclopropane dicarboxylic acid ($[\alpha]_D^{24} = -4.0-2.8°$) is obtained in toluene, but product with $[\alpha]_D^{23} = +8.3$ –9.3° is obtained in DMF.[157] (Another example is shown below in Eq. 5.90.[151]) These effects of the solvent can be explained in terms of the different reaction species produced in each solvent as a result of solvation of the reagent. The solvent can also affect diastereo-differentiation in catalytic reactions, e.g., in the reductive amination of methyl pyruvate (Eq. 5.84, R = Me) with optically active α-methylbenzylamine using Pd-C as a catalyst (Fig. 5.12).[158]

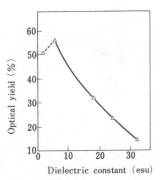

Fig. 5.12. Relationship between the dielectric constant of the solvent and diastereo-differentiation by the catalyst in the reductive amination of methyl pyruvate with α-methylbenzylamine (Pd-C catalyst).[158]

Table 5.30 shows the effect of temperature on the reduction of methyl benzylmethylketone with LiAlH$_4$ or Grignard reagent (Eq. 5.86).[159]

$$\underset{\underset{CH_3}{|}}{C_6H_5CHCOR} \longrightarrow \underset{\underset{CH_3 \ \ OH}{|\ \ \ \ |}}{C_6H_5CH-CHR'} \qquad (5.86)$$

TABLE 5.30
Effect of temperature on diastereo-differentiation in Eq. 5.84[159]

R	R′	X	Reaction temp. (°C)	Product (RS/RR) (SR/SS)
H	CH$_3$	CH$_3$MgBr	−50	2.4
			0	2.3
			35	2.0
CH$_3$	CH$_3$	LiAlH$_4$	−70	5.6
			0	3.0
			35	2.6

As the reaction temperature was reduced, higher diastereo-differentiation was observed. The effect of reaction temperature, however, is often more complex than would be expected from a kinetic viewpoint (see section 7.1). For instance, Fig. 5.13 shows that an increase in the reaction temperature for the reaction shown in Eq. 5.84 (R = COOC$_2$H$_5$) not only greatly decreases the degree of diastereo-differentiation, but also reverses its direction.[160]

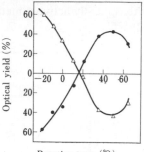

Reaction temp. (℃)

Fig. 5.13.   Effect of temperature on diastereo-differentiating hydrogenation catalyzed by Pd-C.[160] The effective diastereoface differentiation is shown in terms of the optical yield of alanine obtained. ●, Reductive amination with (R)-(+)-methylbenzylamine; △, reductive amination with (S)-(−)-methyl-benzylamine.

The pressure of hydrogen also affects diastereo-differentiation in the reaction shown in Eq. 5.87 (see Table 5.31).[161]

$$\underset{\substack{CH_3}}{\overset{\substack{C_6H_5}}{}}C=N-\underset{\substack{CH_3}}{\overset{\substack{C_6H_5}}{CH}} \xrightarrow{\ H_2/Pd\text{-}C\ } \underset{\substack{CH_3}}{\overset{\substack{C_6H_5}}{}}CHNH\underset{\substack{CH_3}}{\overset{\substack{C_6H_5}}{CH}} \qquad (5.87)$$

TABLE 5.31
Effect of hydrogen pressure on diastereoface-differentiating hydrogenation with Pd-C (Eq. 5.85)

Hydrogen pressure (atm)	*racemic/meso* ratio of products
10	80/20
8	79/21
6	79/21
4	74/26
2	75/25
1	75/25

This phenomenon has been explained in terms of a change in the adsorption state of the substrate with hydrogen pressure, but it seems likely that more complex mechanisms are actually at work in the differentiation process.

From the above discussion, it is clear that diastereo-differentiating reactions are complex, being affected by many factors such as substrate structure, reagent (or catalyst), solvent, reaction temperature, pressure, etc. This must always be considered in attempting to interpret experimental results obtained under various conditions (see Chapter 8).

5.2.2 Diastereoface-differentiating reactions

Diastereoface-differentiating reactions which have been described in the literature will be reviewed briefly below.

A. Reactions of chiral carbonyl compounds

Various diastereoface-differentiating reductions using Grignard reagents, have been described previously (Eq. 5.77). Some analogous results of Grignard reductions are listed in Table 5.32 for the reaction shown in Eq. 5.88.[162a)]

$$
\underset{\overset{|}{CH_3}}{\overset{HO\ \ O}{C_6H_5-\overset{|}{C}-\overset{||}{C}-C_6H_5}} \longrightarrow \underset{\overset{|}{H_3C}\ \overset{|}{CH_3}}{\overset{HO\ \ OH}{C_6H_5-\overset{|}{C}-\overset{|}{C}-C_6H_5}} \tag{5.88}
$$

Diastereoface-differentiating reduction or reductive amination of chiral ketones can be performed by catalytic hydrogenation or by reduction with $LiAlH_4$. An example is provided by the hydrogenation of pyruvoyl-L-alanine isobutyl ester in the presence of benzylamine, using Pd-C as a catalyst.[162b)] This reaction yields 64% diastereomer excess of D-alanyl-L-alanine.

TABLE 5.32
Diastereoface-differentiation by Grignard-type reagents in Eq. 5.86

Reagent	Solvent	*racemic/meso* ratio of product
CH_3Li	Ether	8–11
$(CH_3)_2Mg$	Ether	3.5
	THF	6
$(CH_3)_2MgMgBr_2$	Ether	2.7–3
	THF	10

Eq. 5.89 shows the diastereoface-differentiating formation of 2-amino-3-phenylbutane in 31% diastereomer excess from 2-methylbenzylmethyl-ketone and formamide (Leuckart's reaction).[163)]

$$C_6H_5CHCOCH_3 \xrightarrow{\text{HCOONH}_4} C_6H_5CH-CHCH_3 \qquad (5.89)$$
$$\quad\ \ |\qquad\qquad\qquad\qquad\qquad |\quad\ |$$
$$\quad CH_3 \qquad\qquad\qquad\qquad CH_3\ NH_2$$

Alkoxybutadiene differentiates the diastereofaces of optically active esters of glyoxylic acid in the Diels-Alder reaction (Eq. 5.90).[164)]

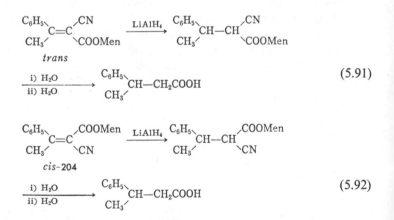

For instance, methoxybutadiene reacts with (−)-menthyl glyoxylate according to Eq. 5.90 to yield the *S* product with 3–6% optical yield at C-6 when methylene chloride is used as a solvent, and the *R* product in similar optical yield when benzene is used.

B.   Reactions of chiral alkenes

(−)-Menthyl *trans*-α-cyanocinnamate can be reduced by LiAlH₄ to yield the (*S*)-(+) product with 4.9% diastereomer excess, as shown in Eq. 5.91, while the cis compound is reduced to (*R*)-(−) form in 14% diastereomer excess (Eq. 5.92).[165)]

These reactions are examples of the chiral factor in the substrate affecting differentiation in opposite ways depending on the geometric structure of the substrate.

Eq. 5.93 shows the formation of acetyl-D-phenylalanyl-D-valine with a maximum optical purity of 45% by the hydrogenation of *N*-acetyldehydrophenylalanyl-D-valine using Pd-C as a catalyst.[166]

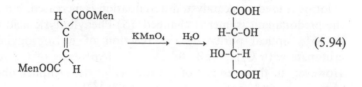

$$C_6H_5CH=C-CONHCHCH\begin{smallmatrix}CH_3\\CH_3\end{smallmatrix} \longrightarrow C_6H_5CH_2CHCONH-CHCH\begin{smallmatrix}CH_3\\CH_3\end{smallmatrix} \quad (5.93)$$

Very few reports of diastereoface-differentiating oxidations of alkenes have been published. Slightly optically active tartaric acid is obtained by the oxidation of dimenthylfumarate with potassium permanganate (Eq. 5.94).[167]

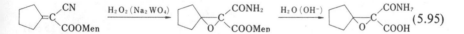

$$\xrightarrow{KMnO_4 \quad H_2O} \quad (5.94)$$

Eq. 5.95 shows the oxidation of an alkene by hydrogen peroxide in the presence of sodium molybdate to give an optically active epoxide with 15% optical purity.[168]

$$\xrightarrow{H_2O_2 (Na_2WO_4)} \quad \xrightarrow{H_2O (OH^-)} \quad (5.95)$$

Diastereoface-differentiation occurs in the reactions of carbenoid reagents such as diazomethane derivatives,[169] dimethyloxosulfonium methylide[170] and Simmons-Smith reagent[171] with menthyl acrylate derivatives. For example, (−)-menthyl-β,β-dimethylacrylate and ethyl diazoacetate give (1*R*,3*R*)-(−)-*trans*-caronic acid in 15% optical purity, as shown in Eq. 5.96.

$$CH_3-\overset{\overset{\displaystyle O}{\|}}{\underset{\underset{\displaystyle CH_3}{|}}{S^+}}-CH_3 I^-$$

dimethyloxosulfonium mothylide

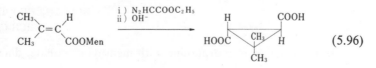

$$\xrightarrow[\text{ii) OH}^-]{\text{i) N}_2\text{HCCOOC}_2\text{H}_5} \quad (5.96)$$

The optical purity of the product in this reaction does not all represent diastereomer excess, but shows that differentiation does occur.

Diastereoface-differentiation occurs in the Diels-Alder condensation of 1,3-dienes with menthyl acrylate. An (R,R)-$(-)$-diol of 1.5% optical purity is obtained as the condensation product of dimenthyl fumarate and butadiene after reduction with LiAlH₄, as shown in Eq. 5.97.[172]

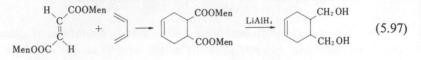

$$(5.97)$$

Diastereoface-differentiating reduction of optically active esters of acrylic acid derivatives can be carried out with Grignard reagents. If cuprous chloride is used as a catalyst, differentiation is increased, but in some cases the predominant isomer is changed. (S)-Phenylbutyric acid was obtained in 16% optical purity by the reaction of di-isopropylidene-D-xylose crotonate with C_6H_5MgBr followed by hydrolysis of the resulting ester. However, in the presence of cuprous chloride, (R)-phenylbutyric acid is obtained in 58% optical purity (Eq. 5.98).[173]

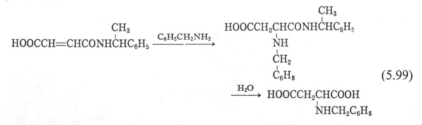

$$(5.98)$$

Diastereoface-differentiation by bromine has already been mentioned, Eq. 5.82 and amines can show differentiation in the same way. For example, N-benzyl-D-aspartic acid is obtained by the addition of benzylamine to (R)-N-α-methylbenzylmaleiamido acid followed by hydrolysis, as shown in Eq. 5.99.[174]

$$\begin{array}{c}
\text{CH}_3\\
|\\
\text{HOOCCH=CHCONHCHC}_6\text{H}_5 \xrightarrow{\text{C}_6\text{H}_5\text{CH}_2\text{NH}_2}
\end{array}
\qquad
\begin{array}{c}
\text{CH}_3\\
|\\
\text{HOOCCH}_2\text{CHCONHCHC}_6\text{H}_5\\
|\\
\text{NH}\\
|\\
\text{CH}_2\\
|\\
\text{C}_6\text{H}_5
\end{array}
$$

$$(5.99)$$

$$\xrightarrow{\text{H}_2\text{O}} \begin{array}{c}\text{HOOCCH}_2\text{CHCOOH}\\ |\\ \text{NHCH}_2\text{C}_6\text{H}_5\end{array}$$

In the reaction of morpholine with menthyl crotonate, diastereoface-differentiating addition occurs (Eq. 5.100). As the chiral center in the cyclohexanone is optically unstable and readily equilibrates, differentiation was confirmed by the removal of chirality in the cyclohexane ring by reduction of the carbonyl group, and the optical activity of the C-2 chiral center alone was measured. That is, differentiation was confirmed by the finding of ($-$) optical activity of the cyclohexylbutyric acid ester.[175]

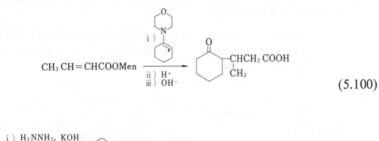

$$(5.100)$$

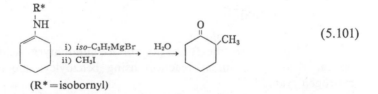

methyl cyclohexylbutyrate

Diastereoface differentiation also occurs in the alkylation of the optically active enamine shown in Eq. 5.102, which is prepared from isobornylalanine and cyclohexanone. Hydrolysis of the initial product gave (S)-2-methylcyclohexanone in 72% optical yield.[176]

$$(5.101)$$

(R* = isobornyl)

The addition reaction of a mercuric salt to an optically active unsaturated acid is also diastereoface-differentiating. The differentiation was detected by conversion of the mercuric compound to the β-methoxy- or β-hydroxy acid or alcohol by treatment with H_2S, $NaBH_4$ or $LiAlH_4$, respectively. For example, Eq. 5.102 shows the reaction of the cyclohexylidene glucose ester of cinnamic acid with mercuric acetate in the presence of nitric acid as a catalyst, yielding β-methoxy-β-phenylpropionic acid in 27% optical purity after reduction with $NaBH_4$.[177]

$$C_6H_5CH=CHCOOR \xrightarrow[\substack{CH_3OH \\ HNO_3}]{Hg(OCOCH_3)_2} \underset{\substack{| \quad | \\ CH_3O \ \ HgOCOCH_3}}{C_6H_5CHCHCOOR} \xrightarrow[\text{ii) } H_2O]{\text{i) } NaBH_4} \underset{\substack{| \\ OCH_3}}{C_6H_5CHCH_2COOH}$$

$$(5.102)$$

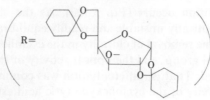

C. Reactions of C=N bonds

The diastereoface of the chiral oxime shown in Eq. 5.103 is differentiated by Grignard reagent, and the resulting ethyleneimine is obtained with 50% diastereomer excess.[178]

$$
\underset{\substack{| \\ CH_3 \ NOH}}{C_6H_5 \ CH{-}CCH_3} \xrightarrow{C_6H_5 \ MgBr} \underset{\substack{| \\ CH_3}}{\overset{C_6H_5}{\underset{\substack{| \\ H}}{C_6H_5 \ CHC{-}CH_2}}} \tag{5.103}
$$

Optically active amino acids can also be obtained by the diastereoface-differentiating addition of HCN to the Schiff base of an optically active amine, followed by hydrolysis. For example, optically active alanine can be obtained in 40–95% optical yield from the Schiff base of optically active α-methylbenzylamine and acetaldehyde, as shown in Eq. 5.104.[179]

$$
\underset{235}{\overset{\substack{CH_3 \\ |}}{C_6H_5CHN{=}CHCH_3}} \xrightarrow{HCN} \underset{\substack{| \\ CN}}{\overset{\substack{CH_3 \\ |}}{C_6H_5CHNHCHCH_3}} \xrightarrow[H^+]{H_2O} \underset{\substack{| \\ COOH}}{\overset{\substack{CH_3 \\ |}}{C_6H_5CHNHCHCH_3}}
$$

$$
\xrightarrow[Pd{-}C]{H_2} \underset{\substack{| \\ COOH}}{NH_2CHCH_3} \tag{5.104}
$$

Eq. 5.105 shows the formation of α-aminobutyric acid in 37.2% optical yield by a similar reaction using benzoylcyanide in place of hydrogen cyanide.[180]

$$
\underset{\substack{| \\ }}{\overset{\substack{CH_3 \\ |}}{C_6H_5CHN{=}CHC_2H_5}} \xrightarrow{C_6H_5COCN} \underset{\substack{| \\ COC_6H_5}}{\overset{\substack{CH_3 \ CN \\ | \ |}}{C_6H_5CHNCHC_2H_5}} \xrightarrow[H^+]{H_2O} \xrightarrow[Pd{-}C]{H_2} \underset{\substack{| \\ NH_2}}{C_2H_5CHCOOH}
$$

$$
\tag{5.105}
$$

Optically active amino acids can also be obtained by the diastereoface-differentiating Grignard reaction of the Schiff base from menthyl glyoxalate with benzylamine derivatives, followed by hydrolysis and hydrogenation. For instance, the use of CH_3MgI, as shown in Eq. 5.106, gives alanine in 53% optical yield.[181]

$$
C_6H_5CH_2N{=}CHCOOMen \xrightarrow{CH_3MgI} \underset{\substack{| \\ CH_3}}{C_6H_5CH_2NHCHCOOMen} \xrightarrow{H_2} \underset{\substack{| \\ CH_3}}{NH_2CHCOOH} \tag{5.106}
$$

Amino acids are also produced by the diastereoface-differentiating addition of $Fe(CO)_9$ to $C=N$ double bonds, followed by alkylation and hydrogenolysis. For example, when the Schiff base of optically active α-methylbenzylamine with glyoxalate is treated with $Fe(CO)_9$, diastereomeric π-complexes are formed disproportionately. Addition of *p*-methoxybenzyl bromide to the complex followed by hydrogenolysis gave *O*-methyltyrosine in 95% optical yield (Eq. 5.107).[182]

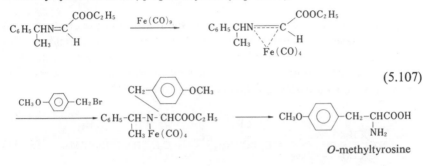

(5.107)

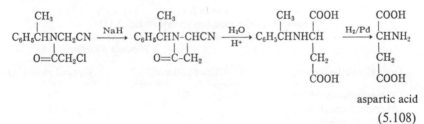

O-methyltyrosine

5.2.3 Diastereotopos-differentiating reactions

Differentiation of diastereotopic hydrogens of a methylene group occurs in various reaction systems. For instance, the reaction of optically active *N*-methylbenzyl-*N*-chloroacetamide acetonitrile with sodium hydride (Eq. 5.108) yields diastereomers of the resulting four-membered lactam disproportionately as a result of differentiation of the diastereotopic hydrogens. Hydrolysis of the lactam followed by hydrogenolysis yields aspartic acid in 21–64% optical yield.[183]

$$
\begin{array}{c}
CH_3 \\
C_6H_5CHNCH_2CN \\
| \\
O{=}CCH_2Cl
\end{array}
\xrightarrow{NaH}
\begin{array}{c}
CH_3 \\
C_6H_5CHN{-}CHCN \\
| \quad | \\
O{=}C{-}CH_2
\end{array}
\xrightarrow[H^+]{H_2O}
\begin{array}{c}
CH_3 \quad COOH \\
C_6H_5CHNHCH \\
| \\
CH_2 \\
| \\
COOH
\end{array}
\xrightarrow{H_2/Pd}
\begin{array}{c}
COOH \\
| \\
CHNH_2 \\
| \\
CH_2 \\
| \\
COOH
\end{array}
$$

aspartic acid

(5.108)

In the reaction of (*S*)-(−)-benzylmethylsulfoxide with methyl iodide in the presence of BuLi (Eq. 5.109), differentiating substitution of diastereotopic hydrogens on the methylene group is observed.[184a]

$$
C_6H_5CH_2\overset{O}{\overset{\uparrow}{S}}CH_3 \xrightarrow{CH_3I,\ C_4H_9Li} C_6H_5CH\overset{O}{\overset{\uparrow}{S}}CH_3
$$
$$
\qquad\qquad\qquad\qquad\qquad | \\
\qquad\qquad\qquad\qquad\quad CH_3
$$

(5.109)

The same type of diastereotopos-differentiation occurs in the halogenation of (R)-$(+)$-ethyl-p-tolylsulfoxide with iodobenzene dichloride (Eq. 5.110).[184b)]

$$ CH_3CH_2\overset{\overset{O}{\uparrow}}{S}\text{-}\langle O \rangle\text{-}CH_3 \xrightarrow{\ C_6H_5ICl_2\ } CH_3\underset{\underset{Cl}{|}}{\overset{\overset{O}{\uparrow}}{C}H}S\text{-}\langle O \rangle\text{-}CH_3 \qquad (5.110) $$

Optically active amines differentiate the diastereotopic electron pairs of the carbene generated from menthyl diazopropionate, yielding diastereomeric amines disproportionately. In the example shown in Eq. 5.111, optically active alanine can be obtained by hydrogenolysis of the initial product with a Pd catalyst, followed by hydrolysis.[185)]

$$ \underset{\underset{COO\text{-}(-)\text{-}Men}{}}{\overset{\overset{CH_3}{}}{N_2C}} + \underset{\underset{}{}}{\overset{\overset{CH_3}{|}}{C_6H_5CH}}NH_2 \xrightarrow{CuCN} C_6H_5\overset{\overset{CH_3}{|}}{C}HNH\overset{\overset{CH_3}{|}}{C}HCOOMen \qquad (5.111) $$

$$ \xrightarrow[Pd]{H_2} CH_3\text{—}\underset{\underset{NH_2}{|}}{C}H\text{—}COOMen \xrightarrow{H_2O} CH_3\text{—}\underset{\underset{NH_2}{|}}{C}H\text{—}COOH $$

Since there are two chiral centers in the reaction system, the configuration of the alanine produced in excess must depend on both centers, as shown in Table 5.33. The results suggest that the chirality of the reagent has a stronger influence on the reaction than that of the substrate.

TABLE 5.33
Results for the reaction shown in Eq. 5.111

	Product alanine	
Optically active amine	Configuration	Optical yield (%)
(S)-methylbenzylamine	R	26
(R)-methylbenzylamine	S	12
(S)-α-naphthylamine	R	26
(R)-α-naphthylamine	S	16

5.2.4   Intramolecular diastereoface- and diastereotopos-differentiating reactions

Several rearrangements of chiral compounds are known which involve the chiral center. These reactions are apparently diastereo-differentiating reaction, but in fact the participation of the chiral center is completely different from that in the other diastereo-differentiating reactions described in this chapter. That is, in general diastereo-differentiating reactions, the chiral center acts only as a reference system, while in the present reactions it is involved directly in the transition intermediate. Therefore, the configuration of the product must depend on the molecular orbital symmetry of the reactant and product, and must be analyzed quantum-mechanically, e.g., by application of the Woodward-Hoffman rules. As mentioned in Chapter 4, we consider that such stereochemical phenomena do not constitute differentiation, but are intrinsic to the reaction character itself.

Many detailed discussions of reactions of this type are available, and only a few limited examples will be mentioned here. Examples include the Claisen rearrangement of allylethers and vinylethers (Eq. 5.112)[186] and the analogous reactions of phenylurethane (Eq. 5.113)[187] and allylenamine (Eq. 5.114).[188]

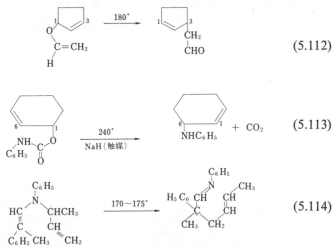

$$(5.112)$$

$$(5.113)$$

$$(5.114)$$

Diastereoface-differentiating hydrogen transfer occurs in the base-catalyzed reactions of 3-alkyl-1-methylindene (Eq. 5.115)[189] and aldimine (Eq. 5.116)[190] which take place with high efficiency.

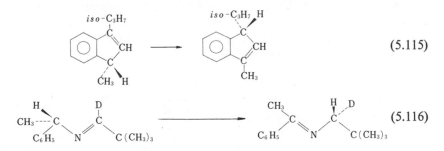

$$(5.115)$$

$$(5.116)$$

A reaction involving acid-catalyzed hydrogen transfer is shown in Eq. 5.117.[191)

$$C_6H_5-\underset{CH_3}{\overset{}{\underset{|}{C}}}HO\!\!\left|\!\!\begin{array}{c}\\H\\CH_3\end{array}\right. \xrightarrow[\text{PPA}]{H^+} C_6H_5\!\!\left|\!\!\begin{array}{c}\\H\\CH_3\end{array}\right.CO\text{-}CH_3 \tag{5.117}$$

(PPA = polyphosphoric acid)

The transfer of chirality from sulfur to carbon can also occur by intramolecular diastereotopos differentiation in sigmatropic rearrangement. This can be seen in the [2,3]-sigmatropic rearrangement of optically active 1-adamantyl allylethylsulfonium fluoroborate in the presence of potassium butoxide in toluene, yielding 1-adamantyl-2-pent-4-enyl sulfide in 94% optical yield,[192) as shown in Eq. 5.118.

$$Ad-\overset{+}{S}\underset{CH_2CH=CH_2}{\overset{CH_2CH_3}{\diagup}} \longrightarrow Ad\text{-}S\text{-}\underset{\underset{\underset{CH_2}{\overset{\diagup}{CH}}}{\overset{|}{CH_2}}}{\overset{CH_3}{\overset{|}{C}}}\text{-}H \tag{5.118}$$

5.3 SPECIAL STEREO-DIFFERENTIATING REACTIONS

When an achiral compound, which is not symmetric across a certain configurational molecular plane, reacts with a reagent or catalyst, as in Eqs. 5.119 and 5.120, it is possible that the reaction may proceed by fundamentally the same differentiating mechanism as in stereo-differentiating reactions to produce diastereomers in different ratios. However, chiral product is not always expected, even when an optically active reagent is used. Such reactions, therefore, do not lie within the strict scope of this book. However, they can, for the above reasons, be considered in a similar manner based on the new concept of "differentiation."

By treatment of π-allyl complex of PdCl₂ and 4-methylenealkylcyclohexane with anions derived from dimethyl malonate in the presence of tri-o-tolylphosphine, the following reaction takes place predominantly to yield diastereomers[193) (Eq. 5.119).

$$\text{[structures]} \xrightarrow[\left(\underset{CH_3}{\underbrace{}}\right)_3 P]{{}^\ominus CH(COOCH_3)_2} \text{[structure]}\text{-}CH(COOCH_3)_2 \tag{5.119}$$

A similar reaction (Eq. 5.120) occurs in the reduction or hydrogenation of 4-substituted cyclohexanones to 4-substituted cyclohexanols. For example, 4-*tert*-butylcyclohexanone is reduced with LiAlH$_4$ to the corresponding *cis* and *trans* alcohols in the ratio 92:8.[194]

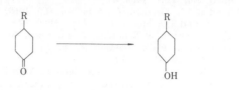 (5.120)

REFERENCES

1. J. D. Morrison and H. S. Mosher, *Asymmetric Organic Reactions*, Prentice-Hall, 1971.
2. Y. Izumi *et al.*, *Kagaku Sosetsu* (Japanese) (ed. Chem. Soc. Japan), vol. 4, Tokyo University Press, 1974.
3. See ref. 1.
4. R. Macleod, F. J. Welch and H. S. Mosher, *J. Am. Chem. Soc.*, **82**, 876 (1960).
5. J. S. Birtwistle, K. Lee, J. D. Morrison and W. A. Sanderson, *J. Org. Chem.*, **29**, 37 (1964).
6. G. Vavon, C. Rivière and B. Angelo, *Compt. Rend.*, **222**, 959 (1946).
7. G. Vavon and B. Angelo, *ibid.*, **224**, 1435 (1947).
8. M. F. Tatibouet, *Bull. Soc. Chim. France*, **1955**, 210.
9. R. Bousett, *ibid.*, **1955**, 210.
10. R. A. Kretchmer, *J. Org. Chem.*, **37**, 801 (1972).
11. G. P. Giacomelli, R. Menicagli and L. Lardicci, *Tetr. Lett.*, **1971**, 4135.
12. W. von E. Doering and R. W. Young, *J. Am. Chem. Soc.*, **72**, 631 (1950).
13. L. M. Jackman, J. A. Mills and J. S. Shannon, *ibid.*, **72**, 4814 (1950).
14. G. Vavon and A. Antonnini, *Compt. Rend.*, **232**, 1120 (1951).
15. O. Cervinka, V. Suchan and B. Masar, *Coll. Czech. Chem. Commun.*, **30**, 1633 (1965).
16. H. C. Brown, N. R. Ayyangar and G. Zweifel, *J. Am. Chem. Soc.*, **86**, 397 (1964).
17. G. Zeifel, N. R. Ayyangar, T. Munekata and H. C. Brown, *ibid.*, **86**, 1076 (1964).
18. W. R. Moore, H. W. Anderson and S. D. Clark, *ibid.*, **95**, 835 (1973).
19. K. R. Varma and E. Caspi, *Tetrahedron*, **24**, 6365 (1968).
20. M. F. Grundon, W. A. Khan, D. R. Boyd and W. R. Jackson, *J. Chem. Soc.* (C), **1971**, 2557.
21. D. R. Boyd, M. F. Grundon and W. R. Jackson, *Tetr. Lett.*, **1967**, 2102.
22. R. F. Borch and S. R. Levitan, *J. Org. Chem.*, **37**, 2347 (1972).
23. H. Pracejus and A. Tille, *Chem. Ber.*, **96**, 854 (1963).
24. K. Harada, *Nature*, **200**, 1201 (1963).
25. A. P. Terentév, R. A. Graheva and T. F. Dendeko, *Dokl. Akad. Nauk SSSR*, **163**, 674 (1965); *Chem. Abstr.*, **63**, 11344b (1965).
26. E. Anders, E. Ruch and I. Ugi, *Angew. Chem.*, **85**, 16 (1973); *Angew. Chem. Intern. Ed. Engl.*, **12**, 25 (1973).
27. L. L. McCoy, *J. Org. Chem.*, **29**, 240 (1964).
28. Y. Inouye, S. Inamasu, M. Ohono, T. Sugita and H. M. Walborsky, *J. Am. Chem. Soc.*, **83**, 2962 (1961).
29. M. H. Palmer and J. A. Reid, *J. Chem. Soc.*, **1962**, 1762.
30. J. Cancell, J. Gabard and J. Jacques, *Bull Soc. Chim. France*, **1968**, 231.

31. S. Mitsui, K. Konno, I. Onuma and K. Shimizu, *Nippon Kagaku Kaishi* (Japanese), **85**, 437 (1964).
32. R. K. Hill and M. Rabinovitz, *J. Am. Chem. Soc.*, **86**, 965 (1964).
33. J. Romeyn and G. F. Wright, *ibid.*, **69**, 697 (1947).
34. F. Montanari, I. Moretti and G. Torre, *Chem. Commun.*, **1969**, 135.
35. D. R. Boyd, *Tetr. Lett.*, **1968**, 4561.
36. D. R. Boyd and R. Graham, *J. Chem. Soc.* (C), **1969**, 2648.
37. F. D. Greene and S. S. Hecht, *Tetr. Lett.*, **1969**, 575.
38. C. R. Johnson and C. W. Schroeck, *J. Am. Chem. Soc.*, **90**, 6852 (1968).
39. G. Bredig and P. S. Fiske, *Biochem. Z.*, **46**, 7 (1912).
40. E. Fischer, *Ber.*, **29**, 3210 (1894).
41. S. Akabori, Y. Izumi, Y. Fujii and S. Sakurai, *Nature*, **178**, 232 (1956).
42. Y. Izumi, M. Imaida, H. Fukawa and S. Akabori, *Bull. Chem. Soc. Japan*, **36**, 21 (1963).
43. W. S. Knowles and M. J. Sabacky, *Chem. Commun.*, **1968**, 1445.
44. W. S. Knowles, M. J. Sabacky B. D. Vinevard, *J. Chem. Soc., Chem. Commun.*, **1972**, 10.
45. E. R. Burnett, M. A. Augiar, C. J. Morrow and C. Phillips, *J. Am. Chem. Soc.*, **93**, 1301 (1971).
46. H. B. Kagan and T. P. Dang, *Chem. Commun.*, **1971**, 481.
47. Y. Ohgo, S. Takeuchi and J. Yoshimura, *Bull. Chem. Soc. Japan.*, **44**, 583 (1971).
48. Y. Ohgo, K. Kobayashi, S. Takeuchi and J. Yoshimura, *ibid.*, **44**, 583 (1971).
49. H. Hirai and T. Furuta, *J. Polymer Sci.*, part B, **9**, 459 (1971).
50. P. Pino, G. Consiglio, C. Botteghi and C. Salomon, *Advan. Chem. Ser.*, **132**, 295 (1974).
51. (*a*) C. Salomon, G. Consiglio, C. Botteghi and P. Pino, *Chimia*, **27**, 215 (1973); (*b*) H. Ogata and Y. Ikeda, *Chem. Lett.*, **1972**, 483; (*c*) M. Tanaka, Y. Watanabe, T. Mitsudo, K. Yamamoto and K. Takegami, *Chem. Lett.*, 1972, 483; (*d*) G. Consiglio, C. Botteghi, C. Salomon and P. Pino, *Angew. Chem. Intern. Ed. Engl.*, **12**, 669 (1973).
52. K. Yamamoto, Y. Uramoto and M. Kumada, *J. Organo-metal Chem.*, **31**, C9 (1971).
53. K. Yamamoto, T. Hayashi and M. Kumada, *ibid.*, **46**, C65 (1972).
54. K. Yamamoto, T. Hayashi, H. Omoto and M. Kumada, *Shokubai* (Japanese), **16**, 8 (1974).
55. J. Furukawa, T. Kakuzen, H. Morikawa, R. Yamamoto and O. Kuno, *Bull. Chem. Soc. Japan*, **41**, 155 (1968).
56. B. Bogdanovic, B. Henc, B. Meister, H. Pauling and C. Wilke, *Angew. Chem. Intern. Ed. Engl.*, **11**, 1023 (1972).
57. (*a*) H. Nozaki, H. Takaya, S. Moriuchi and R. Noyori, *Tetrahedron*, **24**, 3655 (1968); (*b*) T. Aratani, Y. Yoneyoshi and T. Nagase, *Tetr. Lett.*, **1975**, 1707.
58. Y. Tatsuno, A. Konishi, A. Nakamura and S. Otsuka, *J.C.S. Chem. Commun.*, **1974**, 588; A. Konishi, R. Tsujitani, A. Nakamura and S. Otsuka, 16th Symp. Organoradical Reactions, Preprints, p. 69, 1975.
59. L. Lardicci, G. P. Giacomelli, R. Menicagli and P. Pino, *Organometal Chem. Syn.*, **1** (4), 447 (1972).
60. S. Tsuboyama, *Bull. Chem. Soc. Japan*, **35**, 1004 (1962).
61. V. Prelog and M. Wilhelm, *Helv. Chim. Acta*, **37**, 1634 (1954).
62. W. F. Yates and R. L. Heider, *J. Am. Chem. Soc.*, **74**, 4153 (1952).
63. H. Albers and E. Albers, *Z. Naturforsch.*, **9b**, 122, 133 (1954).
64. H. H. Hustedt and E. Pfiel, *Ann.*, **640**, 15 (1961).
65. G. Belluci, C. Giordano, A. Marsili and G. Berti, *Tetrahedron*, **22**, 2977 (1966).
66. S. Inoue, S. Ohashi, A. Tabata and T. Tsuruta, *Makromol. Chem.*, **112**, 66 (1968).
67. Y. Izumi, *Angew. Chem. Intern. Ed. Engl.*, **10**, 871 (1971).
68. (*a*) R. L. Beamer, R. H. Belding and C. S. Fickling. *J. Pharm. Sci.*, **58**, 1142 (1967); (*b*) *ibid.*, **58**, 1419.

69. K. Harada and T. Yoshida, *Naturwissenschaften*, **57**, 306 (1970).
70. R. E. Padgett, Jr. and R. L. Beamer, *J. Pharm. Sci.*, **53**, 689 (1964).
71. R. L. Beamer, C. S. Fickling and J. H. Ewing, *ibid.*, **56**, 1029 (1967).
72. (*a*) Y. Izumi, S. Tatsumi, M. Imaida, A. Fukawa and S. Akabori, *Bull. Chem. Soc. Japan*, **39**, 361 (1966); (*b*) Y. Orito, S. Miwa, S. Imai, J. Synth. *Org. Chem. Japan*. (Yukigosei Kagaku) (Japanese) **34**, 236 (1976); (*c*) L. H. Gross and P. Rys, *J. Org. Chem.*, **39**, 2429 (1974).
73. L. H. Gross and P. Rys, *J. Org. Chem.*, **39**, 2429 (1974).
74. (*a*) E. I. Klabunovskii, V. I. Neupokoev and Yu. I. Petrov, *Itv. Akad. Nauk SSSR Ser. Khim.*, **1970** (12), 2839; (*b*) E. I. Klabunovskii, V. I. Neupokoev and Yu. I. Petrov, *ibid.*, **1971** (9), 2067; (*c*) E. I. Klabunovskii, N. P. Sokolova, A. A. Vendeyapin and Yu. M. Talanov, *ibid.*, **1971** (3), 1803; (*d*) E. I. Klabunovskii, N. P. Sokolova, A. A. Vendeyapin, Yu. M. Talanov, N. D. Zubareva, V. P. Polyakova and N. V. Gorina, *ibid.*, **1972** (10), 2361; (*e*) T. Tanabe and Y. Izumi, *Bull. Chem. Soc. Japan*, **46**, 1550 (1973).
75. Y. Izumi, S. Tatsumi and M. Imaida, *ibid.*, **42**, 2373 (1969).
76. Y. Izumi, S. Yajima, K. Okubo and K. Babievsky, *ibid.*, **44**, 1416 (1971).
77. Y. Izumi, M. Imaida, H. Fukawa and S. Akabori, *Bull. Chem. Soc. Japan*, **36**, 155 (1963).
78. Y. Izumi, S. Akabori, H. Fukawa, S. Tatsumi, M. Imaida, Y. Fukuda and S. Komatu, *Proc. 3rd Intern. Congr. Catalysis*, vol. 2, p. 1364, North-Holland, 1964.
79. (*a*) F. Higashi, T. Ninomiya and Y. Izumi, *Bull. Chem. Soc. Japan*, **44**, 1333 (1971); (*b*) T. Harada, Y. Hiraki, Y. Izumi, J. Muraoka, H. Ozaki and A. Tai, 6th Intern. Congr. Catalysis, 1976. Reprints, in press; (*c*) K. Tanabe and N. Ueda, *Rept. Inst. Chem. Res., Kyoto Univ.* (Japanese), **52**, 616 (1974).
80. Y. Izumi, S. Tatsumi, M. Imaida and K. Okubo, *Bull. Chem. Soc. Japan*, **43**, 566 (1969).
81. Y. Izumi, S. Tatsumi, M. Imaida, Y. Fukawa and S. Akabori, *ibid.*, **39**, 2223 (1966).
82. Y. Izumi, K. Matsunaga, S. Tatsumi and M. Imaida, *ibid.*, **41**, 2515 (1968).
83. Y. Izumi and K. Okubo, *ibid.*, **44**, 1330 (1971); A. Hata and W. Suetaka, *ibid.*, presented for publication.
84. Y. Izumi, T. Tanabe, S. Yajima and M. Imaida, *ibid.*, **41**, 941 (1968).
85. K. Okubo, T. Ninomiya, T. Harada, Y. Igami, *ibid.*, in press.
86. (*a*) Y. Izumi *et al.*, *unpublished;* (*b*) Y. Izumi *et al.*, *unpublished.*
87. (*a*) J. A. Groenewegen and W. M. H. Sachtler, *J. Catal.*, **27**, 369 (1972); (*b*) A. Hata and W. Suetake, *Bull. Chem. Soc., Japan*, **48**, 2428 (1975); (*c*) T. Ninomiya, *Bull. Chem. Soc. Japan*, **45**, 2551 (1972); (*d*) K. Tanabe and N. Ueda, *Rept. Inst. Chem. Res. Kyoto, Univ.* (Japanese), **52**, 616 (1974).
88. T. Harada *et al.*, *unpublished.*
89. E. I. Klabunovskii and Yu. I. Petrov, *Dokl. Akad. Nauk SSSR*, **173**, 1125 (1967)
90. T. Tanabe, *Bull. Chem. Soc. Japan*, **46**, 1482 (1973).
91. (*a*) J. A. Groenewegen and W. M. H. Sachtler, *J. Catal.*, **38**, 501 (1975); (*b*) I. Yasumori, Y. Inouye and K. Okabe, Proc. Intern. Symp. Relations between Heterogeneous and Homogeneous Catalytic Phenomena, 1900.
92. H. L. Cohen and G. F. Wright, *J. Org. Chem.*, **18**, 432 (1953).
93. N. Allentoff and G. F. Wright, *ibid.*, **22**, 1 (1957).
94. T. D. Inch, G. F. Lewis, G. L. Sainsburg and D. J. Sellers, *Tetr. Lett.*, **1969**, 3657.
95. H. J. Bruer and R. Haller, *ibid.*, **1972**, 5227.
96. (*a*) M. Guette and J. P. Gette, *ibid.*, **1971**, 2863; (*b*) S. Sawada, J. Oda and Y. Inouye, *J. Org. Chem.*, **33**, 2141 (1968).
97. J. Grimshaw, R. N. Gourley and P. G. Millar, *J. Chem. Soc.* (C), **1970**, 2318.
98. L. Horner and D. H. Skaletz, *Tetr. Lett.*, **1970**, 3670.
99. L. Horner and D. Degner, *ibid.*, **1968**, 5889.
100. T. Aratani, T. Gonda and H. Nozaki, *ibid.*, **1969**, 2265.
101. P. S. Schwartz and H. E. Carter, *Proc. Natl. Acad. Sci. U.S.A.*, **40**, 499 (1954).

102. R. Altschul, P. Berstein and S. G. Cohen, *J. Am. Chem. Soc.,* **78,** 5091 (1956).
103. W. M. B. Könst, W. N. Speckamp, H. O. Howen and H. O. Huisman, *Rec. Trav. Chim.,* **91,** 861 (1972).
104. M. Pezold and R. L. Shriner, *J. Am. Chem. Soc.,* **54,** 4707 (1932).
105. F. Wuld and T. B. K. Lee, *Chem. Commun.,* **1972,** 61.
106. U. Folli and D. Larossi, *Gazz. Chim. Ital.,* **99,** 1306 (1971).
107. H. Nozaki, T. Aratani and R. Noyori, *Tetr. Lett.,* **1968,** 2087.
108. I. Tomoskozi and G. Janzso, *Chem. Ind.,* **1962,** 2055.
109. D. D. Denney, R. Napier and A. Cammarata, *J. Org. Chem.,* **30,** 3151 (1965).
110. S. I. Goldberg, *J. Chem. Soc.* (D), **1969,** 1409.
111. (*a*) U. Eder, G. Sauer and R. Wiechert, *Angew. Chem. Intern. Ed. Engl.,* **10,** 496 (1971); (*b*) B. Bogdanovic, *ibid.,* **12,** 954 (1973).
112. H. J. Bestmann and J. Lienert, *Chem. Zig. Chem., Appl.,* **94,** 487 (1970).
113. S. J. Cristol, *J. Am. Chem. Soc.,* **71,** 1894 (1949).
114. C. V. Bird, *Tetrahedron,* **18,** 1 (1962).
115. K. Okumoto, E. Minami and H. Shingu, *Bull. Chem. Soc. Japan,* **41,** 1426 (1968).
116. O. Cervinka, *Coll. Czech. Chem. Commun.,* **31,** 1371 (1966).
117. F. Akiyama, K. Sugino and N. Tokura, *Bull. Chem. Soc. Japan,* **40,** 359 (1967).
118. J. D. Morrison, A. Tomash and R. W. Ridgway, *Tetr. Lett.,* **1969,** 565.
119. K. Mislow, R. E. O'Brien and H. Schaeser, *J. Am. Chem. Soc.,* **84,** 1940 (1962).
120. H. C. Brown, N. R. Ayyangar and G. Zweifel, *ibid,* **84,** 4341; **86,** 397 (1964).
121. W. L. Waters, *J. Org. Chem.,* **36,** 1569 (1969).
122. W. L. Waters, W. S. Linn and M. C. Caserio, *J. Am. Chem. Soc.,* **90,** 6741 (1968).
123. K. Balenovic and N. Bregant, *Chem. Ind.,* **1964,** 1577.
124. M. Mikolojczyk, *ibid.,* **1966,** 2059.
125. J. Read and R. A. Story, *J. Chem. Soc.,* **1930,** 2761.
126. N. K. Richtmyer and C. S. Hudson, *J. Am. Chem. Soc.,* **57,** 1716 (1935).
127. N. K. Richtmyer and C. S. Hudson, *ibid.,* **58,** 2540 (1936).
128. (*a*) H. Wuyts, *Bull. Soc. Chim. Belg.,* **30,** 30 (1921); (*b*) Y. Shibata and R. Tsuchida, *Bull. Chem. Soc. Japan,* **4,** 142 (1929).
129. M. Hatano, T. Nozawa, S. Ikeda and T. Yamamoto, *Makromol. Chem.,* **141,** 1 (1971).
130. C. Carlini, D. Politi and F. Ciardelli, *Chem. Commun.,* **1970,** 1260.
131. F. Cramer and W. Dietsche, *Chem. Ber,* **92,** 1739 (1959).
132. S. Akabori and S. Sakurai, *Nippon Kagaku Zasshi* (Japanese), **78,** 1629 (1957).
133. A. N. Kost, R. S. Sagitullin and M. A. Yurooskaja, *Chem. Ind.,* **1966,** 1496.
134. J. P. Vigneron, H. Kagan and A. Horeau, *Tetr. Lett.,* **1968,** 5681.
135. E. J. Corey, R. J. McCally and H. S. Sachdev, *J. Am. Chem. Soc.,* **92,** 2476, 2488 (1970).
136. S. Mitsui and A. Kanai, *Nippon Kagaku Kaishi* (Japanese), **86,** 627 (1965).
137. V. Prelog, E. Philbin, E. Watabe and M. Wilhelm, *Helv. Chim. Acta,* **39,** 1086 (1956).
138. A. McKenzie and P. D. Ritchie, *Biochem. Z.,* **237,** 1 (1931).
139. M. Kawana and S. Emoto, *Bull. Chem. Soc. Japan,* **40,** 2168 (1967).
140. A. McKenzie and E. W. Christie, *Biochem. Z,* **277,** 426 (1935).
141. A. McKenzie and P. D. Ritchie, *ibid.,* **250,** 376 (1932).
142. V. Prelog, *Helv. Chim. Acta,* **36,** 308 (1953).
143. J. A. Marshall and R. D. Carroll, *J. Org. Chem.,* **30,** 2748 (1965).
144. J. A. Reid and E. E. Turner, *J. Chem. Soc.,* **1951,** 3219.
145. H. Matsuo, H. Kobayashi and T. Tatsuno, *Chem. Pharm. Bull.,* **18,** 1693 (1970).
146. C. Moise, J. Tirouflet and D. Sautrey, *C. R. Acad. Sci. Paris, Ser. C,* **271,** 951 (1970).
147. C. Moise, D. Sautrey and J. Tirouflet, *Bull. Soc. Chim. France,* **1971,** 4562.
148. M. S. Biernbaum and H. S. Mosher, *J. Am. Chem. Soc.,* **93,** 6221 (1971).
149. D. J. Abott and J. M. Stirling, *Chem. Commun.,* **1971,** 472.
150. H. C. Brown and H. R. Deck, *J. Am. Chem. Soc.,* **87,** 5620 (1965).

151. J. B. Umland and B. W. Williams, *J. Org. Chem.*, **21**, 1302 (1956).
152. H. C. Brown and D. B. Bigley, *J. Am. Chem. Soc.*, **83**, 3166 (1961).
153. W. Hückel, M. Maier, E. Jordan and W. Seeger, *Ann.*, **616**, 46 (1958).
154. J. P. Charles, H. Christol and G. Solladie, *Compt. Rend.*, **269**, 179 (1969).
155. J. C. Charles, H. Christol and G. Solladie, *Bull. Soc. Chim. France*, **1970**, 4439.
156. S. Yamaguchi, J. A. Dala and H. S. Mosher, *J. Org. Chem.*, **37**, 3174 (1972).
157. Y. Inouye, S. Inamasu, M. Ohno, T. Sugita and H. M. Walborsky, *J. Am. Chem. Soc.*, **83**, 2962 (1961).
158. K. Harada and T. Yoshida, *Bull. Chem. Soc. Japan*, **43**, 921 (1970).
159. Y. Gault and H. Felkin, *Bull. Soc. Chim. France*, **1960**, 1342.
160. K. Harada and T. Yoshida, *Chem. Commun.*, **1970**, 1071.
161. K. Harada and T. Yoshida, *Bull. Chem. Soc. Japan*, **45**, 3706 (1972).
162. (*a*) D. J. Cram and D. R. Wilson, *J. Am. Chem. Soc.*, **85**, 1245 (1963); (*b*) K. Harada and K. Matsumoto, *Bull. Chem. Soc. Japan*, **44**, 1068 (1971).
163. D. J. Cram and J. E. McCarty, *J. Am. Chem. Soc.*, **76**, 5740 (1954).
164. J. Jurezak and A. Zamojski, *Tetrahedron*, **28**, 1505 (1972).
165. D. Cabaret and Z. Welvert, *C. A. Akad. Sci. Paris, Ser. C*, **274**, 1200 (1972).
166. M. Nakayama, G. Maeda, T. Kaneko and H. Katsura, *Bull. Chem. Soc. Japan.*, **44**, 1150 (1971).
167. A. McKenzie and H. Wren, *J. Chem. Soc.*, **91**, 1215 (1907).
168. M. Igarashi and H. Midorikawa, *Bull. Chem. Soc. Japan.*, **40**, 2624 (1967).
169. H. M. Walborsky, T. Sugita, M. Ohno and Y. Inouye, *J. Am. Chem. Soc.*, **82**, 5255 (1960).
170. H. Nozaki, H. Ito, D. Tunemoto and K. Kondo, *Tetrahedron*, **22**, 441 (1966).
171. S. Sawada, J. Oda and Y. Inouye, *J. Org. Chem.*, **33**, 2141 (1968).
172. H. M. Walborsky, L. Barash and I. C. Davis, *ibid.*, **26**, 4778 (1961); *Tetrahedron*, **19**, 2333 (1963).
173. M. Kawana and S. Emoto, *Bull. Chem. Soc. Japan*, **39**, 910 (1966).
174. Y. Liwschitz and A. Singerman, *J. Chem. Soc.* (C), **1966**, 1200.
175. K. Igarashi, J. Oda, Y. Inouye and M. Ohno, *Agr. Biol. Chem.*, **34**, 811 (1970).
176. D. Mea-Jacheet and A. Horeau, *Bull. Soc. Chim. France*, **1968**, 4571.
177. M. Kawana and S. Emoto, *Bull. Chem. Soc. Japan*, **40**, 618 (1967).
178. G. Alvernhe and A. Laurent, *Tetr. Lett.*, **1971**, 1913.
179. K. Harada and T. Okawara, *J. Org. Chem.*, **38**, 707 (1972).
180. K. Harada and T. Okawara, *Bull. Chem. Soc. Japan.*, **46**, 191 (1973).
181. J. C. Fiaud and H. B. Kagan, *Tetr. Lett.*, **1670**, 1813.
182. J. Y. Chernard, D. Commereuc and Y. Chauvin, *Chem. Commun.*, **1972**, 750.
183. T. Okawara and K. Harada, *J. Org. Chem.*, **37**, 3286 (1972).
184. (*a*) K. Nishihata and Nishio, *J. Chem. Soc.* (D) **1971**, 958; (*b*) M. Cinquini, S. Colonna and F. Montanari, *Chem. Commun.*, **1970**, 1441.
185. J. F. Nicoud and H. B. Kagan, *Tetr. Lett.*, **1971**, 2065.
186. R. K. Hill and A. G. Edwards, *ibid.*, **1964**, 3239.
187. M. E. Synerholm, N. W. Gilman, J. W. Morgan and R. K. Hill, *J. Org. Chem.*, **33**, 1111 (1968).
188. R. K. Hill and N. W. Gilman, *Tetr. Lett.*, **1967**, 1421.
189. J. Almy and D. J. Cram, *J. Am. Chem. Soc.*, **91**, 4459 (1969).
190. R. D. Guthrie, W. Meister and D. J. Cram, *ibid.*, **89**, 5288 (1967).
191. R. K. Hill and R. M. Carlson, *ibid.*, **87**, 2772 (1965).
192. B. M. Trost and R. F. Hammen, *ibid.*, **95**, 962 (1973).
193. B. M. Trost and P. E. Strege, *ibid.*, **97**, 2534 (1975).
194. J. C. Richer, *J. Org. Chem.*, **30**, 324 (1965).

CHAPTER 6

Miscellaneous Stereo-Differentiating Reaction Systems

6.1 STEREO-DIFFERENTIATING REACTIONS IN POLYMERIZATION

Polymer synthesis* occurs by the repetition of a reaction or series of reactions in such a way that the product consists of a sequence of monomeric units. In the same way as simple organic chemical reactions, polymerization reactions can be classified into the six categories described in section 4.2.3 as regards stereo-differentiation. The overall ability of the differentiation in polymerization can be specified by the excess of one enantiomeric unit over the other in the total monomeric units in the product polymer. However, a further special feature of the polymer is the stereoregularity of the sequence of monomer units within it, which represents the results of successive differentiating reactions.

6.1.1 Relationship of stereo-differentiation and stereoregulation in polymer synthesis

If a polymer chain consists of n monomeric units (i.e., the degree of polymerization is n) and if there is one chiral center per monomer, then the number of possible diastereomers of the polymer is 2^{n-1}. Thus, many diastereomers will be produced if stereo differentiation at each polymerization step is ineffective. On the other hand, if differentiation is perfect, a polymer containing only one kind of diastereomeric unit will be produced.[†] This process is known as stereoregular polymerization, and the product is

---

* Only linear polymers will be discussed here, since there is no way to determine differentiation in two- and three-dimensional polymers at present.

† Polymers with stereo regular structures can be defined in terms of the repeating unit. For example, the polymers ... D·D·D·D·D ... and ... D·L·D·L·D·L ... can be be written as $-(D \cdot D)_n-$ and $-(D \cdot L)_n-$, respectively, in terms of their diastereomeric units.

a stereoregular polymer. Simple stereoregular polymerization is shown in Eq. 6.1.

$$R \xrightarrow{S} RS_D \xrightarrow{S} RS_DS_D \xrightarrow{S} RS_DS_DS_D\cdots\cdots S_D \qquad (6.1a)$$

$$R' \xrightarrow{S'} RS_L \xrightarrow{S'} RS_LS_L \xrightarrow{S'} RS_LS_LS_L\cdots\cdots S_L \qquad (6.1b)$$

Polymer molecules of the type shown in Eq. 6.1a and 6.1b, in which all the monomeric units have the same configuration, are known as isotactic polymers. Such a polymer is formed when the addition of a new monomeric unit to the reaction terminus of the growing chain polymer produces a new reaction terminus which has the same configuration as the previous one. If the new reaction terminus has the opposite configuration to the previous one, the product will be a syndiotactic polymer in which monomeric units of opposite configuration alternate, as in Eq. 6.2.

$$R \xrightarrow{S} RS_D \xrightarrow{S'} RS_DS_L \xrightarrow{S} RS_DS_LS_D\cdots\cdots S_L \qquad (6.2)$$

If the polymerizations shown in Eqs. 6.1 and 6.2 are carried out with a substrate having a prochiral center ($S = S' = CH_2=CHCH=CHCH_3$) each polymerization step is both enantioface-differentiating. If chiral monomer is used ($S = $ D-propylene oxide, $S' = $ L-propylene oxide) then each polymerization step is enantiomer-differentiating.

Since these definitions depend on the sequence of monomeric units in the polymer, and are independent of the formation of optically active polymer, further examples will not be given here. Detailed discussions of this type of polymerization have been given widely in the literature.

6.1.2 The stereo-differentiating process in polymerization

Stereo-differentiation during polymerization by means of an optically active initiator or catalyst is essentially similar to that in ordinary organic reactions. As mentioned above, the differentiating ability of the initiator or catalyst can be evaluated from the excess of one enantiomeric unit over the other in the product polymer. We may distinguish two general types of polymerization; one in which the active reagent is the terminus of the growing polymer (the monomer simply acts as a substrate) and another in which more than one species acts as a reagent, as in condensation polymerization. Various types of stereo-differentiating polymerization reactions are categorized in Table 6.1.

TABLE 6.1
Classification of stereo-differentiating

Location of chirality relating to the differentiation	Classification according to Marckwald	Type of differentiation	
	Asymmetric reaction	Enantioface-differentiating	
	Asymmetric reaction	Enantiotopos-differentiating	
	Kinetic resolution	Enantiomer-differentiating	
Chiral center in initiator or catalyst	Asymmetric reaction	Enantiotopos-differentiating	Enantio-differentiating
	Asymmetric reaction	Enantiotopos-differentiating	
	Kinetic resolution	Enantiomer-differentiating	
	Asymmetric reaction	Diastereoface-differentiating	
Chiral center in monomer	Asymmetric reaction	Diastereoface-differentiating	Diastereo-differentiating
	Kinetic resolution	Diastereoface-differentiating	
None	Not classified	No differentiation	

† Chiral center introduced from the substrate or by differentiation; *
Chiral center which does not contribute to differentiation; *
Chiral factor; *

polymerization reactions

Reaction model†

Substrate	Product	Remarks
CH_3 $\overset{\ast}{C}H=CH-CH=CH_2$	CH_3 $\left(\overset{\vert}{\underset{H}{\overset{\ast}{C}}}-CH=CH-CH_2\right)_n$	The main chain chiral center is produced by enantioface differentiation by the chiral catalyst.
$\qquad\qquad COOR$ $NH_2-(CH_2)_n-\overset{\vert}{\underset{\vert}{C}}H$ $\qquad\qquad COOR$	$\qquad\quad H$ $\left(NH-(CH_2)_n-\overset{\vert}{\underset{COOR}{\overset{\ast}{C}}}-CO\right)_n$	One enantiotopic carboxymethyl group of the amino malonic ester is polymerized by enantiotopos differentiation by the chiral catalyst.
$O\ \ CH_3\qquad CH_3\ O$ $\overset{\ast}{C}-\overset{\vert}{C}-H\quad H\overset{\ast}{C}-C_O$ $\overset{\vert}{C}-NH\qquad NH-\overset{\vert}{C}$ $O\qquad\qquad\quad O$	$\qquad H$ $\left(CO-\overset{\ast}{\underset{CH_3}{C}}-NH\right)_n$	One enantiomer is polymerized by enantiomer differentiation by the chiral catalyst.
R $CH_3-\overset{\ast}{C}H$ $\qquad CH=CH-CH=CH_2$	R $CH_3-\overset{\ast}{C}H$ $\left(\overset{\vert}{\underset{H}{\overset{\ast}{C}}}-CH=CH-CH_2\right)_n$	The main chain chiral center is produced by the chirality contributing to the differentiation that lies in the initiator or catalyst, independently of the chiral center of the monomer.
R $H\overset{\ast}{C}-R'\qquad COOR'$ $NH_2-CH-(CH_2)_n-\overset{\vert}{C}H$ $\qquad\qquad\qquad COOR'$	R $H-\overset{\ast}{C}-R'\qquad H$ $\left(NH-CH-(CH_2)_n-\overset{\ast}{\underset{COOR''}{C}}CO\right)$	
$R\qquad\qquad R$ $O\ H\overset{\ast}{C}-R'\ \ H\overset{\ast}{C}-R'\ O$ $\overset{\ast}{C}-\overset{\ast}{C}H\quad H-\overset{\ast}{C}—C_O$ $\overset{\vert}{C}-NH\qquad NH-\overset{\vert}{C}$ $O\qquad\qquad\qquad O$	R $H-\overset{\ast}{C}-R'$ $\left(CO-\overset{\vert}{\underset{H}{C}}-NH\right)_n$	
Ph $H-\overset{\ast}{C}-CH_3$ $\overset{\vert}{C}H=CH-CH=CH_2$	Ph $H-\overset{\ast}{C}-CH_3$ $\left(\overset{\ast}{\underset{H}{C}}CH=CH-CH_2\right)_n$	The chiral center in the monomer acts as the chiral factor in the reaction.
Ph $R-\overset{\ast}{C}H\qquad\quad COOR'$ $NH_2-\overset{\vert}{C}H-(CH_2)_n-\overset{\vert}{C}H$ $\qquad\qquad\qquad\quad COOR'$	Ph $R\overset{\ast}{C}H$ $\left(NH-\overset{\ast}{\underset{H}{C}}-(CH_2)_n\overset{\ast}{C}HCO\right)_n$ $\qquad\qquad\qquad\qquad COOR$	Differing reactivities of the diastereomers produce the differentiation.
$R\qquad\qquad R$ $H-\overset{\ast}{C}-R\quad H-\overset{\ast}{C}-R'$ $\overset{\vert}{CO}-\overset{\vert}{C}H\quad H\overset{\vert}{C}-CO_O$ $O\overset{}{}\quad CO-NH\ ,\ \ HN-CO$	$\qquad H$ $\left(CO-\overset{\ast}{C}-NH\right)_n$ $R-\overset{\vert}{C}H$ $\qquad R'$	

A.   Addition and ring-opening polymerizations

Enantioface-, enantiomer- and diastereoface-differentiating reactions have been identified as possible mechanisms of differentiation among addition and ring-opening polymerizations.

It should be noted that the differentiating effects of optically active initiators are different from those of optically active catalysts, in that the differentiating effect of the initiator decreases as the reactive end of the polymer chain becomes more distant from it (due to the addition of further monomer units), whereas that of a catalyst is constant at each step of monomer addition. In the former case, if the initial differentiating ability is α ($\alpha < 1$) then when the degree of polymerization is n, the apparent differentiating ability at that step will be α^n. Polymerization reactions with an optically active initiator (R^*) and an optically active catalyst (C^*) are shown in Eqs. 6.3 and 6.4, respectively.

$$R^* \xrightarrow{S_1} R^*S_1^* \xrightarrow{S_2} R^*S_1^*S_2^* \xrightarrow{S_3} R^*S_1^*S_2^*S_3^* \xrightarrow{S_n} R^*S_1^*S_2^*S_3^* \ldots S_n^* \quad (6.3)$$

$$C^* \xrightarrow{S_2} C^*S_1^* \xrightarrow{S_2} C^*S_2^*S_1^* \xrightarrow{S_3} C^*S_3^*S_2^*S_1^* \xrightarrow{S_n} C^*S_n^* \ldots S_3^*S_2^*S_1^* \quad (6.4)$$

a.   Enantioface-differentiating polymerization

The polymer of methyl sorbate ($CH_3CH=CHCH=CHCOOCH_3$) has been obtained in 6 % optical yield as a result of the enantioface-differentiating polymerization of methyl sorbate with ($-$)-2-menthylethylester-C_4H_9Li as a catalyst.[1]

A very special enantioface-differentiating polymerization was carried out by the radiation of γ-ray to the lathrate compound of 1,3-pentadiene and R-($-$)-perhydrotriphenylene in chloroform by Farina et al (Eq. 6.5).[2] The above polymerization gave optically active trans-1,4-isotactic polypentadiene o, $[\alpha]_D + 2.5° \pm 0.5°$. This reaction system must bring an useful information on the independence of the differentiating process from the reaction process.

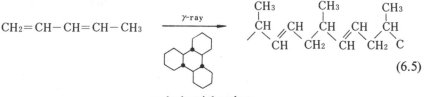

$$CH_2=CH-CH=CH-CH_3 \quad \xrightarrow{\gamma\text{-ray}}$$

(6.5)

perhydrotriphenylene

b.   Enantiomer-differentiating polymerization

Polyleucine having the same configuration as the initiator has been obtained by the enantiomer-differentiating polymerization of DL-leucine-*N*-carbonate anhydride using leucine methylester, as shown in Eq. 6.5.[3]

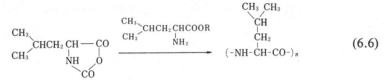

$$(6.6)$$

In addition, (+)-borneol-diethyl zinc catalyzes the enantiomer-differentiating polymerization of DL-propylene oxide to give the polymer shown in Eq. 6.6, which contains more D-monomer units than L-monomer units.[4,5]

$$H_2C-CH-CH_3 \quad \longrightarrow \quad (-O-CH_2-\overset{CH_3}{\underset{|}{CH}}-)_n \qquad (6.7)$$

c. Diastereoface-differentiating polymerization

Although diastereoface differentiation has not been clearly demonstrated in polymerization reactions, Eq. 6.7 shows a possible example.[6]

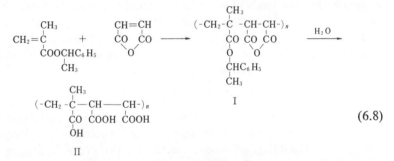

$$(6.8)$$

The polymer I is obtained by radical polymerization of a mixture of optically active α-methylbenzyl methacrylate and maleic anhydride. However, the removal of the optically active α-methylbenzyl groups by hydrolysis leaves a polymer (II) which still shows optical activity. This polymer has chiral centers at C-1, C-2 and C-3 (*). It is therefore possible that the optical activity arises from the C-1 center, although this has not yet been determined.

B. Polycondensation and polyaddition

Polycondensation and polyaddition reactions have the feature that the structures of both the reagent (or catalyst) and the substrate change during the polymerization process. Few examples of these reactions are known to be stereo-differentiating. One example is provided by the polyaddition reaction between ethylene dimethacrylate and a dithiol using (*S*)-butyl-ethyleneimine as a catalyst, which yields an optically active polymer (Eq. 6.8).[7]

$$CH_2=\overset{CH_3}{\underset{|}{C}}COO(CH_2)_2OCO\overset{CH_3}{\underset{|}{C}}=CH_2+HS(CH_2)_4SH \longrightarrow$$

$$(-S-(CH_2)_4SCH_2\overset{CH_3}{\underset{|}{C}}HCOO(CH_2)_2OCO\overset{CH_3}{\underset{|}{C}}HCH_2-)_n \quad (6.9)$$

6.2 STEREO-DIFFERENTIATION IN ENZYME REACTIONS[8]

Naturally occurring optically active compounds are in most cases produced by highly specific enzyme-controlled differentiating reactions in living systems. Enzyme reactions are often regarded as a special type of reaction, but there is no essential difference between an enzyme reaction and an ordinary stereo-differentiating reaction, so that enzyme reactions can be classified into the same six categories mentioned in the previous section. The main characteristic difference between enzymic reactions and ordinary stereo-differentiating reactions is that enzymic reactions often couple with other kinds of stereo-differentiating processes. Since the comprehensive review[8] on the enzymic reactions of prochiral compounds eddited by the conventional system has been published, the few enzymic reactions will be discussed from the view point of the stereo-differentiation as an example in this section.

6.2.1 Aconitase

Aconitase is one of the enzymes of the TCA cycle and catalyzes the dehydration of isocitric acid to aconitic acid, successively differentiating the pro-R CH_2COOH of isocitric acid and the pro-R hydrogen of the CH_2COOH group, as shown in Fig. 6.1. Thus the dehydration is coupled with enantiotopos- and diastereotopos-differentiating reactions.[†]

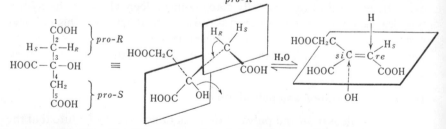

Fig. 6.1. The dehydration of isocitric acid to aconitic acid by aconitase: coupled enantiotopos and diastereotopos differentiation.

† As the dehydration is achieved after enantiotopos differentiation of the pro-R CH_2COOH, the isocitric acid bound to the enzyme can be regarded as enantiomeric at this stage, and thus it is reasonable to regard the two hydrogens on C-2 as enantiotopic.

6.2.2 Alcohol dehydrogenase

Alcohol dehydrogenase catalyzes the reduction of acetaldehyde to ethanol in the presence of reduced nicotinamide-adenine dinucleotide (NADH). Since this enzyme catalyzes the enantioface-differentiating reduction of acetaldehyde using the pro-R hydrogen at C-4 of NADH (Fig. 6.2), the reaction is coupled with enantioface- and enantiotopos-differentiating reactions.

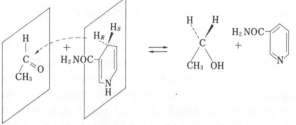

Fig. 6.2. The enantioface-differentiating reduction of acetaldehyde by alcohol dehydrogenase with NADH.

6.2.3 Enantiomer-differentiating reactions

Optical resolution of amino acids can be carried out on an industrial scale using various enzymes. The use of acylase, papaine and amino acid oxidase will be discussed below.

An acylase catalyzes the enantiomer-differentiating hydrolysis of N-acyl-DL-amino acids to give the L-amino acid as shown in Eq. 6.10, while the N-acyl-D-amino acid is recovered unchanged in this enzyme system. On an industrial scale, the enzyme is used in the solid phase, supported on an ion exchange resin.

$$\text{L-}\begin{matrix}\text{R-CHCOOH}\\|\\\text{NHCOCH}_3\end{matrix} \xrightarrow{+H_2O} \text{L-}\begin{matrix}\text{R-CHCOOH}\\|\\\text{NH}_2\end{matrix} \qquad (6.10)$$

Papaine was used for the optical resolution of amino acids before the development of acylase for this purpose. It catalyzes the enantiomer-differentiating amidation of N-acyl-L-amino acids, conventionally benzoyl-amino acid, with aniline, yielding the acyl-L-amino anilide which is insoluble and can therefore be easily collected. The reaction is shown in Eq. 6.11.

$$\text{L-}\begin{matrix}\text{R-CHCOOH}\\|\\\text{NHCOC}_6\text{H}_5\end{matrix} \xrightarrow{+C_6H_5NH_2} \text{L-}\begin{matrix}\text{R-CHCONHC}_6\text{H}_5\\|\\\text{NHCOC}_6\text{H}_5\end{matrix} \qquad (6.11)$$

L-Amino acid oxidase yields the α-ketoacid from an L-amino acid by enantiomer-differentiating oxidation of the DL-amino acid (Eq. 6.12), and can therefore be used for the small-scale preparation of D-amino acids. As this enzyme has an extremely high differentiating ability, it can be used to determine the optical purity of amino acids (see Chap. 8).

$$\text{L-} \underset{\underset{\text{NH}_2}{|}}{\text{R-CHCOOH}} \xrightarrow{+O_2} \text{R-COCOOH} \tag{6.12}$$

6.3 SO-CALLED ABSOLUTE ASYMMETRIC SYNTHESIS

The synthesis of optically active compounds by the stereo-differentiating reactions described above involves the presence of an optically active compound in the reaction system, and since optically active compounds are intimately associated with living systems, there has been considerable interest in processes by which chirality could have been introduced into living systems. In principle, two kinds of processes are known. One involves the accidental resolution of an optically active compound by preferential crystallization. An example is provided by the crystallization of methylethylarylaniline iodide, which is easily racemized in solution. The R or S compound is obtained exclusively, for instance, when solutions of (R,S)-methylethylarylaniline iodide are stored for long periods and crystallize out.[9] Under the influence of these optically active crystals, stereo-differentiating reactions might occur. Thus, many studies have been made on the synthesis of optically active compounds under the influence of dextrorotatory or levo-rotatory crystals of inorganic compounds.[10] The mechanisms of such reactions will be the same as those of ordinary stereo-differentiating reactions.

The other kind of process which has been considered involves the action of a chiral physical force, such as circularly polarized light or γ-rays.[11] In reactions performed by such a chiral physical force, which are classified as "absolute asymmetric reactions" in the narrow sense, one would expect enantioface, enantiotopos and enantiomer differentiation to be possible, but in spite of considerable efforts, only enantiomer-differentiating reactions have been observed. These can be divided into configurational and conformational enantiomer-differentiating reactions, and will be described in more detail below.

6.3.1 Configurational enantiomer-differentiating reactions

It was suggested by van't Hoff in 1894 that the synthesis of optically active compounds might be achieved by means of circularly polarized light. Cotton irradiated copper tartrate with circularly polarized light at its absorption wavelength, but failed to detect any differentiating reaction.

The reason for this was inadequate energy of the irradiating light.[12] Thirty years later, Kuhn et al.[13] succeeded in performing the first enantiomer-differentiating reaction with circularly polarized light in the ultraviolet region.

The rate of reaction induced by light is proportional to the extinction coefficient ε. Since circular dichroism arises from a difference, $\Delta\varepsilon$, in the absorption of left- and right-circularly polarized light, left- or right-rotatory polarized light should interact preferentially with one enantiomer of substances exhibiting circular dichroism (See Chapter 9). If the racemic substrate is decomposed under irradiation by circularly polarized light and the reaction is stopped before completion the relation between the optical rotatory power of unreacted substrate and the anisotropy factor, g, is as follows:

$$[\alpha]_D = \frac{g}{2} \times ([\alpha]_D)_0 \cdot (1 - x) \ln \frac{1}{1 - x} \qquad (6.13)$$

$[\alpha]_D$: optical rotatory power of unreacted substrate
$([\alpha]_D)^0$: optical rotatory power of substrate (100% optical purity)
x: rate of reaction of substrate

$g = \dfrac{\varepsilon_l - \varepsilon_r}{\varepsilon}$: anisotropy factor of substrate

If we take $g = 0.02$ to 0.03 for α-azidopropionic acid, good correspondence is obtained between the calculated results and the experimental results for the enantiomer-differentiating reaction of α-azidopropionic amide with circularly polarized light. There was a good correlation between the conversion of substrate and the $[\alpha]_D$ of recovered substrate, and furthermore $[\alpha]_D$ values of recovered substrates under the action of levo- and dextro-rotatory light were identical except for their sign. Thus Kuhn[13] confirmed the occurrence of enantiomer differentiation by circularly polarized light both experimentally and theoretically, verifying van't Hoff's prediction. This type of reaction is not a synthetic reaction, but a kind of optical resolution, since the optical activity of the unreacted substrate depends on the transformation of one enantiomer in the racemic mixture.

Stevenson et al.[14] discovered that the irradiation of $(\pm)$-chromium oxalate aqueous solution with circularly polarized light gradually gave the solution optical rotatory power (increasing to a constant value) without affecting the chromium oxalate content. This complex is interconvertible between the $(+)$ and $(-)$ forms through an excited state which can be produced by irradiation with light. When circularly polarized light is used, one enantiomer is excited and converted to the racemic modification. As a

Circularly polarized light[†1]	Configurational Substrate
RCL and LCL ($\lambda = 280$ nm)	$(\pm)$-CH$_3$CHBrCOOC$_2$H$_5$
RCL and LCL ($\lambda = 290$ nm)	$(\pm)$-CH$_3$CHCON(CH$_3$)$_2$ with N$_3$ substituent
RCL and LCL ($\lambda = 600$–700 nm)	$(\pm)$
CL[†2] ($\lambda = 360$ nm)	
LCL ($\lambda = 330$ nm)	$(\pm)$
RCL and LCL ($\lambda = 313$ nm)	$(\pm)$
RCL and LCL ($\lambda = 540$ nm)	(a) chromium (III) trioxalate
	(b) chromium (II) dioxalate

[†1] RCL, right-circularly polarized light; LCL, left-
[†2] Direction of polarization not specified

TABLE 6.2
enantiomer-differentiating reactions

Operational results	Ref.
A 5–8% solution of the substrate in methanol was irradiated with CL. After 50% conversion, unchanged substrate was recovered; RCL and LCL gave $[\alpha]_D$ values of $+0.05°$ and $-0.05°$, respectively.	(12)
The substrate had a large g value at 290 nm. A 2.5% solution of the substrate in hexane was irradiated. After 40% conversion, unchanged substrate was recovered; RCL and LCL gave $[\alpha]_D$ values of $+0.78°$ and $-1.04°$, respectively.	(13)
A 7% solution of the substrate in ethyl acetate was irradiated. The optical activity of the reaction system was measured periodically. RCL and LCL gave optimum activities of $+0.21°$ and $-0.21°$ at 546 nm. The reaction products were not separated, and prolonged irradiation destroyed the optical rotatory power.	(15)
The substrate (0.6 g in 1700 ml of ethanol) was irradiated for 24–36 hr. After 37% conversion (forming the pyridine derivative) the solvent was removed and the residue was dissolved in acetic acid. The optical activity, $[\alpha]_D$, was $-0.022°$.	(16)
The substrate (1.5 g in 6 ml of methylene chloride) was irradiated for 120 hr. The recovered substrate gave $[\alpha]_{385}$ of $-0.13°$ and $[\alpha]_{345}$ of $-0.43°$ (in $CHCl_3$).	(17)
The substrate (43 mg) was irradiated in 140 ml of benzene for 110 hr. RCL gave $[\alpha]_{589}^{23}$ of $+2.6°$ and $[\alpha]_{436}^{23}$ of $+7.5°$, while LCL gave $[\alpha]_{589}^{23}$ of $-1.1°$ and $[\alpha]_{436}^{23}$ of $-5.2°$.	(18)
The substrate (1.8×10^{-3} M aqueous solution) was irradiated. The optical activity increased with irradiation time, reaching equilibrium at 4 hr.	(19)
The equilibrium optical activities with RCL were $[\alpha]_D$ of $+0.06°$ (for irradiation at 1.4°C) and $[\alpha]_D$ of $+0.03°$ (at 13°C) for substrate (a), and corresponding values of $+0.02°$ and $+0.01°$ for substrate (b). LCL gave equal but opposite optical activities.	

circularly polarized light; CL, circularly polarized light

result, the other enantiomer, which is scarcely affected, accumulates in the solution until the isomer ratio reaches an equilibrium value. Stevenson *et al.* named this reaction *partial photoresolution*. It is a special case of a combined enantiomer-differentiating reaction and equilibrium reaction. Well-established enantiomer-differentiating reactions which involve the use of circularly polarized light are summarized in Table 6.2.

6.3.2   Conformational enantiomer-differentiating reactions

Conformational enantiomer differentiating reactions can be carried out with circularly polarized light if right- or left-circularly polarized light preferentially excites one of the conformational enantiomers, causing it to react with some reagent to yield a configurational enantiomer as a product.

Karagounis *et al.*[20] attempted to react the triphenylmethyl radical derivative shown below with bromine or chlorine under the influence of

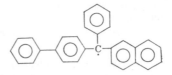

triphenylmethyl radical derivative

circularly polarized light. They found that the optical rotatory power of the reaction system was $+0.07°$ and $-0.06°$ with right- and left-circularly polarized light after one hour, but the optical activity had disappeared after 2 hr. They explained these results in terms of an equilibrium between planar and trigonal pyramidal structures for the radical, suggesting that one enantiomer of the trigonal pyramidal structure is preferentially excited and thus reacts with the halogen to a greater extent. However, these results are questionable for the following reasons. Firstly, the optically active halogenated triarylmethane was not isolated. Secondly, reinvestigation of the reactions failed to confirm the initial results, and finally, the triarylmethyl radical is now known to have a propeller-like structure[21] in which the central carbon is planar, not a trigonal pyramidal structure. However, as radical which has strongly electronegative ligands is known to maintain a trigonal structure,[22] and also the propeller-like structure is chiral, the possibility cannot be completely ruled out that the reaction proceedes through the conformational enantiomer differentiation of the chiral radicals by circular polarized light.

It was found that helicene is produced by the irradiation of the diarylethylene shown in Eq. 6.14 in the presence of a small amount of iodine.[23]

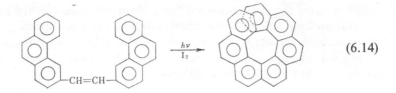

$$\text{(6.14)}$$

Martin *et al.*[24] succeeded in resolving the optical isomers of helicene, finding that it has extremely strong optical rotatory power, with $[\alpha]_D = 6200 \pm 200°$. Subsequently, Kagan *et al.*[18,25] and Buchardt *et al.*[26-28] independently succeeded in using circularly polarized light to produce enantiomer-differentiating syntheses of helicenes, as shown in Table 6.3.

The mechanism of the cyclization and dehydrogenation reactions is shown in Eq. 6.15, and three possible types of differentiation can be distinguished.

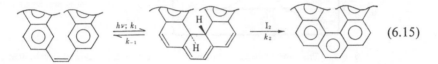

$$\text{(6.15)}$$

First, differentiation may take place by the enantiomer-differentiating decomposition of the helicene once produced, so that the remaining helicene becomes optically active. The second possibility is that differentiation takes place as a result of a reversible cyclization process (first step in Eq. 6.15) with enantiomer-differentiating ring opening of the enantiomeric intermediates under the influence of circularly polarized light as the reverse reaction. In this case, the remaining intermediate will become optically active under stationary conditions, and if dehydrogenation (second step in Eq. 6.15) then occurs with retention of configuration, optically active helicene will be obtained as a product. For this process to occur, the reaction rates should be in the relation $k_1, k_{-1} > k_2$. The third possibility is conformational enantiomer differentiation of the starting material, which has helical chirality, since the terminal aromatic rings of the diarylethylene cannot lie in the same plane. In this case, one conformational enantiomer of diarylethylene will be activated more than the other, and hence, one configurational enantiomer of the intermediate will be produced preferentially under the influence of circularly polarized light. If the configuration of the intermediate is retained during dehydrogenation (second step in Eq. 6.15), optically active helicene will be produced.

Kagan *et al.* ruled out the first possibility, since the direction of optical rotation of the helicene obtained on irradiation of racemic helicene with circularly polarized light is opposite to that produced with diarylethylene using the same polarized light. Buchardt *et al.* reached the same conclusion from studies of the relationship between the optical yield and the anisotropy factor (g) of the helicene produced in the reaction, as shown in Fig. 6.3. The sign of the anisotropy factor changes at 335 nm, while that of the optical yield changes at 305 nm.

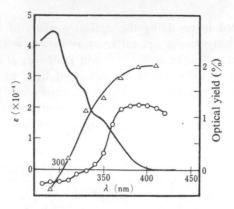

Fig. 6.3. The optical yield ($\triangle$) of helicene produced, g value of octahelicene ($\bigcirc$) and the ultraviolet absorption spectrum ($-$) of octahelicene. [From Bernstein, Calvin and Buchardt (26)].

As for the second possibility, Muszkat *et al.*[29] found that the kinetic constants for the reaction shown in Eq. 6.16 were in the relation k_1, $k_{-1} < k_2$.

$$\text{stilbene} \underset{k_{-1}}{\overset{k_1}{\rightleftharpoons}} \text{dihydrophenanthrene} \overset{k_2}{\longrightarrow} \text{phenanthrene} \tag{6.16}$$

By analogy, one would expect a similar relationship of the kinetic constants in Eq. 6.15, which would rule out the second possibility. Furthermore, several transition points are found in the long wavelength region for dihydrophenanthrene, and again by analogy one would expect the sign of the optical yield to change in this region for reaction 6.15; Fig. 6.3 shows that this is not the case. This possibility also fails to account for the differences of optical yield between compounds I and II in Eq. 6.17 (see Table 6.3).

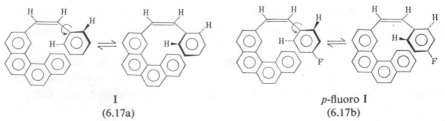

I
(6.17a)

p-fluoro I
(6.17b)

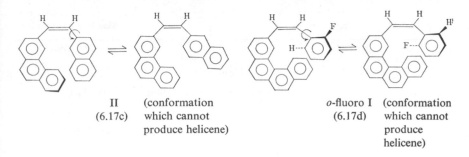

II (6.17c)	(conformation which cannot produce helicene)	*o*-fluoro I (6.17d)	(conformation which cannot produce helicene)

As shown in Eq. 6.17, the differences in optical yields can be accounted for in terms of the third possibility. Thus, the cyclization of a nonsymmetric 1,2-diarylethylene by photoreaction will proceed through the excited state of lowest energy, and the excitation will be localized in the aryl group which can take the lowest energy state[30] (benzophenanthryl and phenanthryl groups in compounds I and II (Eq. 6.17), respectively). In these excited states, the phenyl group of I and the naphthyl group of II can rotate 180° around the bond to the ethylene carbon. Now the phenyl group of I is symmetric with respect to this axis and both rotational states can act as precursors of helicene, but the original conformation and that produced by 180° rotation are in an enantiomeric relation, so that racemization occurs before cyclization. However, in the cyclization of II, the structure produced by rotation of the naphthyl group cannot act as a precursor of helicene, and cyclization can only occur in the excited state in which the original structure is retained. The higher optical yield of II than I can thus be explained by supposing that the proportion of helicene enantiomers produced reflects the proportion of the corresponding enantiomers in the excited reaction species. The remaining possibility that the results might be due to a higher anisotropy factor of II than of I was ruled out by Buchardt *et al.* They also provided further support for the above view by investigating derivatives of compound I with *ortho* and *para* fluorine substituents in the phenyl ring (shown in Eq. 6.17b and d) under the same conditions. If the third possible mechanism of differentiation, described above, is correct, a higher optical yield should be obtained with the *ortho*-fluoro compound than with the *para*-fluoro one, which can racemize by rotation of the *p*-fluorophenyl ring in the same way as compound I. On the other hand, the introduction of a fluorine atom should not change the anisotropy factor, g, markedly, so if differentiation is controlled by g, all three compounds should produce similar optical yields of helicenes. The results (shown in Table 6.3) supported the former view.

Other results in Table 6.3 are less clear-cut, but it was clearly shown that the reaction of 1,2-diarylethylene according to Eq. 6.15 involves enantio-differentiation by circularly polarized light. The relation between

TABLE 6.3
Absolute asymmetric

$$R_1 \diagdown \quad \diagup R_2$$
$$C=C$$
$$H \diagup \quad \diagdown H$$

R₁	R₂		Light source		Product
			Wavelength	Polarization	
(structure)	(structure)		350 370 350 370 390	RCL RCL RCL LCL	Hexahelicene
(structure)	(structure)		350 390 350	RCL LCL	Hexahelicene
(structure)		X=F	370	RCL LCL	2-Fluorohexahelicene
		X=Cl	370	RCL LCL	2-Chlorohexahelicene
		X=Br	370	RCL LCL	2-Bromohexahelicene
(structure)		X=F	370	RCL LCL	4-Fluorohexahelicene
		X=Cl	370	RCL LCL	4-Chlorohexahelicene
		X=Br	370	RCL LCL	4-Bromohexahelicene
(structure)	(structure)		370 400 335	RCL LCL RCL LCL RCL LCL	Heptahelicene
(structure)	(structure)		390 370	RCL LCL	Heptahelicene
(structure)	(structure)		290–370 290–370 390	RCL LCL	Octahelicene
(structure)	(structure)		290–310 290–310	RCL LCL	Octahelicene
(structure)	(structure)		290–370 390	LCL	Nonahelicene

†¹ To amplify the effect, samples obtained by irradiation with right- and left-circularly polarized light (RCL and LCL, respectively) were placed in the sample and reference phials of the ORD spectrometer. The maximum difference (α) in

Time of irradiation (hr)	Optical rotatory power		ORD or CD measurements	Ref.
	$[\alpha]_{589}$ (concentration)	$[\alpha]_{436}$ (concentration)		
6	−1.8(1.29)	−7.6(1.29)		(18)
24		−8.0(0.59)		(28)
6	+1.9(0.99)	+8.4(0.99)		(18)
24		+7.6(0.57)		(28)
0.5			0.0007°/500–400 nm†1	(26)
6	−7.5(2.08)	−30.0(2.08)		(18)
0.5			0.0086°/500–370 nm†1	(26)
6	+7.9(0.77)	+30.5(0.77)		(18)
24		−10.5(0.55) +10.7(0.54)		(28)
24		−16.4(0.66) +15.8(0.65)		(28)
36		−23.8(0.59) +23.2(0.59)		(28)
36		−31.5(0.41) +34.0(0.38)		(28)
72		−45.8(0.36) +46.9(0.26)		(28)
68		−38.7(0.35) +34.2(0.21)		(28)
4			(−)0.74%†2 (+)0.76%	
24			(−)1.29%†2 (+)1.26%	(28)
24			(−)0.47%†2 (+)0.47%	
1.5			0.0132°/520–385 nm†1	(26)
24			(−)0.57%†2 (+)0.62%	
	At 80% conversion	{−21 (1.22) +20.3(1.12)		(18)
2			0.0320°/500–410 nm†1	(26)
	At 45% conversion	−7.6(0.25)		(25)
	At 100% conversion	0		(25)
	At 75% conversion	+3.7(0.188)		(25)
2	At 50% conversion	+30.4(0.65)		(25)
			0.0144°/500–415 nm†1	(26)

optical rotatory power between wavelengths λ_1 and λ_2 is shown as $\alpha/\lambda_1 \sim \lambda_2$.

†2 Optical yield calculated from the g ratio between the sample and an optically pure helicene

the helicity of helicene and the optical rotatory power was determined by Lightner et al.,[31] as shown in Eq. 6.18.

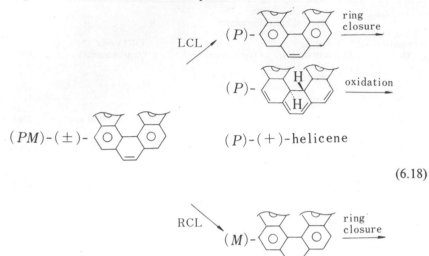

(6.18)

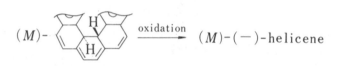

6.3.3   Enantioface-differentiating reactions

Following the use of circularly polarized light by Kuhn to carry out enantiomer-differentiating reactions, several attempts were made to obtain enantioface-differentiation with achiral reagents under the influence of circularly polarized light. However, these all involved very low optical rotatory power of the reaction mixture, and in no case was an optically active product isolated. In these experiments, circularly polarized light was expected to cause enantioface-differentiating excitation of sp^2-prochiral compounds, but such effects would not be expected on the basis of present theories. Although circularly polarized light might differentiate independently produced racemic enantiomers or enantiomeric excited complexes, these reactions would actually be enantiomer-differentiating reactions.

Recently, Boldt[17] reexamined the reactions shown in Table 6.4 from experimental and theoretical viewpoints. He pointed out that the circularly polarized light used in experiments 1 and 2 had insufficient energy to excite the ethylene bond, but could excite chlorine or bromine to their radicals. Even if Cl and Br radicals could produce chiral effects, the photoquantum yield of the reaction, $\phi \gg 1$, rules out the possibility in this case, so that the circularly polarized light can only participate in the reaction as an initiator of the chain reaction, as ordinary light does. Reinvestigation of reactions 3 and 4 revealed that reaction 3 did not occur under the reported conditions, and reaction 4 failed to produce any optically active product.

Boldt reinvestigated reactions 6 and 7 under the assumption that differentiation could only occur under conditions where the photoquantum yield is less than 1. However, no optically active product could be obtained in reaction 6. Though slight optical activity was produced in the case of reaction 7, it was insufficient to distinguish between the alternative possibilities of enantioface differentiation or enantiomer differentiation of the preformed *RS*-product by circularly polarized light.

Thus, although the possibility of enantioface differentiation by circularly polarized light cannot be completely ruled out on the basis of present analytical methods, it seems unlikely that such differentiation can occur.

6.4 ENANTIO-DIFFERENTIATING RACEMIZATION

The synthesis of amino acids is very important commercially, but general synthetic methods yield racemic mixtures, and a catalyst which could convert the DL-form to D- or L-form would be of great value industrially. In living organisms, racemase operates on the two forms at different rates,[37,38] and synthetic catalysts have also been prepared.[39] However, none of them actually work as a catalyst for the conversion of DL-form to D- or L-form.

We will first consider the energetics of racemization. As the substrate has the same internal energy as the product, we have $\Delta H = 0$, and this reaction is therefore controlled by entropy effects. Now in the racemization of pure L (or D) enantiomer, ΔS corresponds to the entropy of mixing, if the reaction system is treated as an ideal solution, and is given by Eq. 6.19.

$$\Delta S = -R(X_A \ln X_A + X_B \ln X_B) \tag{6.19}$$

Here, X_A and X_B are the molar ratios of A and B, respectively, in the reaction system. As these will both take values of 0.5 for the racemate, the entropy change per mole is given by Eq. 6.20.

$$\Delta S = -R(0.5 \ln 0.5 + 0.5 \ln 0.5) = 1.37 \text{ cal/deg} \cdot \text{mole} \tag{6.20}$$

TABLE 6.4
Enantioface-differentiating"

Reaction No.	Chiral force†	Reagent	Substrate	Product(s)

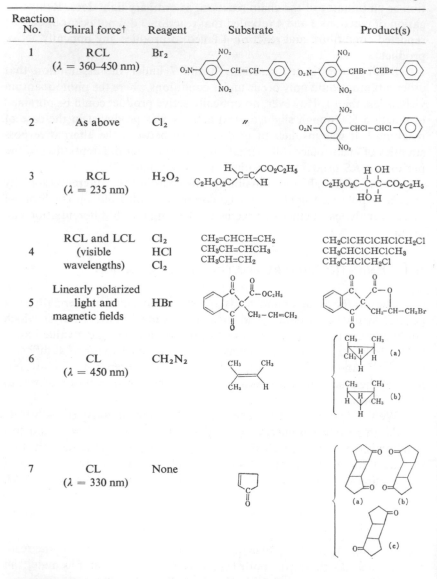

† RCL, right-circularly polarized light; LCL, left-circularly polarized light; CL,

reactions (see text)

Results	Ref.
The optical rotatory power of the reaction system reached a maximum, 0.04°, after irradiation with CL for 390 min. It subsequently decreased and fell to zero. No optically active compound could be isolated.	(32)
As above, the optical rotatory power reached a maximum, 0.03°, after 45 min, then decreased to zero. No optically active product could be isolated.	(33)
The optical rotatory power of the product reached a maximum of 0.073° (580 nm) after irradiation for 119 hr. However, in another run it did not exceed 0.03°.	(34)
Optical rotatory powers of $(+)$ and $(-)$ 0.04–0.05° were observed on irradiation in the gas phase with RCL and LCL, respectively.	(35)
Slight optical activity (± 0.15–0.25°) was reported.	(36)
Cyclobutanes were obtained in 10% yield with respect to diazomethane on irradiation with CL for 130 hr, but no significant optical activity was observed. The 2-methyl-2-butene solution was subjected to react with diazomethane introduced by argon stream.	(17)
The cyclopentanone dimer produced by irradiation with CL at 330 nm was purified by silica gel chromatography. The optical activity of the racemate (a and b) $[\alpha]_{385}^{20}$, was $-0.20 \pm 0.04°$. As irradiation of the racemic dimer with CL gave a product with $[\alpha]_{385}^{20}$ of $-0.13 \pm 0.04°$, the activity presumably arises by enantiomer-differentiating decomposition of the product by CL.	(17)

circularly polarized light

Thus, the driving force of racemization, ΔF, at 25 °C is as follows.

$$\Delta F = -T\Delta S = -410 \, \text{cal/mole} \qquad (6.21)$$

Since ΔF is less than zero, the reaction goes to completion, i.e., it reaches an equilibrium state.

If an enzyme or catalyst which can differentiate the D or L enantiomers is used, the reactions L → DL and D → DL will proceed at different rates, though the final product (racemate) is the same in each case.* Thus, the chiral properties of the catalyst can change the kinetic but not the thermodynamic parameters of the reaction. This means that the reaction cannot proceed in the direction DL → L (or D) ($\Delta F > 0$) with any type of catalyst if the reaction is performed in a closed system.* In the racemate at equilibrium, L- and D-forms are still being interconverted by catalytic differentiation, but at an equal rate, so that the bulk properties of the system remain constant.

Lambert et al.[38] studied the racemization of D- and L-alanine using racemase obtained from *Escherichia coli* and defined the enantiomer-dif-

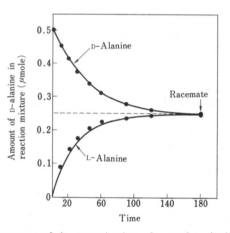

Fig. 6.4.   Time course of the racemization of L- and D-alanine by alanine racemase. [From Lambert and Neuhaus (38).]

---

* Strictly speaking, in the final stages of the reaction, the amounts of D- and L-forms bound to the chiral catalyst will not be equal, so that the ratio is not exactly 50: 50. This is so because the reaction is a kinetic resolution process.

* At least two chemical steps are necessary to obtain a pure enantiomer from the racemate against the entropy of mixing. That is, energy must be supplied to compensate for the increasing ΔF due to the entropy of mixing, so that an open energy system must be used. Optical resolution is thus energetically different from the process described in the text.

ferentiating ability of enzymes. This enzyme catalyzes the process D → DL faster than L → DL, as shown in Fig. 6.4.

If the mechanism of this reaction is given by Eq. 6.22, the rate of decrease of L-isomer is given by Eq. 6.23.

$$E + L \underset{k_2}{\overset{k_1}{\rightleftarrows}} X \underset{k_4}{\overset{k_3}{\rightleftarrows}} E + D \tag{6.22}$$

$$v = -\frac{d[L]}{dt} = \frac{d[D]}{dt} = \frac{(k_1 k_3 [L] - k_2 k_4 [D])[E_0]}{k_1 [L] + k_2 + k_3 + k_4 [D]} \tag{6.23}$$

Here, [L] and [D] are the concentrations of the L- and D-isomers, respectively, and $[E_0]$ is the total concentration of enzyme in the system. As $v = 0$ in the equilibrium state, and the thermodynamic condition is [L]/[D] = 1, we can derive Eqs. 6.24 and 6.25.

$$k_1 k_3 [L] - k_2 k_4 [D] = 0 \tag{6.24}$$

$$\frac{k_1 k_3}{k_2 k_4} = \frac{[L]}{[D]} = 1 \tag{6.25}$$

Eq. 6.23 can now be converted to the Michaelis-Menten form as follows.

$$v = \frac{(V_L/K_L)[L] - (V_D/K_D)[D]}{1 + [L]/K_L + [D]/K_D} \tag{6.26}$$

$V_L = (v_L)_{max} = k_3[E_0]$: reaction rate as [L] → ∞
$V_D = (v_D)_{max} = k_2[E_0]$: reaction rate as [D] → ∞
$K_L = \dfrac{k_2 + k_3}{k_1}$: Michaelis constant for L
$K_D = \dfrac{k_2 + k_3}{k_4}$: Michaelis constant for D

The Michaelis constants for the reactions L → DL and D → DL can be calculated from the results in Fig. 6.4 as $K_L = 9.7 \times 10^{-4}$ mole/l and $K_D = 4.6 \times 10^{-4}$ mole/l, respectively, and the maximum velocities as $V_L = 2.22$ mole/hr and $V_D = 0.95$ mole/hr, respectively. From these results, the ratios of the rate constants for L → DL and D → DL can be calculated as $k_1/k_4 = 1/2$ and $k_3/k_2 = 2/1$ approximately. Thus the recemization of L-amino acid occurs twice as rapidly as that of D-amino acid in the initial period. Fig. 6.4 shows good correspondence between the experimental results (black dots) and the calculated values (lines) obtained by the integration of Eq. 6.26 using the values of K_L, V_L, K_D and V_D obtained above. It is clear that racemase does in fact have enantiomer-differentiating ability, like other enzymes.

We will next consider the equilibrium state. The rate of formation of the enzyme-substrate complex (ES complex, X) of L is greater than that of D due to the enantiomer-differentiating ability of the enzyme. However, the rate of release of L from the ES complex is slower than that of D. Thus, the equilibrium depends on the diastereomer-differentiating decomposition of the ES complex, and even though enantiomer differentiation is occurring at equilibrium, the ratio of D to L remains 1.

Since the racemase reacts with both L and D enantiomers at rates appropriate for analytical studies, kinetic studies and studies of the differences between the ES complexes can give useful information in fundamental studies of the stereo-differentiating reaction in enzymatic systems.

The authors[39,40] have synthesized (S)-$(-)$-2-hydroxy-2'-nitro-5-nitroso-6,6'-dimethylbiphenyl, which has racemization activity, in order to compare its activity with that of the racemase enzyme.

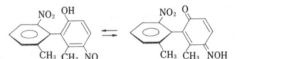

(6.27)

2-hydroxy-2'-nitro-5-nitroso-6,6'-dimethylbiphenyl

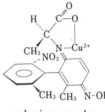

L-alanine complex

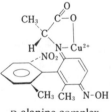

D-alanine complex

In the presence of Cu^{2+} under basic conditions, this compound forms a Schiff base with amino acids, and the Schiff base chelates Cu^{2+} (see Eq. 6.27). On the basis of molecular models, it is expected that the complex with L-alanine should be formed more readily than that with D-alanine. In fact, the compounds catalyze the racemization of L-alanine more rapidly than that of D-alanine, as expected (Fig. 6.5). Fig. 6.5 shows that the time course of racemization by this catalyst is considerably different from that with the enzyme, racemase. In particular, there is an induction period, which suggests that the catalytic species is not one of the complexes shown in Eq. 6.27, but some other species produced from them. On the basis of

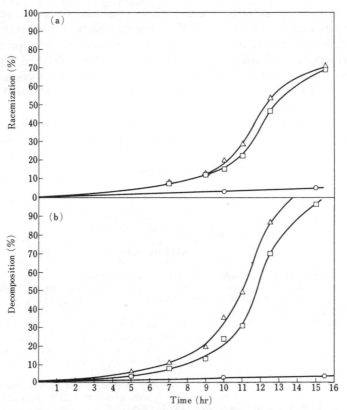

Fig. 6.5. (a) Time course of the enantiomer-differentiating racemization of alanine by (S)-(−)-2-hydroxy-2′-nitro-5-nitroso-6,6′-dimethylbiphenyl. (b) Time course of decomposition of (S)-2-hydroxy-2′-nitro-5-nitroso-6,6′-dimethylbiphenyl. △, racemization of L-alanine; □, racemization of D-alanine.

these and other observations, the catalytic scheme shown in Eq. 6.28 is proposed.

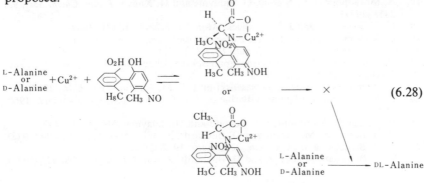

As shown, the catalytic species, X, is produced from the complex. It appears that the enantiomer-differentiating ability of the catalyst plays a part in the formation of the complexes shown. With respect to steric interactions, formation of the L-alanine complex is favored. Thus, if the species X is produced in proportion to the amount of complex in the system, the rate of racemization of L-alanine should be higher than that of D-alanine at the initial stage.

If a catalyst can be prepared in which the intermediate complex is stable, comparison with the enzyme intermediate ES will be possible. It is interesting that the catalytic chemical process and the stereo-differentiating process occur independently, and further studies may clarify relationship between these two processes.

REFERENCES

1. P. Pino, *Fortschr. Hochpolymer Forsch.*, **4**, 393 (1965).
2. M. Farina, G. Audisio and G. Natta, *J. Am. Chem. Soc.*, **89**, 5071 (1967).
3. H. C. Bücherer and H. C. Elias, *Makromol. Chem.*, **169**, 145 (1973).
4. S. Inoue, T. Tsuruta and J. Furukawa, *ibid.*, **53**, 215 (1962).
5. N. Nakaniwa, I. Kameoka, R. Hirai and J. Furukawa, *ibid.*, **155**, 197 (1972).
6. N. Beredjick and C. Schuerch, *J. Am. Chem. Soc.*, **78**, 2646 (1956).
7. S. Inoue, S. Ohashi, A. Tabata and T. Tsuruta, *Makromol. Chem.*, **112**, 66 (1968).
8. T. W. Goodewin, *Essays in Biochemistry*, vol. 9, p. 103, Academic Press, 1973.
9. E. Havinga, *Biochim. Biophys. Acta*, **13**, 171 (1954).
10. E. L. Klabunovskii, *Absolute Asymmetric Synthesis and Asymmetric Catalysis in Origins of Life on Earth*, Rept. Intl. Symp., Moscow, p. 158, Pergamon, 1959.
11. T. L. V. Ulbricht, *Quart. Rev.*, **13**, 48 (1959).
12. A. Cotton, *Ann. Chim. Phys.*, **8**, 360 (1896).
13. W. Kuhn and E. Braun, *Naturwissenschaften*, **17**, 227 (1929); W. Kuhn and E. Knopf, *ibid.*, **18**, 183 (1930).
14. K. L. Steveson and J. F. Verdieck, *J. Am. Chem. Soc.*, **90**, 2974 (1968).
15. S. Mitchell, *J. Chem. Soc.*, **1930**, 1829.
16. J. A. Berson and E. Brown, *J. Am. Chem. Soc.*, **77**, 450 (1955).
17. P. Boldt and W. Thielecke, *Chem. Ber.*, **104**, 353 (1971).
18. A. Moradpour, J. F. Nicoud, G. Balavine and H. Kagan, *J. Am. Chem. Soc.*, **93**, 2353 (1971).
19. K. L. Stevenson and J. F. Verdieck, *Mol. Photochem.*, **1**, 271 (1969).
20. G. Karagounis and G. Drikos, *Naturwissenschaften*, **21**, 607 (1933).
21. Per Anderson, *Acta Chem. Scand.*, **19**, 622 (1965).
22. H. Kawamura, *Kagaku Sosetsu* (Japanese) (ed. Chem. Soc. Japan), vol. 1, p. 186, Tokyo University Press, 1973.
23. M. Flammang-Barbieux, J. Nasellski and R. M. Martin, *Tetr. Lett.*, **1967**, 743.
24. R. M. Martin, M. Flammang-Barbieux, J. P. Cosyn and M. Gelbcke, *ibid.*, **1968**, 3507.
25. H. Kagan, A. Mordpour, J. F. Nicoud and G. Balavine, *ibid.*, **1971**, 2479.
26. W. J. Bernstein, M. Calvin and O. Buchardt, *J. Am. Chem. Soc.*, **94**, 494 (1972).
27. W. J. Bernstein and M. Calvin, *Tetr. Lett.*, **1972**, 2195.
28. W. J. Bernstein, M. Calvin and O. Buchardt, *J. Am. Chem. Soc.*, **95**, 527 (1973).

29. K. A. Muszkat and E. Fischer, *J. Chem. Soc.* (B), **1967**, 662.
30. G. S. Hammond and S. P. Van, *Mol. Photochem.*, **1**, 99 (1969).
31. D. A. Lightner, D. T. Helefinger, T. W. Powers, G. W. Frank and K. N. Trueblood, *J. Am. Chem. Soc.*, **94**, 3492 (1972).
32. T. L. Davis and R. Heggie, *ibid.*, **57**, 377 (1935).
33. T. L. Davis and R. Heggie, *ibid.*, **57**, 430 (1935).
34. T. L. Davis and J. Ackerman, *ibid.*, **67**, 486 (1945).
35. M. Betti and E. Lucchi, *Atti Conger. Int. Chim., X. Congr.*, Roma, vol. II, 1938.
36. D. Radulescus and V. Moga, *Bull. Soc. Chim. Romania*, **A1**, 18 (1939); *Chem. Abstr.*, **37**, 4070.
37. H. Hiyama, S. Fukui and K. Kitahara, *J. Biochem.* (*Tokyo*), **64** (1), 99 (1968).
38. M. P. Lambert and F. C. Neuhaus, *J. Bact.*, **110**, 978 (1972).
39. K. Hirota and Y. Izumi, *Bull. Chem. Soc. Japan*, **44**, 2287 (1971).
40. K. Hirota and Y. Izumi, *unpublished*.

CHAPTER 7

Mechanisms of Stereo-Differentiating Reactions

The mechanisms of stereo-differentiating reactions can be described in terms of stereochemical models or from a kinetic viewpoint. The former approach is based on experimental rules or on a mathematical description utilizing parameters known to be relevant to differentiation. The kinetic approach is based on the fundamental principle that differentiation is due to differences of activation energies between two diastereomeric reaction intermediates. However, it was recently noticed that differentiation is not always performed in the transition process of a reaction but may also occur in an independent process. In the last part of this section, we shall discuss the assignment of differentiating processes to particular reaction pathways.

In kinetic studies and studies using stereochemical models, the degree of stereo-differentiation obtained under very specific reaction conditions is used simply as a parameter for the stereo-differentiating ability of the reagent or catalyst. However, as described in Chapter 8, the degree of stereo-differentiation does not give a simple indication of the nature of the stereo-differentiating ability. Therefore, this approach is effective only in reactions involving limited kinds of substrates and limited reagents and catalysts under very limited reactions conditions, and in these cases only is it possible to predict the degree of stereo-differentiation satisfactorily.

7.1 THE KINETIC APPROACH

In order to produce stereoisomers in different proportions in one reaction system, it is essential that competitive reactions take place which produce stereoisomers. One possibility involves thermodynamically controlled reactions. The ratio of stereoisomers produced is determined by the equilibrium constant, which depends on the free energy difference between the two stereoisomers produced.

The other possibility is a kinetically controlled reaction in which the rates of formation of stereoisomers depend on the equilibrium concentrations of activated complexes at the transition state. The equilibrium concentration of each activated complex is dependent on the free energy difference between the substrate and the active complex (activation free energy).

If the chiral factor[†] in the reaction system acts on the thermodynamic equilibrium, the reaction is called an asymmetric transformation.[†] If the chiral factor acts on the equilibrium concentrations of activated complexes, the reaction is called a stereo-differentiating reaction.

As a basis for an understanding of stereo-differentiating reactions, we will first consider asymmetric transformations.

The epimerization of sugars is a typical example of asymmetric transformation. An equilibrium mixture of (S, R) and (R, R) glucose is obtained in the ratio of 36/64 when glucose is dissolved in water (see Fig. 7.1).

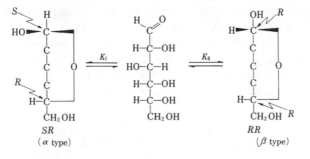

Fig. 7.1. Epimerization of glucose.

Designating the free energies of the diastereomers as G_{RR}^0 and G_{SR}^0 and the concentrations of the diastereomers at equilibrium as c_{RR} and c_{SR}, we obtain the following relation.

$$-RT \ln \frac{K_{II}}{K_I} = -RT \ln \frac{c_{RR}}{c_{SR}} = G_{RR}^0 - G_{SR}^0 \qquad (7.1)$$

The free energy difference between the diastereomers at 20°C cal-

---

[†] As the product obtained by an asymmetric transformation is readily converted to another stereoisomer, and there is no positive differentiating process from the viewpoint of synthetic chemistry, we do not include asymmetric transformation in the category of stereo-differentiating reactions.

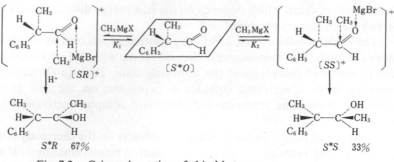

Fig. 7.2. Grignard reaction of chiral ketones.

culated for $c_{SR} = 0.36$ and $c_{RR} = 0.64$ is 0.37 kcal/mole. This difference of free energy acts as the driving force of this reaction.

The Grignard reaction of chiral ketones (Fig. 7.2) can be taken as an example of a stereo-differentiating reaction.[†] The equilibrium concentrations of the activated diastereomeric $[SR]^{\neq}$ and $[SS]^{\neq}$ relate to the activation free energies of the reactions, as shown in Eqs. 7.2 and 7.3, respectively.

If the intermediates $[SR]^{\neq}$ and $[SS]^{\neq}$ are transformed into products SR and SS with the same transmission coefficients, Eqs. 7.2 and 7.3 reduce to Eq. 7.4, and since the reaction orders producing SR and SS are the same, Eq. 7.4 reduces to Eq. 7.5. That is, the ratio of produced diastereomers is dependent only on the difference of activation energies between the diastereomeric intermediates, $\Delta\Delta G^{\neq}$.

$$-RT \ln K_1^{\neq} = -RT \ln \frac{[S^*R]^{\neq}}{[S^*O][CH_3MgX]} = \Delta G_{SR}^{\neq} \qquad (7.2)$$

$$-RT \ln K_2^{\neq} = -RT \ln \frac{[S^*S]^{\neq}}{[S^*O][CH_3MgX]} = \Delta G_{RR}^{\neq} \qquad (7.3)$$

$$-RT \ln \frac{K_1^{\neq}}{K_2^{\neq}} = -RT \ln \frac{[S^*R]^{\neq}}{[S^*S]^{\neq}} = \Delta G_{SR}^{\neq} - \Delta G_{SS}^{\neq} = \Delta\Delta G^{\neq} \qquad (7.4)$$

Eq. 7.5 can be replaced by Eq. 7.6 if $\Delta\Delta G^{\neq}$ is calculated from the optical purity of the product, P, instead of the ratio of diastereomers.

$$-RT \ln \frac{c_{SR}}{c_{SS}} = -RT \ln \frac{[S^*R]^{\neq}}{[S^*S]^{\neq}} = \Delta\Delta G^{\neq} \qquad (7.5)$$

[†] O and S in $[SO]$ in Fig. 7.2 represent a chiral factor and a prochiral center, respectively. The chiral factor is indicated by S^* or R^*, while the chiral catalyst is represented as C^*.

$$-RT \ln \frac{c_R}{c_S} = -RT \ln \frac{1+P}{1-P} = \Delta\Delta G^{\neq} \tag{7.6}$$

The relation between the optical purity of the product and the free energy difference between the diastereomeric reaction intermediates at 20 °C is shown in Fig. 7.3. It is remarkable that the free energy difference between enantiomeric intermediates necessary to obtain a product with 98 % optical yield is only 2.66 kcal/mole.

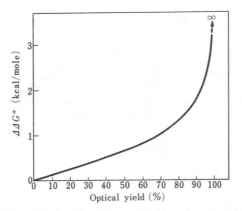

Fig. 7.3. Relationship between $\Delta\Delta G^{\neq}$ and optical yield.

The presence of a chiral factor is essential for the production of diastereomeric intermediates which have different activation free energies, in both an asymmetric transformation and a stereo-differentiating reaction.

The process of stereo-differentiation is illustrated in Fig. 7.4.[1] Fig. 7.4 (A and B) shows reactions which produce diastereomers with different free energies in the ground state of the product. These reactions are classified as A and B type reactions according to whether the sign of the difference in activation energies corresponds to the sign of the difference of free energies in the products or not. Diastereoface- and diastereotopos-differentiating reactions (Eq. 7.7),[2] and enantioface-, enantiotopos- and enantiomer-differentiating reactions which produce diastereomers as

[1] All kinetically controlled differentiating reactions can be classified into the types of reactions shown in Fig. 7.4, except for stereo-differentiating reactions caused by a chiral physical force, i.e., so-called absolute asymmetric reactions. Stereo-differentiating reactions caused by a chiral physical force are different from kinetically controlled reactions (see section 6.3).

[2] The reaction of RO can be treated in the same manner as that of SO, but for the sake of simplicity, we will describe the reaction of only one of the diastereomeric substrates in this section.

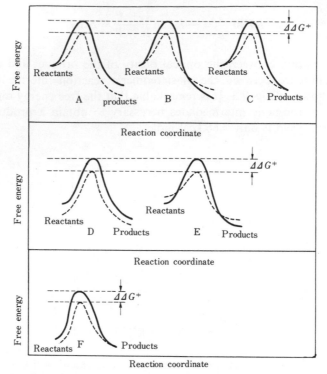

Fig. 7.4.   Free energy-reaction coordinate diagram (see the text).

products (Eqs. 7.8 and 7.9) are included in these reactions. Enantioface-differentiating (Eq. 7.10), enantiotopos-differentiating (Eq. 7.10), and enantiomer-differentiating (Eq. 7.11), reactions with self-immolative reagents,[†] enantioface- and enantiotopos-differentiating catalytic reactions (Eq. 7.12) and enantiomer-differentiating catalytic reactions (Eq. 7.13) are included in the type of reaction shown in Fig. 7.4(C).

$$S^*O \rightarrow S^*S + S^*R \qquad (7.7)$$

$$O + S^* \rightarrow SS^* + RS^* \qquad (7.8)$$

$$\left.\begin{matrix} S \\ R \end{matrix}\right\} + S^* \rightarrow SS^* + RS^* \qquad (7.9)$$

$$O + S^* \rightarrow [SS^* + RS^*] \rightarrow S + R + S^*(\text{or } SO + RO) \qquad (7.10)$$

---

† In a reaction in which the chiral factor is decomposed during the reaction and is thus not retained in the structure of the product, the reagent involved is described as self-immolative.

$$\left.\begin{array}{c} S \\ R \end{array}\right\} + S^* \rightarrow [SS^* + RS^*] \rightarrow S + R + S^*(\text{or } SO + RO) \qquad (7.11)$$

$$O \xrightarrow{C^*} S + R \qquad (7.12)$$

$$\left.\begin{array}{c} S \\ R \end{array}\right\} \xrightarrow{C^*} S + R \qquad (7.13)$$

As the diastereomers have different free energies, a diastereomer-differentiating reaction is similar to an ordinary competitive reaction between two different substances. Generally in a diastereomer-differentiating reaction, the substrates, reaction intermediates and products all have different free energies, as shown in Fig. 7.4(D and E). However, in the special case shown in Eq. 7.14, in which one of the two chiral centers in the substrate is destroyed in the course of the reaction, the products from both diastereomers have the same free energies, as shown in Fig. 7.4(F).

$$\left.\begin{array}{c} SS \\ RR \end{array}\right\} \rightarrow SO + RO \qquad (7.14)$$

Eq. 7.6 can be rewritten as Eq. 7.15 in terms of $\Delta S^{\neq}$ and $\Delta H^{\neq}$.

$$\ln \frac{c_R}{c_S} = \frac{\Delta\Delta S^{\neq}}{R} - \frac{\Delta\Delta H^{\neq}}{RT} \qquad (7.15)$$

In a system in which the entropy term is negligible, it can be expected on the basis of Eq. 7.15 that a fall in the reaction temperature should lead to an increase in the degree of differentiation. However, this situation is rare. On the other hand, cases in which the direction of differentiation is reversed with increase or reduction of the reaction temperature due to a large entropy term $\Delta\Delta S^{\neq}$ are frequent. Furthermore, it is very rare for a linear relation to exist between $\ln (c_R/c_S)$ and $1/T$, though a reaction which satisfied Eq. 7.15 should give such a relation. Very complicated relationships between c_R/c_S and reaction temperature are common in stereo-differentiating reactions. For instance, if the $\Delta\Delta G^{\neq}$ which controls a stereo-differentiation depends on a difference of molecular interaction, such as steric interaction or ion-dipole and dipole-dipole interactions between the substrate and the catalyst or the reagent, a small change of reaction conditions, such as temperature or solvent, might be expected to have a great effect on $\Delta\Delta G^{\neq}$. Moreover, there is no evidence that the rate-determining process necessarily coincides with the process of differentiation.

Though we shall consider the mechanism of stereo-differentiating reactions on the basis of a new concept in the following section, we should mention here the main features of a kinetic study. The result of a kinetic study is obtained as an activation energy; that is, the result is obtained as a difference of free energies between the substrate and the reaction intermediate at the transition state. The schemes in Fig. 7.4 do not represent complete real reaction processes but simply show the relation between the energy levels of the substrate, activated complex and product. Thus, a kinetic study of a reaction mechanism only compares the free energy levels at the entry and exit points of a "black box", without telling us anything about the mechanism within the box.

7.2 THE USE OF STEREOCHEMICAL MODELS

7.2.1 Empirical stereochemical models (experimental rules for stereo-differentiating reactions)

At the beginning of the 1950's, the first attempts to predict and explain stereo-differentiation were made by Prelog[1,2] and Cram,[3] who used a stereo model of the transition state of a differentiating reaction in order to explain the experimental results. Although a natural result of the historical development of stereo-differentiating reactions, both experimental rules relate only to diastereoface-differentiating reactions. They are often used to explain the mechanisms and to predict the results of diastereoface-differentiating reactions at present. Cornforth[4] also presented a supplementary experimental rule.

There is no experimental rule for enantioface-differentiating reactions which can be applied as widely as the rules of Prelog and Cram. That is, there is no general mode of interaction between a chiral factor and the enantioface of a substrate such as exists in diastereoface-dffierentiating reactions. However, as enantioface-differentiating reactions with a reagent proceed through diastereomeric intermediates, Prelog's and other rules are often applied to explain or predict the results of enantioface-differentiating reactions.

As the structure of a reaction intermediate in the transition state in enantio-differentiating reactions with homogeneous catalysts such as organometallic complexes can be postulated more easily than those with heterogeneous catalysts, experimental rules based on stereomodels for reactions with organometallic complexes are often proposed, but these are not general experimental rules.

In the explanation of reaction mechanisms by means of empirical models, in other words when an experimental rule is used, some care is required. Thus, the experimental rule is built up from the correlation between a single specific factor among various factors affecting the reaction.

Accordingly, the experimental rule often cannot be applied to reactions performed under conditions different from those typically used. The reader will find examples of serious conditions in section D. Such contradictions may also be ascribed to the look of proper evaluation of differentiating ability for reagents and catalysts (see section 8.2.1).

A. Prelog's rule

Prelog's rule applies to the diastereoface-differentiating reaction of chiral esters of α-keto acids with Grignard reagents, based on reexamination and arrangement by Prelog of extensive studies by McKenzie *et al.* It was proposed that the α-carbonyl, the carbonyl in the ester moiety, the chiral center and the substituent R_L (or R_S)[†] form a diastereo-zeroplane, as shown in Fig. 7.5, and the reagent attacks the α-carbonyl from the side on which R_S (or R_M) is present. This assumption is known as Prelog's rule.

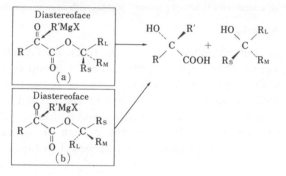

Fig. 7.5. Diastereoface differentiation of α-keto acids by Grignard reagents.

B. Cram's rule

In 1952, Cram presented an experimental rule for the diastereoface-differentiating reaction of ketones with a chiral center at the vicinal carbon with organometallic compounds or metal hydride complexes. This rule consists of two parts; one is the fundamental rule and the other applies to

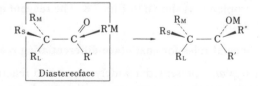

Fig. 7.6. Diastereoface differentiation of ketones by organometallic compounds.

[†] R_L, R_M and R_S represent the largest, medium-sized and smallest substituents of the chiral moiety, respectively.

special cases. The fundamental rule corresponds quite closely to Prelog's rule; that is, the diastereo-zeroplane is supposed to consist of the carbonyl group, chiral center and R_L, and the attacking reagent performs diastereoface-differentiation based on the location of R_S (see Fig. 7.6).

Since the fundamental rule does not apply to the reaction of a substrate containing a chiral center having a substituent with a high coordination ability for metals (e.g., hydroxy or amino group), Cram proposed a diastereo-zeroplane consisting of a chelate ring, as shown Fig. 7.7, for such cases and assumed that the reagent acts from the side of R_S.

Fig. 7.7.   Diastereoface differentiation of ketones having a substituent with high coordination activity toward metals with organometallic reagents.

C.   Cornforth's rule

Cornforth *et al.* presented a rule for the reaction of a substrate which has a chiral center linked with a polar atom or group (e.g., halogen); this rule is the so-called "dipolar model". That is, as the bond between a strongly electronegative atom or group and the carbon of the chiral center is easily polarized, such an electronegative atom or group takes a conformation as far as possible from the oxygen of the carbonyl; the electronegative atom, chiral center, and carbonyl group constitute the

Fig. 7.8.   "Dipole model" for diastereoface-differentiating reactions with organometallic reagents.

diastereo-zeroplane, as shown in Fig. 7.8. The reagent attacks from the R_S side of the plane.

D.   Experimental rules for enatioface-differentiating reactions

Mosher *et al.*[5] presented a widely supported reaction mechanism for enantioface-differentiating reactions with chiral Grignard reagents in 1950, as shown in Fig. 7.9, in order to account for the observations of Whitmore *et al.*[6] Although, as shown in the following examples, various reactions can be described reasonably, there are inconsistencies between

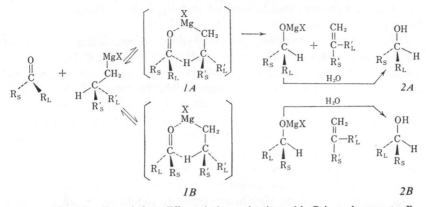

Fig. 7.9. Enantioface-differentiating reduction with Grignard reagents. R_L, R'_L are substituents of large steric volume; R_S, R'_S are substituents of small steric volume.

different reactions and it is unclear whether such reactions are typical.

The mechanism of the enantioface-differentiating reaction of carbonyl compounds with optically active Grignard reagents involves differentiation by the competitive reaction of two six-membered cyclic transition states, *1A* and *1B*, in Fig. 7.9. In this mechanism, the diastereo-zeroplane is supposed to be constituted by the chelate ring of the transition state. Since R_L and R'_L of *1A*, which are largest substituents, are on opposite sides of the diastereo-zeroplane, *1A* is more stable than *1B*, in which both R_L and R'_L are located on the same side, and in which considerable steric hindrance would be expected. Thus we would except that *1A* should have a lower energy than *1B*.

For example, in the enantioface-differentiating reaction of alkyl-phenylketones ($R_L = C_6H_5$, $R_S = CH_3$, C_2H_5, $n\text{-}C_3H_7$) with Grignard reagent A (Fig. 7.10) the alcohols *2A* and *2B* were obtained in 18–72%

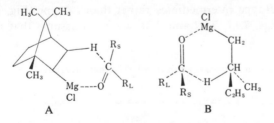

Fig. 7.10. Grignard reagents used as enantioface-differentiating reagents (see the text).

optical yields in excess of the *R* isomer.[7] In these reactions, if a *gem*-dimethyl group R'_L is present, the structure A is favorable, corresponding to *1A* and the product should have an *R* configuration. This accounts well for the observed results.

As a further example, the enantioface-differentiating reaction of alkyl-*tert*-butyl ketones (R_L = *tert*-C_4H_9, R_S = CH_3, C_2H_5, *n*-C_3H_7, *n*-C_4H_9, *iso*-C_3H_7, *iso*-C_4H_9, *cyclo*-C_6H_{11}) with Grignard reagent prepared from (+)-1-chlor-2-methylbutane will be described.[8,9] (*S*)-Alcohols are obtained predominantly in all cases. Thus, when *tert*-butyl is the largest group and compound B (Fig. 7.9) is taken as the model of the transition state, a good correlation is obtained with the observed results. These views are confirmed by the reaction of alkyl cyclohexylketones (R_L = *cyclo*-C_6H_{11}, R_S = CH_3, C_2H_5, *n*-C_3H_7, *iso*-C_3H_7, *n*-C_4H_9, *iso*-C_4H_9, *tert*-C_4H_9) with the Grignard reagent B (Fig. 7.10), that is, in all cases except for *tert*-butylcyclohexylketone, (*S*)-alcohols are obtained. Accordingly, in this case the cyclohexyl group should be the largest of all. Furthermore, as shown in Table 7.1, in the reaction of phenylketones with Grignard reagent B, the products change their configurations between *tert*-butyl- and trimethylsilicylketones.[10,11] These results can be explained readily, since the phenyl group is larger than the alkyl groups (CH_3, C_2H_5, *iso*-C_3H_5, *tert*-C_3H_9) and smaller than trimethylsilicyl, triphenylmethyl and triphenylsilicyl groups.

However certain results have been found that conflict with the above discussions in the reactions of phenylketones and in reactions with Grignard reagents having phenyl substituents. Several examples are shown in Eqs. 7.16 through 7.24. These schemes are based on the structures of the products preferentially obtained. Comparing Eqs. 7.16 and 7.17, if the phenyl group is taken as being larger than the *tert*-butyl group, Eq. 7.16 can be explained. However, Eq. 7.17 cannot be explained if the phenyl group in the Grignard reagent is taken as being larger than the methyl group. In Eqs. 7.18 and 7.19, if we take the phenyl groups as being larger than the trifluoromethyl group, the preferred intermediates in these reactions should be *1B*-type intermediates rather than *1A*-type (Fig. 7.9). In order to explain Eqs. 7.17, 7.19 and 7.21, we must assume that the two

TABLE 7.1

Reduction of phenyl ketones with Grignard reagent

$$
\begin{array}{c}
R \\
| \\
C=O \\
| \\
Ph
\end{array}
\; + \;
\begin{array}{c}
CH_2MgCl \\
| \\
H\!-\!C\!-\!C_2H_5 \\
| \\
CH_3
\end{array}
\longrightarrow \longrightarrow
\begin{array}{c}
R \\
| \\
H\!-\!C\!-\!OH \\
| \\
Ph
\end{array}
\; + \;
\begin{array}{c}
CH_2 \\
\| \\
C\!-\!C_2H_5 \\
| \\
CH_3
\end{array}
$$

R	CH_3	C_2H_5	*iso*-C_3H_5	*tert*-C_3H_9	$(CH_3)_3$Si	$(C_6H_5)_3$C	$(C_6H_5)_3$Si
Optical (%)	4	6	24	17	[3]†	[8]†	[11]†

† [ ] indicates product of opposite configuration

phenyl groups are located on the same side of the diastereo-zeroplane in the favored intermediate, but the results of Eqs. 7.20 and 7.22 contradict this assumption.

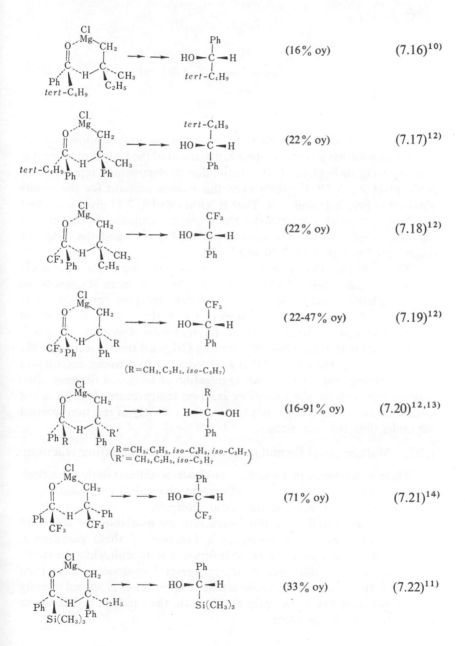

$(16\% \text{ oy})$ $(7.16)^{10)}$

$(22\% \text{ oy})$ $(7.17)^{12)}$

$(22\% \text{ oy})$ $(7.18)^{12)}$

$(22\text{-}47\% \text{ oy})$ $(7.19)^{12)}$

$(16\text{-}91\% \text{ oy})$ $(7.20)^{12,13)}$

$(71\% \text{ oy})$ $(7.21)^{14)}$

$(33\% \text{ oy})$ $(7.22)^{11)}$

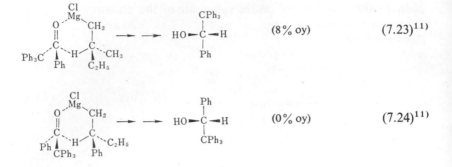

$$(8\% \text{ oy}) \qquad (7.23)^{11)}$$

$$(0\% \text{ oy}) \qquad (7.24)^{11)}$$

However, if we assume that the favored structural relation between the trifluoro-methyl group and the phenyl group of the Grignard reagent is that involving their greatest separation, due to electrostatic repulsion, we can explain Eq. 7.21. However even this cannot account for the results obtained in Eqs. 7.23 and 7.24. That is, although Eq. 7.23 gives a product in 8% optical yield, unexpectedly the racemic compound is obtained in Eq. 7.24, which involves the ketone used in Eq. 7.23 and the Grignard reagent used in Eqs. 7.19, 7.20 and 7.22.

The reason for this complexity seems to be the use of the excessively simple empirical models *1A* and *1B* (Fig. 7.9). It is more reasonable to assume different reaction species for different reaction conditions. For example, as Grignard reagents coordinate with the solvent, the reaction species must actually be different in different solvent systems. This was clearly shown in enantioface-differentiating Grignard reactions in optically active solvents by Morrison.[15] It is also necessary to consider aggregation of the Grignard reagent, since the aggregation of Grignard reagents often depends markedly on the solvent or reaction temperature. Thus, it is not feasible to use model compounds to describe a variety of reactions carried out under different conditions.

7.2.2. Mathematical formulation of stereo-differentiating reactions

There is no doubt that enantio- or diastereo-differentiating reactions involve interaction between a chiral reagent or catalyst and substrate or between a reagent or catalyst and chiral substrate.

If parameters indicating this interaction are available, the degree of differentiation may be estimated as a function of these parameters. Although the formulation of a suitable function is difficult while the mechanisms of differentiation remain unclear, several approaches have been attempted. In this section, two representative approaches proposed recently will be discussed briefly in order to illustrate the fundamental concepts used in treating this problem.

A. Stereochemical analog model

In diastereo- or enantio-differentiating reactions, where the chiral factor in the substrate or reagent does not participate directly in the reaction on the reaction center, the formation of isomers occurs at the reaction center under the indirect influence of the chiral factor.

Ugi, Ruch and co-workers described such a competing situation as qualitatively identical bonding behavior.[16] They introduced the concept that, in the absence of qualitative differences in the bonding behavior for stereoisomeric products or transition complexes, the prediction of the isomer ratio can be achieved by mathematical formulae based on analogies with eigenstates. That is, the states of stereoisomers may be considered to be analogous as regards electron distribution and the motion of atomic nuclei. Therefore stereoisomers are expected to possess qualitatively the same molecular skeletons, but to differ as regards ligand permutations[†] on a common molecular skeleton. The free energy of formation of stereoisomeric species is determined by the molecular skeleton, its bonds with ligands and the nature of the ligands. Since the contributions due to the mutual interactions of ligands are relatively small, the free energies of stereoisomeric species are functions of properties of the ligands. Their values may thus be interchanged or may remain unchanged depending on whether a permutation of the ligands corresponds to a change of configuration or a conformational interconversion in the same isomer. Using this concept, the relative free energies of stereoisomers or isomeric transition complexes can be expressed by a function $Q(L_1 L_2 \ldots L_n)$, where $L_1 L_2 \ldots L_n$ represents the nature of the ligands. When the relationship introduced in Eq. 7.1 is taken into account, the relative concentrations of isomers can be expressed as $\ln (c_I/c_{II}) = Q(L_1 L_2 \ldots L_n)$, where c_I and c_{II} are the concentrations of the isomeric species I and II. Thus, if the function is formulated, the efficiency of a stereo-differentiating reaction or asymmetric transformation can be predicted.

The function $Q(L_1 L_2 \ldots L_n)$ may be a basis of representation of the group of permutations and its value will vary with the ligand assortment. Ugi and Ruch replaced this function with a function $F(\lambda_i)$ in which λ_i represent ligand parameters of the ligand L_i. The set value of λ_i is not determined *a priori*, but is chosen appropriately so that $F(\lambda_i)$ becomes

---

[†] We previously described in Chapter 2 how a molecule can be divided into a molecular skeleton and an assortment of ligands. Stereoisomers are constituted from ligands such that either all the ligands or all those in a given assortment of ligands are bound to the same atomic neighbors in an equivalent manner. These ligands are said to be constitutionally equivalent, and permutations of these ligands are known as constitution-preserving permutations.

a smooth function of λ_i. Such treatments of reactions involving stereo-isomers are called "stereochemical analog models" by Ugi and Ruch.[†]

Since the ratio of stereoisomers is determined by the enantio- or diastereo-differentiating ability of the reagent and a reference chiral system, the function $F(\lambda_i)$ may be approximated to $\rho \cdot \chi(\lambda_i)$, in which ρ represents a reaction parameter[†] and $\chi(\lambda_i)$ is a symmetry function derived from the skeletal structure of the chiral center. An illustration of this treatment will now be given with the reaction system shown in Eq. 7.25, in which the chiral center $C^{(2)}$ is epimerized under thermodynamically controlled conditions. In this case, the ratio of diastereomers depends on the nature and relative position of the three ligands bonded to the chiral center $C^{(1)}$.

$$
\begin{array}{c}
\underset{c}{\overset{a}{\diagdown}}\text{C}^{(2)}\text{--X--}\underset{3\,L_3}{\overset{1\,L_1}{\diagup}}\text{C}^{(1)}\text{--}2\,L_2 \\
b \\
C_R^{(2)} \sim X \sim C_R^{(1)}
\end{array}
\;\rightleftharpoons\;
\begin{array}{c}
\underset{b}{\overset{a}{\diagdown}}\text{C}^{(2)}\text{--X--}\underset{3\,L_3}{\overset{1\,L_1}{\diagup}}\text{C}^{(1)}\text{--}2\,L_2 \\
c \\
C_S^{(2)} \sim X \sim C_R^{(1)}
\end{array}
\qquad (7.25)
$$

Suppose that the configuration of $C^{(1)}$, in which the distribution of ligands is such that the ligand numbers and skeletal numbers coincide, is R. Then the two diastereomers can be expressed by $C_R^{(2)} \sim X \sim C_R^{(1)}$ and $C_S^{(2)} \sim X \sim C_R^{(1)}$, and their free energies of formation are given by G and G', respectively. If the configuration of $C^{(1)}$ is S the free energies of formation for $C_R^{(2)} \sim X \sim C_S^{(1)}$ and $C_S^{(2)} \sim X \sim C_S^{(1)}$ are G' and G, respectively, as shown in Table 7.2.

TABLE 7.2
Free energies of diastereomers

configuration of $C^{(1)}$	R		S	
configuration of $C^{(2)}$	Isomer produced	Free energy of formation	Isomer produced	Free energy of formation
R	$C_R^{(1)} \sim X \sim C_R^{(2)}$	G	$C_S^{(1)} \sim X \sim C_R^{(2)}$	G'
S	$C_R^{(1)} \sim X \sim C_S^{(2)}$	G'	$C_S^{(1)} \sim X \sim C_S^{(2)}$	G

If the concentration of the isomer with R configuration is expressed by c_R and that of the isomer with S configuration is expressed by c_S, the ratio of c_R and c_S can be expressed as follows.

$$
\ln \frac{c_R}{c_S} = \frac{-(G - G')}{RT} = \frac{-\Delta G}{RT}
$$

where $\Delta G = G - G'$.

[†] Ugi *et al.* refer to ρ as a reaction parameter. However, it corresponds to the differentiating ability of the reagent or catalyst in our new approach.

In the same way, if the configuration of $C^{(1)}$ is S, the above equation becomes

$$\ln \frac{c_R}{c_S} = \frac{-(G' - G)}{RT} = \frac{\Delta G}{RT}$$

In these two equations, ΔG can be expressed as $F(\lambda_i)$, a function of the ligand parameters λ_i, and the reaction parameter ρ, which depends on the nature of the reaction. That is,

$$\ln \frac{c_R}{c_S} = \rho F(\lambda_i)$$

As mentioned in section 7.2, the configuration of the chiral center $C^{(1)}$ is determined by the permutation of ligands on the skeleton. In this case, the number of ligands is three, so the distribution of ligands corresponds to the permutation group $\mathscr{S}_3$. Since the point group to which the skeleton of the chiral center $C^{(1)}$, belongs is C_{3v}, the relations $C_{3v} \rightsquigarrow \mathscr{S} = \mathscr{S}_3{}^{\dagger 1}$ and $\mathscr{D}(\mathscr{D} \subset \mathscr{C}_{3v}) \rightsquigarrow \mathscr{N}(\mathscr{N} \subset \mathscr{S}_3)$ hold.[2]

The operation $s \in \mathscr{N}$ on the ligands of $C^{(1)}$ does not change the configuration of $C^{(1)}$ (R), but $s' \in \mathscr{N}'(\mathscr{N}'$ is a coset of $\mathscr{N}$ in $\mathscr{S}_3$) changes the configuration of $C^{(1)}$ to S. Thus, the relation between $\ln (c_R/c_S)$ and G is summarized in Table 7.3.

TABLE 7.3
Relationship between $\ln (c_R/c_S)$ and ΔG with change of permutation

	$s \in \mathscr{N}$	$s' \in \mathscr{N}'$
$\ln \dfrac{c_R}{c_S}$	$\dfrac{-\Delta G}{RT}$	$\dfrac{\Delta G}{RT}$

The function $F(\lambda_i)$ presenting $\ln (c_R/c_S)$ must have the same value and sign for the operation with $s \in \mathscr{N}$ and have same value and opposite sign for the operation with $s' \in \mathscr{N}'$. In addition, as $c_S/c_R = 0$ in the case of an achiral molecule, the value of $F(\lambda_i)$ should be zero. (For instance if $\lambda_1 = \lambda_2$ in the molecule shown in Eq. 7.25, $F(\lambda)$ must be zero.)

The properties of $F(\lambda_i)$ can be summarized as follows: (1) $F(\lambda_i)$ must be a base of the chirality representation[3] of $\mathscr{S}_3$. (2) $F(\lambda_i)$ must be zero in the case of achiral molecules.

[1] See Chapter 2, section 2.2 and Appendix B.
[2] See Chapter 2, Eq. 2.6.
[3] The irreducible representation of $\mathscr{S}$ (see section 9.3.2), in which the character $(s \in \mathscr{N})$ is $+1$ and $(s' \in \mathscr{N}')$ is -1 is called the chirality representation.

As an approximate formula for $F(\lambda_i)$, Ruch and Ugi introduced the function $\chi(\lambda_i)$, the lowest ordered polynomial of λ_i.[17] The formula for $\chi(\lambda_i)$ is determined by the skeletal symmetry and number of ligands.[†1] For a chiral center $C^{(1)}$ in which the skeletal symmetry is C_{3v} and the number of ligands is three, the following equation can be applied.

$$\chi = (\lambda_1 - \lambda_2)(\lambda_2 - \lambda_3)(\lambda_3 - \lambda_1) \qquad (7.26)$$

Thus the function expressing the ratio of R and S isomers at $C^{(2)}$ can be formulated as follows

$$\ln \frac{c_R}{c_S} = \rho \cdot \chi \qquad (7.27)$$

To assign the direction of the reaction, a new parameter $\delta = \pm 1$ is introduced; $\delta = +1$ means that reaction occurs with the R configuration of $C^{(1)}$ and $\delta = -1$ means that reaction occurs with the S configuration of $C^{(1)}$.

Eq. 7.27 can be rewritten as follows.

$$\delta \ln \frac{c_R}{c_S} = \rho \cdot \chi \qquad (7.28)$$

The relation shown in Eq. 7.28 is derived for a thermodynamically controlled reaction, but if we suppose that c_R and c_S correspond to the equilibrium concentrations of the transition complexes, $c_R^{\neq}$ and $c_S^{\neq}$, respectively, the equation can also be applied to kinetically controlled reactions. In such cases, Eq. 7.28 can be expressed as

$$\ln \frac{c_R^{\neq}}{c_S^{\neq}} = \frac{-(\Delta G_R^{\neq} - \Delta G_S^{\neq})}{RT} = \frac{-\Delta\Delta G^{\neq}}{RT} \qquad (7.29)$$

Since the parameters ρ and λ_i are arbitrary numbers, they are not determined *a priori* from physical or chemical properties of the ligands and reaction center. Ugi and co-workers arbitrarily chose $\lambda_H = 0$, $\lambda_{CH_3} = 1$ and obtained the other parameters λ_i and ρ experimentally using Eq. 7.28. As there are two unknown parameters in one reaction, λ_i and ρ must be determined from the results of several reactions in a given series so that the all parameters are self-consistent.

Ugi and co-workers carried out three representative reactions[†2]

[†1] Ruch and Ugi called these functions chirality functions. The functions corresponding to various skeletal structures and numbers of ligands are described in ref. 4 in Chap. 2.

[†2] Eq. 7.30 is a diastereoface-differentiating reaction; Eq. 7.31 is an enantioface-differentiating reaction; and Eq. 7.32 is an asymmetric transformation.

TABLE 7.4

Application of Eq. 7.28 (see the text)

Type of reaction	Ligands			Ligand parameters			Observed value	Value calculated from Eq. 7.28
	L_1	L_2	L_3	λ_1	λ_2	λ_3		
Eq. 7.30 ($\rho = 0.313$)	H	CH_3	tert-C_4H_9	0	1	1.49	c_{RR} (mole %) 62.0	62.4
	H	CH_3	C_6H_5	0	1	1.23	51.5	55.0
	H	CH_3	Mesityl	0	1	1.58	65.0	65.4
	H	CH_3	1-Naphthyl	0	1	1.29	56.0	56.3
	H	CH_3	$(C_6H_5)_3C$	0	1	1.75	74.5	71.3
	H	CH_3	2,4,6-Tricyclohexylphenyl	0	1	2.10	83.0	83.2
	H	C_6H_5	$(C_6H_5)_3C$	0	1.23	1.75	63.5	68.4
Eq. 7.31 ($\rho = 0.512$)	H	CH_3	C_2H_5	0	1	1	c_{RS} (mole %) 50.0	50.0
	H	CH_3	iso-C_3H_7	0	1	1	50.0	50.0
	H	C_2H_5	iso-C_3H_7	0	1	1	52.0	50.0
	H	C_2H_5	tert-C_4H_9	0	1	1.25	54.0	54.0
	H	iso-C_3H_7	Mesityl	0	1	1.79	56.0	67.3†
	H	CH_3	tert-C_4H_9	0	1	1.25	55.0	54.0
	H	C_2H_5	C_6H_5	0	1	1.60	61.0	62.0
	H	CH_3	C_6H_5	0	1	1.60	64.0	62.0
	H	CH_3	Mesityl	0	1	1.79	68.0	67.3
	H	C_2H_5	Mesityl	0	1	1.79	66.0	67.3

(Eqs. 7.30,[16] 7.31[18] and 7.32[16]) with various types of ligands and determined ρ and λ_i.

For Eqs. 7.30 and 7.31, ρ and λ_i are determined from Eq. 7.28 with $\chi = (\lambda_1 - \lambda_2)(\lambda_2 - \lambda_3)(\lambda_3 - \lambda_1)$. For Eq. 7.32, Eq. 7.33 is used for the determination of λ_i.

The calculated values using the resulting values of ρ and λ_i and experimental values of $\ln(c_{RR}/c_{SR})$ from Eqs. 7.30 and 7.31 are listed in Table 7.4.

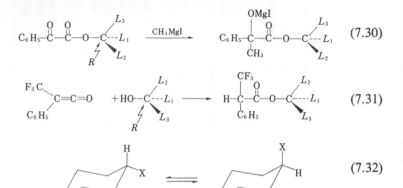

$$\ln \frac{c_6}{c_5} = \rho(\lambda_X - \lambda_H) = \frac{-\Delta G}{RT} \tag{7.33}$$

The values of λ_X corresponding to each ligand X, obtained for Eqs. 7.30, 7.31 and 7.32 are listed in Table 7.5.

It is to be expected that the calculated value of $\ln(c_{RR}/c_{RS})$ with λ_3 agrees with experimental values in the system of $L_1 = H$, $L_2 = CH_3$ and $L_3 = X$ in Eqs. 7.30 and 7.31 because λ_3 is chosen to agree with the experimental value. However, it is noteworthy that good agreement is ob-

TABLE 7.5
λ values determined from three different reactions

	Eq. 7.30 ($\rho = 0.313$)	Eq. 7.31 ($\rho = 0.512$)	Eq. 7.32 ($\rho = 1$)
C_2H_5	—		—
iso-C_3H_7	—		—
tert-C_4H_9	1.49	1.25	1.45
C_6H_5	1.23	1.60	1.24
Mesityl	1.58	1.79	1.55
1-Naphthyl	1.29	—	1.28
2,4,6-Tricyclohexylphenyl	2.10	—	2.03

tained between the calculated and experimental values even in the system of $L_1 = H$, $L_2 = X$, $L_3 = Y$ (Table 7.4, lines 7, 10, 11, 12, 13 and 17).

Originally, λ_i should be equal for identical ligands even in different systems, but they are rather different in Eqs. 7.30 and 7.31. As the difference is not controlled by ρ, some problems still remain with this formulation. The other problem is the difficulty of applying λ and ρ to the elucidation of the reaction mechanism, due to the lack of physical significance of λ and ρ. Further improvements of this formulation will be necessary.

B. Model based on the rotational potential around the chiral center

This model was formulated by Salem[19] to predict the ratio of diastereomers produced in diastereoface-differentiating reactions. Since this model is based on a geometrical model of the reaction, the function of ligands in the substrate and reagent are explained in terms of observable physical and chemical quantities such as temperature, interatomic forces, bond lengths, etc. We will consider a diastereo-zeroplane[†1] as a geometric model of a diastereo-differentiating reaction, as shown in Fig. 7.11.

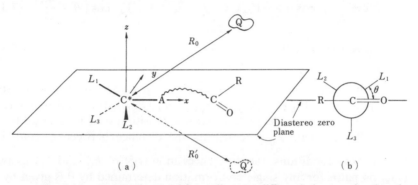

Fig. 7.11. Geometric model for a diastereo-differentiating reaction (see the text).

The substrate is placed in such a way that the origin of coordinates is at the chiral center C^* and the x coordinate lies along the C^*–A bond. Thus, it is assumed that the three ligands, L_1, L_2, L_3, of the chiral center rotate about the x coordinate. Suppose that the positions of the achiral reagent in the transition state of a diastereo-differentiating reaction are defined by Q (x, y, z) and Q' $(x, y, -z)$. The points Q and Q' eventually become mirror images with respect to the enantio-zeroplane.[†2] The

---

[†1] Salem does not use the new terminology, such as diastereo zeroplane, differentiating reaction, etc. However, we have used the new terminology here for convenience.

[†2] We ignore minor differences in the positions of Q and Q' due to the differences of interaction between Q and Q' with $C^*L_1L_2L_3$.

distances between Q and C* and Q' and C* are defined R_0 and R'_0, respectively.[†]

With this geometrical model, the maximum and minimum values of rotational potential $V^{\neq}(\theta)$ can be described as a mutual interaction of A and L_1, L_2, L_3 at the transition state.

The potential energy will contain a threefold component $V^{\neq}_{\text{III}}(\theta)$ of periodicity $2\pi/3$ and a singlefold component $V^{\neq}_{\text{I}}$ of periodicity 2π.

$$V^{\neq}(\theta) = V^{\neq}_{\text{I}}(\theta) + V^{\neq}_{\text{III}}(\theta) \tag{7.34}$$

The origin of θ is chosen when L_1 lies in the diastereo-zeroplane *cis* to the carbonyl group: the position of the ligand L_1 with rotation of θ from the plane is shown in Fig. 7.11(b).

If the reagent Q has different interactions with L_1, L_2 and L_3, and the potential of each interaction is expressed by V_{L_1}, V_{L_2} and V_{L_3}, $V^{\neq}_{\text{I}}(\theta)$ and $V^{\neq}_{\text{III}}(\theta)$ can be expressed as follows.

$$V^{\neq}_{\text{I}}(\theta) = \frac{1}{2}V_{L_1}\cos\theta + \frac{1}{2}V_{L_2}\cos\left(\theta + \frac{2\pi}{3}\right) + \frac{1}{2}V_{L_3}\cos\left(\theta + \frac{4\pi}{3}\right) \tag{7.35}$$

$$V^{\neq}_{\text{III}}(\theta) = \frac{1}{2}V_{\text{III}}\cos 3\theta \tag{7.36}$$

If the interactions between Q and L_1, L_2, L_3 are described by entry force potentials, $W_{L_1}(R_{QL_1})$, $W_{L_2}(R_{QL_2})$, $W_{L_3}(R_{QL_3})$, the total interaction energy between the reagent and C* at the transition state is assumed to be $\sum_i W_{L_i}(R_{QL_i})$, where R_{QL_i} represents the distance between Q and L_i. Under these conditions, the differences in energy of $E^{\neq}_{Q}$ and $E^{\neq}_{Q'}$ as two reaction paths for any single conformation determined by θ is given by

$$E^{\neq}_{Q} - E^{\neq}_{Q'} \equiv \Delta E(\theta) = \sum_i W_{L_i}(R_{QL_i}) - \sum_i W_{L_i}(R_{Q'L_i}) \tag{7.37}$$

$$(i = 1, 2, 3)$$

If the rotation of ligand L_1, L_2, L_3 is fast relative to the time spent by Q at the transition state, the transition state Q will be influenced by all the conformations obtained by the rotation of ligand L_1, L_2, L_3 and $\Delta\bar{E}$ can be expressed by the average of $\Delta E(\theta)$ for all these conformations.

$$\Delta\bar{E} = \frac{\int_0^{2\pi}(\Delta E(\theta)\exp(V^{\neq}(\theta)/kT))\,d\theta}{\int_0^{2\pi}(\exp(V(\theta)/kT))\,d\theta} \tag{7.38}$$

[†]　This corresponds to the distance Q–C* in the transition state complex.

The ratio of the rates of Q and Q' is

$$k_Q/k_{Q'} = \exp(-\Delta \bar{E}/kT) \tag{7.39}$$

If, on the other hand, the rotation of ligands L_1, L_2, L_3 is slow relative to the time spent by Q in the transition state, we have to consider the interaction between Q and ligands L_1, L_2, L_3 with respect to each species of rotation θ. Thus the value of $k_Q/k_{Q'}$ becomes the sum of the values for each conformation.

$$\frac{k_Q}{k_{Q'}} = \frac{\int_0^{2\pi} \exp(-(\sum_i W_{L_i})(Q, \theta)/kT) \exp(-V^{\neq}(\theta)/kT) \, d\theta}{\int_0^{2\pi} \exp(-(\sum_i W_{L_i})(Q', \theta)/kT) \exp(-V^{\neq}(\theta)/kT) \, d\theta} \tag{7.40}$$

In general, the rotational energy barrier $V^{\neq}$ is far smaller than that of the transition state. We will first consider the system expressed by Eq. 7.38.[†1] To obtain $\Delta \bar{E}$ from Eq. 7.38, we must derive descriptive formulae for $\Delta E(\theta)$, $V_{III}^{\neq}(\theta)$ and $V_I^{\neq}(\theta)$. The Taylor expansion of $W_{L_i}(R_{QL_i})$[†2] yields an approximate expression of $\Delta E(\theta)$ as follows

$$\Delta E(\theta) = \frac{-2Z}{R_0}\left(\sum_i \frac{2}{\sqrt{3}} r_i \sin\left(\theta + \frac{2\pi(i-1)}{3}\right) F_{L_i}\right) \tag{7.41}$$

where $F_{L_i} = (\partial W_{L_i}/\partial R)_{R_0}$, $i = 1, 2, 3$ and r_i represents the distance from C* to L_i. $V^{\neq}(\theta)$ may be chosen in several ways, according to the reaction model selected.

(1) The case where $V_I^{\neq}$ is relatively small compared with kT. The exponential of $V^{\neq}/kT$ can be expressed using Eq. 7.34 as follows.

$$\exp(-V^{\neq}/kT) = \exp(-V_I^{\neq}/kT) \exp(-V_{III}^{\neq}/kT) \tag{7.42}$$

[†1] In a system with a high steric hindrance, the ratio of diastereomers should be derivable from Eq. 7.40. Although Salem has solved Eq. 7.40, the same results can be obtained from Eq. 7.38 under conditions of high rotational hindrance.

[†2] If we assume $R_{QL_i} = R_0$, $W_{L_i}(R_{QL_i})$ can be approximated by the first term of the Taylor expansion with respect to the coordinates (x_i, y_i, z_i) of ligand L_i as follows.

$$W_{L_i}(R_{QL_i}) = W_{L_i}(R_0) - x_i\left(\frac{\partial W_{L_i}}{\partial x}\right)_{XYZ} - y_i\left(\frac{\partial W_{L_i}}{\partial y}\right)_{XYZ} - z_i\left(\frac{\partial W_{L_i}}{\partial z}\right)_{XYZ} \tag{1}$$

$$W_{L_i}(R_{Q'L_i}) = W_{L_i}(R_{0'}) - x_i\left(\frac{\partial W_{L_i}}{\partial x}\right)_{XY-z} - y_i\left(\frac{\partial W_{L_i}}{\partial y}\right)_{XY-z} - z_i\left(\frac{\partial W_{L_i}}{\partial z}\right)_{XY-z} \tag{2}$$

If the energy W_{L_i} is regarded as a function of distance, $R_0 = (x^2 + y^2 + z^2)^{1/2}$, the differentials of x and y of W_{L_i} at the points Q (X, Y, Z) and Q' $(X, Y, -Z)$ will be equal. For the z differentials of W_{L_i}, the equation $(\partial W_{L_i}/\partial z)_{XYZ} = -(\partial W_{L_i}/\partial z)_{XY-z}$ holds. Thus,

$$\Delta E(\theta) = -2\sum_i z_i\left(\frac{\partial W_{L_i}}{\partial z}\right)_{XYZ} \tag{3}$$

Since W_{L_i} only depends on z_i and since $R_0 = (x^2 + y^2 + z^2)^{1/2}$, the relation

The first term of this equation can be approximated as follows when $V_I^{\neq} \ll kT$.

$$\exp\left(-V_I^{\neq}/kT\right) \doteqdot 1 - \frac{1}{kT}\left\{\frac{V_{L_1}}{2}\cos\theta + \frac{V_{L_2}}{2}\cos\left(\theta + \frac{2\pi}{3}\right)\right.$$
$$\left. + \frac{V_{L_3}}{3}\cos\left(\theta + \frac{4\pi}{3}\right)\right\} \qquad (7.43)$$

Bringing in Eq. 7.36, we have

$$\exp\left(-\frac{V_{III}}{2kT}\cos 3\theta\right) = I_0\left(\frac{V_{III}}{2kT}\right) + 2\sum_{l=1}^{\infty} I_l\left(\frac{V_{III}}{2kT}\right)\cos 3l\theta \qquad (7.44)$$

Where $I_l(V_{III}/2kT)$ is the Bessel function of order l. Substituting Eqs. 7.41 and 7.43, and the first term of Eq. 7.44 into Eq. 7.38, $\Delta\bar{E}$ can be expressed as

$$\Delta\bar{E} = \sqrt{\frac{2}{3}}\cdot\frac{Z}{R_0}\frac{1}{kT}\left\{\frac{V_{L_1}}{2}(r_2 F_{L_2} - r_3 F_{L_3}) + \frac{V_{L_2}}{2}(r_3 F_{L_3} - r_1 F_{L_1})\right.$$
$$\left. + \frac{V_{L_3}}{2}(r_1 F_{L_1} - r_2 F_{L_2})\right\} \qquad (7.45)$$

The ratio of diastereomers $c_Q/c_{Q'}$ is thus given by

$$\log\frac{c_Q}{c_{Q'}} = -\frac{\Delta\bar{E}}{kT} = \frac{\text{const}}{(kT)^2} \qquad (7.46)$$

(*contined from p. 199*)
$(\partial W_{L_i}/\partial z)_{XYZ} = Z/R_0 \cdot (\partial W_{L_i}/\partial R)_{R_0} \equiv (-Z/R_0)\cdot F_{L_i}$ is obtained. Substituting into Eq. 7.37, we obtain

$$\Delta E(\theta) = \frac{2Z}{R_0}(\sum_i z_i F_{L_i}) \qquad (i = 1, 2, 3) \qquad (4)$$

Now r_1, r_2 and r_3 are the distances $C^*–L_1$, $C^*–L_2$ and $C^*–L_3$, respectively. Thus, z_1, z_2 as z_3 are given as functions of distance as follows.

$$z_1 = -\frac{2\sqrt{3}}{3}r_1\sin\theta$$
$$z_2 = -\frac{2\sqrt{3}}{3}r_2\sin\left(\theta + \frac{2\pi}{3}\right) \qquad (5)$$
$$z_3 = -\frac{2\sqrt{3}}{3}r_2\sin\left(\theta + \frac{4\pi}{3}\right)$$

Substituting into Eq. 4, we obtain Eq. 6 as a general form.

$$z_i = -\frac{2\sqrt{3}}{3}r_i\sin\left(\theta + \frac{2\pi(i - 1)}{3}\right) \qquad (6)$$

With Eq. 4, this yields Eq. 7.41 in the text.

According to Eq. 7.46, the logarithm of the diastereomer ratio is inversely proportional to the square of reaction temperature.

Rewriting Eq. 7.45 in comparison with the expression of Ugi *et al.* (Eq. 7.28), we obtain Eq. 7.47, which also satisfies $\mathscr{S}_3$ symmetry.

$$\log\frac{c_{RR}}{c_{RS}} = \rho[\lambda_1(\lambda_2^2 - \lambda_3^2) + \lambda_2(\lambda_3^2 - \lambda_1^2) + \lambda_3(\lambda_1^2 - \lambda_2^2)] \quad (7.47)$$

Since Eq. 7.45 is a function involving three physical quantities V, r and F, it is essentially different from Eq. 7.28.

The controversial feature of Eq. 7.45 is that the stereo-differentiation is determined only by $V_{\mathrm{III}}^{\neq}$. If L_1, L_2 and L_3 are of the same size stereochemically, no stereo-differentiation is expected according to this equation. To avoid this problem, introduction of three center forces, a second-order approximation of $\Delta\bar{E}$ such as $W_{L_1L_2}(R_{QL_1}, R_{QL_2}, R_{QL_1L_2})$ has been proposed by Salem.

(2) The case where one substituent L_1 is very large, i.e.,

$$(V_{L_1} \gg V_{L_2}, V_{L_3}, kT).$$

Since the rotation of ligands L_1, L_2, L_3 is completely blocked by L_1, the average energy should be determined by that of this rotation.

$$\Delta\bar{E} = \Delta E(180°) = 2\sqrt{\frac{2}{3}\frac{Z}{R_0}}(r_2E_{L_2} - r_3F_{L_3}) \quad (7.48)$$

In this case the logarithm of the diastereomer ratio varies as the inverse of the temperature.

(3) Cases such that $V_{L_1} \gg V_{L_2}$, V_{L_3} and $V_I^{\neq} \fallingdotseq kT$. The first and second terms of Eq. 7.42 in the case of $V_I^{\neq} \fallingdotseq kT$ can be expressed as follows

$$\exp(-V_I^{\neq}/kT) = \exp\left(\frac{-V_{L_1}}{2kT}\cos\theta\right) = I_0(\alpha) + 2\sum_I^\infty I_I(\alpha)\cos l\theta \quad (7.49)$$

$$\alpha = \frac{-V_{L_1}}{2kT}$$

$$\exp(-V_{\mathrm{III}}^{\neq}/kT) = \exp\left(\frac{-V_{\mathrm{III}}}{2kT}\cos3\theta\right) = I_0(\beta) + 2\sum_I^\infty I_I(\beta)\cos 3l\theta \quad (7.50)$$

$$\beta = \frac{-V_{\mathrm{III}}}{2kT}$$

Integration of Eq. 7.38 after substitutions with Eqs. 7.49 and 7.50 gives

$$\Delta \bar{E} = \frac{-2\sqrt{(2/3)/(2/R_0)}(r_2 F_{L_2} - r_3 F_{L_3})[I_0(\beta)I_1(\alpha) + I_1(\beta)\{I_2(\alpha) + I_4(\alpha)\} + \ldots]}{[I_0(\beta)I_0(\alpha) + I_1(\beta)I_3(\alpha) + \ldots]}$$

(7.51)

The term $I_m(\alpha)$ can be approximated to $I_0(\alpha)$ and $I_1(\alpha)$, because the value of $I_m(\alpha)$ rapidly decreases with increasing m.

If the terms $m = 1$ and 0 are used, Eq. 7.51 is reduced to

$$\Delta \bar{E} = 2\sqrt{\frac{2}{3}}(r_2 F_{L_2} - r_3 F_{L_3})\frac{I_1(\alpha)}{I_0(\alpha)} \qquad \alpha = -\frac{V_{L_1}}{2kT}(V_{L_1} \gg V_{L_2}, V_{L_3}) \quad (7.52)$$

The logarithm of the diastereomer ratio is then given by

$$\log \frac{c_Q}{c_{Q'}} = -\frac{\Delta E_{\max}}{kT}\frac{I_1(\alpha)}{I_0(\alpha)}$$

(7.53)

where $\Delta E_{\max}$ is the maximum value of Eq. 7.48.

The values of $I_1(\alpha)/I_0(\alpha)$ for given α can be obtained from tables. The values of $\ln (c_Q/c_{Q'})$ are proportional to $1/T$.

Experimental verification of this model has not been obtained, and its practical significance is not clear at present. Since reliable data on the temperature dependence of the diastereomer ratio are not readily available, the feature of this model, that the differentiation is related to $1/T$ when $V_I^{\neq} > kT$ and to $1/T^2$ when $V_I^{\neq} < -kT$, would be difficult to prove experimentally.

7.3   ILLUSTRATION BASED ON THE NEW CONCEPT OF "DIFFERENTIATION"

The present section will illustrate the importance of our new concept of "differentiation" in the elucidation of the mechanisms of stereo-differentiating reactions, as well as in the classification of stereochemical reactions.

As mentioned previously, the kinetic and stereochemical descriptions of differentiating reactions make the underlying assumption that differentiation takes place within the same process as the reaction. That is, the process of differentiation has never been considered independently of the chemical reaction process. It is not possible to discuss the mechanism of stereo-differentiating reactions fully without the concept that the differentiation and reaction processes are independent, even with the aid of conventional stereochemical models.

The origin of differentiation used to be ascribed to differences in the free energies of competitive transition states in the course of the differentiating reaction. In fact, this is correct insofar as the degree of stereo-differentiation is considered from the kinetic viewpoint, because the difference in activation free energies can be obtained from the ratio of isomers by calculation.

However it is not necessary to suggest that the origin of differentiation lies in the rate-determining step in a reaction process. It seems more reasonable to consider that differentiation may occur in a process other than the rate-determining step in the reaction. A comparative study of reaction rates and the degree of stereo-differentiation would clarify not only the process of differentiation but also the overall mechanism of the reaction itself.

By comparative studies of the degree of enantioface-differentiation and the rate of reaction in the catalytic hydrogenation of acetoacetate on Raney nickel modified with optically active substances, we have reached the conclusion that enantioface differentiation must be performed independently of the hydrogenation, for the following reasons.[20]

Table 7.6 shows the optical yield of methyl-3-hydroxybutyrate (MHB) and the initial hydrogenation rate of the enantioface-differentiating hydrogenation of methyl acetoacetate (MAA) with MRNi catalysts prepared and modified under the same conditions except for the modifying reagent. As shown in the table, although the rates depend on the type and number of functional groups of the modifying reagent, MRNi's in which the modifying reagents are homologs have the same reaction rate. However, we found great differences of differentiating abilities among MRNi modified with homologs.

TABLE 7.6

Comparison of reaction rate with optical yield in the enantiface-differentiating hydrogenation of MAA

Modifying reagent	Rate of hydrogenation[†]	Optical yield (%)
D_s-tartaric acid	13.1	26.2
(L)-Ala	9.8	0.0
(L)-Val	10.2	13.3
(L)-Leu	10.3	4.3
(L)-Ile	9.8	8.1
(L)-Glu	4.3	11.6
(L)-Orn	3.3	6.5
(L)-Lys	3.1	6.4

† Determined from the volume of hydrogen consumed in the first hr

TABLE 7.7

Comparison of reaction rate with optical yield in the enantioface-differentiating
hydrogenation of homologous esters of acetoacetic acid

Substrate	Rate	Optical yield (%)
Methyl acetoacetate	10.7	11.7
Ethyl acetoacetate	9.7	16.4
n-Propyl acetoacetate	8.7	14.9
n-Butyl acetoacetate	8.4	11.1

Table 7.7 shows the initial hydrogenation rate and optical yield of
products in the enantioface-differentiating hydrogenation of homologous
esters of acetoacetic acid with L-Val-MRNi. As shown in the table,
although the rates decreased regularly with increase in the alkyl chain
length of the ester group, there was no correlation between rate and optical
yield.

Table 7.8 shows the temperature dependencies of the degree of
differentiation and the reaction rate of MRNi. An Arrhenius plot of the
reaction rates of any sort of MRNi gives a straight line with an apparent
activation energy of 10.5 ± 0.5 kcal/mole in all cases (Fig. 7.12).

A similar value of activation energy was also obtained for the hydrogenation of homologous esters of acetoacetic acid (Fig. 7.13).

Since the order of the reaction rate is 0.2–0.3 with respect to the concentration of substrate, and the activation energy is larger than for a
diffusion-controlled reaction, it is unlikely that the rate of hydrogenation
is controlled by diffusion of the substrate or hydrogen. Thus, the rate-
determining step of hydrogenation seems to be the surface reaction.

TABLE 7.8

Dependence of reaction rate and optical yield on reaction temperature

Modifying reagent	Temp. (°C)	Reaction rate[†]	Optical yield (%)
D_s-tartaric acid	70	13.1	26.2
	50	6.6	37.8
	30	2.7	22.6
(L)-Val	70	10.2	13.3
	60	6.2	8.7
	40	2.6	2.6
(L)-Glu	70	4.3	11.6
	50	2.0	1.3
	30	0.85	0.0

† Determined from the amount of hydrogen absorbed by the system in the first hr

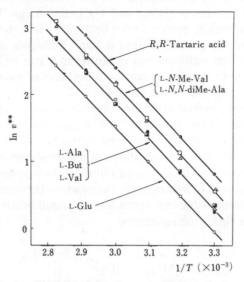

Fig. 7.12. Arrhenius plots of reaction rate in the hydrogenation of MAA over various MRNi; Pressure of hydrogen: 1 atm.

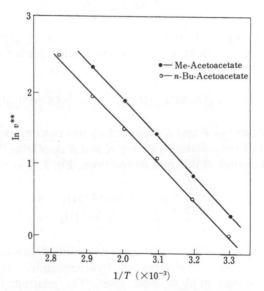

Fig. 7.13. Arrhenius plots of reaction rates in the hydrogenation of homologous esters of acetoacetic acid over Valines MRNi; Pressure of hydrogen: 1 atm.

In spite of the good linearity of the Arrhenius plots of reaction rate, the relation between $\ln (c_S/c_R)$ and $1/T$ is not linear. The accuracy of the apparent activation energy (± 0.5 kcal/mole) is not sufficient to permit discussion of the possibility that a small difference of activation energy may lead to stereo-differentiation. However, it can be assumed that differentiation is not solely dependent on differences of activation energy in the rate-determining step of the reaction.

Since the adsorption of the substrate is not a rate-determining step, the substrates on the surface of MRNi are present in the diastereomeric states, $S\text{-}M^*_{ads}$ and $R\text{-}M^*_{ads}$, produced by interaction with the chiral modifying reagent M*. The concentrations of the substrates in the two diastereomeric adsorption states are in equilibrium. In a system which contains a sufficient amount of substrate O, the relation between the concentrations of the substrate in the two states and the equilibrium constant K will be given by the following equations.

$$[M^*]_{ads} + [O] \underset{K_2}{\overset{K_1}{\rightleftarrows}} \begin{array}{c} [R\text{-}M^*]_{ads} \\ \updownarrow \\ [S\text{-}M^*]_{ads} \end{array} \tag{7.54}$$

$$\frac{[S\text{-}M^*]_{ads}}{[R\text{-}M^*]_{ads}} = \frac{K_2}{K_1} = K \tag{7.55}$$

Taking k_R and k_S as the rate constants of hydrogenation for $[R\text{-}M^*]_{ads}$ and $[S\text{-}M^*]_{ads}$, respectively, the reaction velocities for R and S, v_R and v_S, are given by Eq. 7.56.

$$v_R = k_R[R\text{-}M^*]_{ads}[H]_{ads}, \quad v_S = k_S[S\text{-}M^*]_{ads}[H]_{ads} \tag{7.56}$$

Since the ratio of R and S produced by this reaction is proportional to the velocities of the reactions yielding R and S insofar as the ratio of R and S does not change at different conversions, Eq. 7.57 is obtained.

$$\frac{c_S}{c_R} = \frac{v_S}{v_R} = \frac{k_S[S\text{-}M^*][H]}{k_R[R\text{-}M^*][H]} = \frac{k_S}{k_R} \times K \tag{7.57}$$

As was seen in Table 7.6, catalysts modified with homologs give nearly the same reaction velocities in hydrogenation, since $v_R + v_S = v =$ constant should hold in these cases. The relation, $v_R + v_S =$ constant means that $[R\text{-}M^*]_{ads} + [S\text{-}M^*]_{ads} = [R\text{-}M^* + S\text{-}M^*]_{ads} =$ const. or $k_S = k_R = k =$ const. with each catalyst system.

The former relation arises since the total amount of adsorbed species is determined by the active surface area of catalyst, not by the interaction of the modifying reagent with the substrate, but no clear explanation for the latter relation can be found. The values of apparent activation energy of hydrogenation are the same in all cases, regardless of differences in enantio-differentiating ability, and the mechanism of hydrogen addition is not affected by changing the modifying reagent or ester group of the substrate, or by changes in the orientation of adsorbed substrate. If we suppose that the rate constant of hydrogenation is not influenced by environmental factors, $k_S \fallingdotseq k_R$ holds. Eq. 7.57 can then be transformed to Eq.

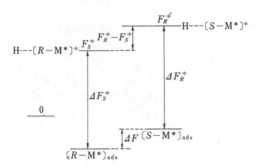

Fig. 7.14. Energy diagram of stereo-differentiation.

7.58 (see Fig. 7.14 for the reaction scheme based on the above discussion and nomenclature).[†]

$$\frac{c_S}{c_R} \fallingdotseq K \tag{7.58}$$

[†] Eq. 7.58 seems to conflict with the Curtin-Hammett principle. However, detailed consideration of the Curtin-Hammett principle shows that there is no contradiction between Eq. 7.58 and the principle. That is, the Curtin-Hammett principle states that the ratio of products obtained in competitive reactions depends on the difference between the activation energies of the intermediates at the transition state. In considering the ratio of products in Fig. 7.14, k_R, k_S and K are expressed as follows

$$\left. \begin{array}{l} k_R = \kappa \dfrac{kT}{h} \exp\left(-\Delta F_R^{\ddagger}/RT\right) \\[2mm] k_S = \kappa \dfrac{kT}{h} \exp\left(-\Delta F_S^{\ddagger}/RT\right) \\[2mm] K = \exp\left(\Delta F/RT\right) \end{array} \right\} \tag{1}$$

where κ is the transmission coefficient and k is the Boltzmann constant. From Eqs. 1 and 7.57, Eq. 2 can be obtained.

$$\frac{v_S}{v_R} = \frac{\exp\left(-\Delta F_S^{\ddagger}/RT\right) \cdot \exp\left(\Delta F/RT\right)}{\exp\left(-\Delta F_R^{\ddagger}/RT\right)} = \exp\left(\Delta F_R^{\ddagger} + \Delta F - \Delta F_S^{\ddagger}\right)/RT \tag{2}$$

As shown in Fig. 7.14, $\Delta F_R^{\ddagger} + \Delta F - \Delta F_S^{\ddagger} = F_R^{\ddagger} - F_S^{\ddagger}$. The ratio of R and S pro-

Thus, the degree of differentiation is expected to be controlled by the equilibrium between the concentrations of diastereomeric adsorption states, independently of the hydrogenation process.[20b]

We will next discuss the reaction mechanism of enantioface-differentiating reactions with optically active metal complexes from the viewpoint of the new concept.

Although the results of enantioface-differentiating reactions with optically active Wilkinson complexes are described in section 5.1.1.B.a, the reaction mechanism of enantioface differentiation with a Wilkinson complex will be reconsidered here from the viewpoint of the new concept of "differentiation".

The most extensively studied reaction mechanism of hydrogenation of alkenes with a Wilkinson complex is shown in Fig. 7.15. Horner *et al.*

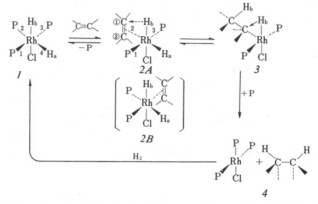

Fig. 7.15. The hydrogenation of alkenes with Wilkinson-type complex catalysts.

proposed *2B* as a reaction intermediate, but *2A* was proposed more recently based on a ^{31}P-NMR study by Meakin *et al.*[21] According to this mechanism, the hydrogenation proceeds via the following steps.

That is, (1) the production of *2A* by the addition of alkene to *1*, (2) the addition of H_b *cis* in local relation to Rh to produce the σ-complex *3*, (3) the addition of H_a with retention of configuration. Thus, the overall addition of hydrogen is *cis*.

Since the ratio of enantiomers produced by 100% *cis* addition of

duced corresponds to the ratio of v_R and v_S. Thus, when we substitute these into Eq. 2, we have

$$c_S/c_R = \exp{(F_R^{\ddagger} - F_S^{\ddagger})/RT} \qquad (3)$$

In conclusion, the reaction mechanism based on the new concept does not conflict with the Curtin-Hammett principle that the amount of each product formed in a competitive reaction depends on the difference between the free energies of the intermediates at the transition state.

hydrogen will be determined by the ratio of the hydrogens introduced from the *si* and *re* enantiofaces, the ratio of enantiomers produced is not affected whether $C^{(1)}$ is located on the same side or on the opposite side as C1.

The optical yield of product depends on the ratio of the concentrations of the diastereomeric intermediates of *3*, since the reaction *3* → *4* (Fig. 7.15) can be assumed to proceed as an irreversible process at high velocity; the configuration of the diastereomeric intermediates must therefore be determined in compounds *2A* and *3*. Accordingly, there must be a differentiating process at step *1* ⇄ *2* or *2* ⇄ *3*. The differentiation of the catalyst should be due to enantioface differentiation during the coordination of the complex with the alkene, if step *1* ⇄ *2* is the rate-determining step. On the other hand, if the rate-determining step is at *2* ⇄ *3*, the enantioface-differentiation of the catalyst should depend on both the equilibrium concentrations of the diastereomeric intermediates *2A* and/or the ratio of velocities of H_b transfer to the diastereomeric coordination compounds.

According to the study of Siegel,[22] the rate-determining step of the reaction is step *1* → *2* in Fig. 7.15, so we can expect from the above discussion that the differentiating process should be considered independently from the hydrogenating process, and that the optical yield of product depends on the relative rate of formation of diastereomeric coordination compounds of the alkene.

Hydroformylation with Wilkinson complexes is known to be analogous to hydrogenation. Although the reaction proceeds by *cis* addition of H and CO to the alkene, the reaction species of complex containing the optically active phosphine ligand is greatly affected by the ratio of phosphine to rhodium and by the pressure of CO. The optical yield of this reaction shows complicated variations with type of ligand, the reaction temperature and the reaction pressure. Thus, examination of the mechanism of differentiation in hydroformylation with Wilkinson complexes is more difficult than in the case of hydrogenation.[23]

However, since the reaction is a *cis* addition, the mechanism of differentiation can be determined to some extent by experiments on the reactions of various substrates using the same catalyst under the same reaction conditions.

Since the process of hydroformylation is known to occur as shown in Fig. 7.16, differentiation must be performed at step *1* ⇄ *2*, *2* ⇄ *3* or *3* ⇄ *4*.

The possibility of differentiation at step *3* ⇄ *4* in Fig. 7.16 can be excluded since the complex catalyzes the decarbonylation of aldehyde in the reverse reaction with retention of configuration,[24] and 2-methylbutanols with opposite configuration are obtained from 1-butene and

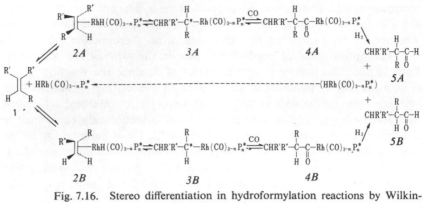

Fig. 7.16.   Stereo differentiation in hydroformylation reactions by Wilkinson-type complex catalysts.

cis-2-butene with the same catalyst.[25] Thus, differentiation is achieved at either *1* ⇄ *2* or *2* ⇄ *3*, and the optical yield of the product depends on the concentrations of *3A* and *3B* in Fig. 7.16. However, it is difficult to suggest which is the differentiating step (*1* ⇄ *2* or *2* ⇄ *3*) at present.[†]

Elucidation of the energy difference between the diastereomeric π-complexes formed competitively in this reaction process and identification of the differentiating process at the transition state of the π- to σ-complex thus represent most important problems for future study in relation to the mechanism of enantioface-differentiating reactions with optically active metal complex catalysts.

Studies on the differentiating process and the factors contributing to differentiation are essential if we are to increase the differentiation of catalysts or reagents and also for the development of new catalysts or reagents with superior differentiation based on theoretical considerations. The importance of the new concept of "differentiation" for basic studies of stereo-differentiating reactions is clear.

[†] Pino *et al.*[26] have simulated the reaction intermediate using a stable platinum complex and postulated that the equilibrated concentration of diastereomeric π-complex, *2A* and and *2B*, plays an important role in the differentiation. Since the enantiofaces of the olefin represent a diastereomeric π-complex, the differentiation can be considered as enantioface-differentiation based on our concept.

Although justification for a direct comparison of the natures of Rh- and Pt-complexes is still in doubt, this approach could be important in clarifying the mechanisms of differentiation reactions.

REFERENCES

1. V. Prelog, *Helv. Chim. Acta*, **36**, 308 (1953).
2. V. Prelog and W. Dauben, XII Intern. Congr. Pure Appl. Chem., 1951. Abstracts, p. 401.
3. D. J. Cram and F. A. Abd Elhafez, *J. Am. Chem. Soc.*, **74**, 5828, 5851 (1952); D. J. Cram and K. R. Kopecky, *ibid.*, **81**, 2748 (1959).
4. J. W. Cornforth, R. H. Cornforth and K. K. Mathew, *J. Chem. Soc.*, **1959**, 112.
5. H. S. Mosher and E. M. La Combe, *J. Am. Chem. Soc.*, **72**, 3994 (1950).
6. F. C. Whitmore, R. S. George, *ibid.*, **64**, 1239 (1942).
7. G. Vavon and B. Angelo, *C. R. Acad. Sci., Paris*, **222**, 959 (1946); *ibid.*, **224**, 1435 (1947).
8. W. M. Foley, F. J. Welch, E. M. La Combe and H. S. Mosher, *ibid.*, **81**, 2779 (1959).
9. E. P. Burrows, F. J. Welch and H. S. Mosher, *ibid.*, **82**, 880 (1960).
10. R. Macleod, F. J. Welch and H. S. Mosher, *ibid.*, **82**, 876 (1960).
11. M. S. Biernbaum and H. S. Mosher, *J. Org. Chem.*, **36**, 3168 (1971).
12. J. D. Morrison and H. S. Mosher, *Asymmetric Organic Reactions*, Prentice-Hall, 1971.
13. J. S. Birtwistle, K. Lee, J. D. Morrison, W. A. Sanderson and H. S. Mosher, *J. Org. Chem.*, **29**, 37 (1964).
14. D. Dull, *Ph.D. Thesis, Stanford Univ.*, 1967; *Chem. Abstr.*, **70**, 3115 (1969).
15. J. D. Morrison and R. W. Ridgway, *Tetr. Lett.*, **1969**, 569.
15. E. Ruch and I. Ugi, *Topics in Sterochemistry*, vol. 4, p. 99, Interscience, 1969.
17. E. Ruch and A. Schönhofer, *Theoret. Chim. Acta.*, **10**, 91 (1968).
18. E. Anders, E. Ruch and I. Ugi, *Angew. Chem. Intern. Ed. Engl.*, **12**, 25 (1973).
19. L. Salem, *J. Am. Chem. Soc.*, **95**, 94 (1973).
20. (*a*) H. Ozaki, A. Tai and Y. Izumi, *Chem. Lett.*, *1974*] 935 & (*b*) T. Harada, Y. Hiraki, Y. Izumi, J. Muraoka, H. Ozaki and A. Tai, 6th Int. Congr. on Catalysis, Preprints, 1976.
21. P. Meakin, J. P. Jesson and C. A. Tolman, *J. Am. Chem. Soc.*, **94**, 3240 (1972).
22. S. Siegel and D. W. Ohrt, *Tetr. Lett.*, **1972**, 5155.
23. M. Tanaka, Y. Watanabe, T. Mitsudo and Y. Takegami, *Bull. Chem. Soc. Japan*, **47**, 1968 (1974).
24. J. Tsuji and K. Ohno, *Tetr. Lett.*, **1967**, 2173.
25. G. Consiglio, C. Botteghi, C. Salomon and P. Pino, *Angew. Chem. Intern. Ed. Engl.*, **12**, 669 (1973).
26. P. Pino, C. Consiglio, C. Botteghi and C. Salomon, *Advan. Chem. Ser.*, **273**, 295 (1974).

CHAPTER 8

Methods for the Study of Stereo-Differentiating Reactions

Since the stereo-differentiating ability of a reagent or catalyst is reflected in the ratio of enantiomers or diastereomers produced, there are special considerations in the study of stereo-differentiating reactions beyond the conventional methods of synthetic organic chemistry. In order to obtain reliable experimental results and to determine the mechanism of differentiation, it is important to understand the basic experimental techniques and methods of data analysis relating to stereo-differentiating reactions.

8.1 REQUIREMENTS FOR EXPERIMENTAL CONDITION FOR STEREO-DIFFERENTIATING REACTIONS

Stereo-differentiating reactions suitable for experimental study should possess the characteristics outlined below.

8.1.1 Substrate

A. Substrates for enantio-differentiating reactions

(1) The substrate must give a product which can be easily purified without any change in the ratio of enantiomers during purification. Since the crystallization of an optically unpurified sample often changes the ratio of enantiomers, a compound which can be distilled or transformed to a volatile derivative is to be preferred as the product of an enantio-differentiating reaction.

(2) The substrate should give a product such that the optical and chemical purities can be determined accurately.

(3) The substrate should give a product, the enantiomer ratio of which can be easily determined by conventional analytical methods. Although

the polarimetric method was formerly the most familiar and widely used, chromatographic and NMR spectrometric methods have now become conventional. Thus, substrates giving a product which can be transformed into diastereomers are also suitable, in addition to those having a large specific rotation and a high solubility in solvents for optical rotatory power measurement.

(4)   The substrate should give a product which does not racemize under the reaction conditions used.

Thus, a substrate which gives a volatile liquid product having high optical rotatory power is most desirable.

B.   Substrates for diastereo-differentiating reactions

(1)   The substrate should be a chiral compound. However, if the diastereomers produced can be analyzed for diastereomer ratio, the substrate need not to be an optically active compound.

(2)   If it is necessary to measure the ratio of diastereomers produced by determining the optical activity of a compound derived from the newly produced chiral moiety, the following features are required in the substrate.

(i)   Racemization of the newly produced chiral moiety should not take place during the chemical procedures necessary to separate it from the chiral factor.

(ii)   The products formed from the newly produced chiral moiety and those formed from the chiral factor should be readily separable.

(iii)   The products formed from the newly produced chiral moiety should have the properties required of a product of an enantio-differentiating reaction, as described in the previous section.

8.1.2   Reagent and catalyst

A.   Fundamental requirements for the reagent or catalyst in an enantio-differentiating reaction

(1)   The reagent or catalyst must be chiral, and must be optically active except in special cases, for instance where the reaction gives a product with a diastereomeric intermediate, and the purpose of the investigation is limited to estimation of the degree of enantio-differentiation.

(2)   The optically active reagent or catalyst should be easily separable from the reaction product.

(3)   The reagent or catalyst should be detectable with high sensitivity, for instance by chromatography or a colorimetric method. This is very important in reactions with reagents or catalysts having high optical activ-

ity, but which produce a compound with very low optical activity.
(4)   The reagent and catalyst should be optically stable.

B.   Reagent and catalyst in diastereo-differentiating reactions

It is not necessary to use a chiral compound as the reagent or catalyst for diastereo-differentiating reactions. Any usual reagent or catalyst can be used.

8.1.3   Reaction conditions and subsequent isolation and purification of the reaction product

Reliable results can be obtained under the following reaction conditions.

(1)   The chiral factor in the substrate and the newly produced chiral center should not be racemized during the reaction. In addition, the newly produced chiral center should not be racemized or the ratio of diastereomers should not be changed during treatment of the reaction product. If necessary, it should be checked whether the chiral factor or newly produced chiral center is racemized or whether the stereoisomer ratio changes under the reaction and post-treatment conditions, using an authentic sample of known optical purity.

(2)   As the ratio of enantiomers and especially the ratio of diastereomers can easily be changed by recrystallization under usual conditions, distillation may be the preferable method for purification of the products of stereo-differentiating reactions in most cases.

8.1.4   Confirmation of the experimental accuracy

It is most important to check the reliability of the results in studies of stereo-differentiating reactions. When a stereo-differentiating reaction is performed with a new system, the following points should be checked.

(1)   The absence of contamination of the optically active chiral factor should be confirmed.

(2)   No change should occur in the ratio of stereoisomers during the post-treatment.

(3)   No racemization of the newly produced chiral center should occur during the reaction or post-treatment.

(4)   Errors may arise in enantiomer-differentiating reactions during the preparation of analytical samples, e.g. during derivation to diastereomers (as a post-treatment) for chromatographic and NMR analysis.

(5)   Errors may arise in the measurement of optical rotation (as a check of reliability, the observed value of optical rotation, α, or the concentration of the solution used for measurement should always be specified).

8.1.5   Confirmation of stereo-differentiation

When stereo-differentiation is achieved by a new process, the differentiation should be confirmed using the chiral factor having the opposite optical rotatory power. Results obtained with chiral factors having the opposite configurations should show the same value of optical rotatory power, but with opposite sign. This is very important in cases where the optical activity of the product is low.

8.2   EVALUATION OF STEREO-DIFFERENTIATING ABILITY

Conventionally, the efficiency of stereo-differentiation of a reagent or catalyst is represented an asymmetric yield, which is based on the ratio of stereoisomers produced. However, this efficiency is not related simply to the differentiating ability of the reagent or catalyst: stereo-differentiating ability consists of very complex factors. In this section, therefore, the nature of the stereo-differentiating ability will be discussed first, before describing methods for evaluation of stereo-differentiating reactions.

8.2.1   Stereo-differentiating ability and the degree of differentiation

A linear relationship does not hold between the logarithms of enantiomer ratio and reaction temperature in most stereo-differentiating reactions; also, the direction of differentiation is often reversed by a change in the reaction temperature, as found in Figs. 5.9 and 5.13. These facts indicate that the differentiation is not controlled by a single, simple factor but by more than two factors.

Stereo-differentiating ability has been evaluated from the disproportionality of stereoisomers produced under specific reaction conditions. However, the above facts cast doubts on whether such a simple approach can reveal the true nature of the stereo-differentiating ability of the reagent or catalyst. That is to say, if we discuss the results of Fig. 5.9, the following inconsistencies arise. Alanine(Ala)–, α-aminobutyric acid(But)–, valine (Val)– and glutamic acid(Glu)–MRNi hydrogenate methyl acetoacetate to methyl 3-hydroxylbutyrate at 60°C with optical yields in the order, Val–MRNi > Glu–MRNi > But–MRNi > Ala–MRNi, where the latter shows no differentiating ability. However, in hydrogenation at 30°C, the order of differentiating ability is essentially reversed, i.e., Ala–MRNi > But–MRNi > Val–MRNi > Glu–MRNi, where both the latter two catalysts show almost negligible optical yields. Moreover, the direction of differentiation of Ala–, But– and Val–MRNi at 60°C is opposite to that at 30°C. Thus, the degree of differentiation obtained under certain reaction conditions does not give, and cannot be used as a representative measure of the differentiating ability of the catalyst.

It is necessary therefore that the differentiating abilities of reagents and catalysts be evaluated comprehensively from results obtained under a variety of reaction conditions, although at present the authors have not developed a formula for the method of quantitative representation. Nevertheless, in the case of comparing the differentiating abilities of Ala–, But–, Val– and Glu–MRNi, which show a similar mode of temperature dependence, the temperature (T_0) at which the change in direction of differentiation occurs might give an improved index on which to base the comparison than the optical yields at a specific temperature. The nature of the differentiating ability of the catalysts could also be presented more qualitatively by utilizing an additional index such as the tangent of the slope of the curves in the above-mentioned figure.

A meaningful illustration of the stereo-differentiating reaction using a stereomodel would be given only in limited reactions performed at a reaction temperature remote from T_0. Since T_0 may shift with a change in reaction conditions other than temperature, such as pressure, solvent, etc., such conditions must also be incorporated into studies of the reaction mechanism. Part of the reason for the contradictions arising in the explanation of reaction mechanisms using the empirical model described in section 7.2 may be attributable to this cause.

Establishment of a reliable qualitative evaluating method for the stereo-differentiating ability thus represents the most important and pressing problem to be resolved in giving a mathematical estimation of stereo-differentiation, although the usefulness of the stereo-differentiating ability in practical applications of stereo-differentiating reactions can be assessed adequately from the two viewpoints of the degree of differentiation and the practicability of the reaction conditions employed. In this book, therefore, the degree of differentiation (such as the optical yield) is used in a distinct manner from the stereo-differentiating ability.

8.2.2   Terms expressing the proportion of stereoisomers

Since the degree of stereo-differentiation is expressed in a relation that incorporates terms for the proportion of stereoisomers, these terms will be described next.

A.   Optical purity

The proportion of enantiomers is customarily expressed as an optical purity, P (%), due to the historical background of the chemistry of chiral compounds and for convenience in the presentation of results obtained by polarimetry.

The specific optical rotation $[\alpha]$ is calculated from Eq. 8.1, where α is the angle of rotation (degrees), $[c]$ is the concentration of the sample solution (g/ml) and d is the length of the light path (dm).

$$[\alpha]_\lambda^T = \alpha \times \frac{1}{d \times [c]} \qquad (8.1)$$

Here, T is the temperature of measurement and λ is the wavelength of light used for measurement (when the sodium D line is used ($\lambda = 589$–589.6 nm), the subscript D is used in place of a numerical value λ).

When the optical rotation of a liquid sample is measured, the density ρ is used instead of $[c]$ and $[\alpha]$ is calculated with Eq. 8.2.

$$[\alpha]_\lambda^T = \alpha \times \frac{1}{d \times [\rho]} \qquad (8.2)$$

The D and L forms of a chiral molecule (represented here by D and L) have the same optical rotation but with opposite sign. Thus, the optical rotation of mixtures of enantiomers depends on the number ratio of enantiomer molecules, so that the optical rotation can be taken as a measure of the enantiomer ratio.

The largest optical rotation is seen with samples consisting only of L or D molecules; it decreases in mixtures, disappearing completely in mixtures containing equal amounts of D and L molecules. The optical purities of pure L or D samples and of an equal mixture of them are therefore 100% and 0%, respectively. The optical purity (P) of an arbitrary mixture of enantiomers can be calculated from Eq. 8.3.

$$P(\%) = \frac{[\alpha]}{[\alpha]_{max}} \times 100 = \frac{|[D] - [L]|}{[D] + [L]} \times 100 \qquad (8.3)$$

Here $[\alpha]$ is the specific optical rotation of the sample, $[\alpha]_{max}$ is the specific optical rotation of pure L or D compound and $[L]$ and $[D]$ are the concentrations of L and D compounds, respectively, in the sample.

B.   Enantiomer excess

It was recently found that $[\alpha]/[\alpha]_{max}$ is not always proportional to the value of $(|[D] - [L]|)/([D] + [L])$.[1] The real value of $(|[D] - [L]|)/([D] + [L])$ is often called the enantiomer excess (e.e.). However, since such a lack of proportionality is not widely encountered, for practical purposes, optical purity is employed in conventional expressions. We will use the term optical purity to indicate values obtained from optical rotation, and enantiomer excess for values obtained by analytical methods other than the optical method.

C. Enantiomer and diastereomer ratios

The ratio of enantiomers or diastereomers is expressed as r or d respectively.

D. Diastereomer excess

The diastereomer excess (d.e., %) is the term for diastereomers corresponding to the enatiomer excess. It is expressed as follows:

$$\text{d.e.} = \frac{1 - d}{1 + d} \times 100 \qquad (8.4)$$

8.2.3 Determination of the proportion of stereoisomers

Accurate determination of the ratio of enantiomers produced is a very important prerequisite for accurate determination of the degree of differentiation. For reliable evaluation of the results, suitable procedures for analysis and for pretreatment of the sample before analysis are required. In this section, therefore, the analytical methods available for determination of the stereoisomer ratio are described, together with suitable pretreatments in each case.

A. Enantiomer

a. Polarimetric determination

Polarimetry has been the conventional and longest-known method for the determination of enantiomer excess.

Principles of polarimetry

Though there are two types of polarimeter, the conventional manual polarimeter and the automatic polarimeter, both are based on the same optical principles. Fig. 8.1 schematically illustrates the structure of a

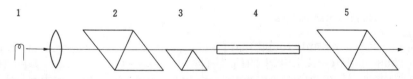

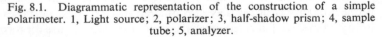

Fig. 8.1. Diagrammatic representation of the construction of a simple polarimeter. 1, Light source; 2, polarizer; 3, half-shadow prism; 4, sample tube; 5, analyzer.

manual polarimeter. The optical rotation of a sample can be measured by determination of the rotatory angle of the polarized light passing through the sample. However, in practice it is difficult to locate the angle at which the transmitted light intensity is maximum or minimum with high accuracy. The half-shadow prism makes it possible to overcome this difficulty. This prism rotates the angle of polarization of half the light source by a few degress, θ. Fig. 8.2 shows the relationship between the intensity of transmitted light from the light source and that passed through the half-shadow prism with respect to the analyzer angle. The two beams

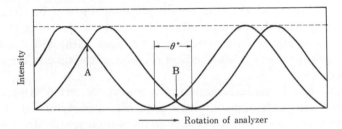

Fig. 8.2. Representation of the relationship between light intensity and the rotation of the analyzer.

have the same intensity at the crossing points A and B. If the analyzer angle changes to either side of A or B, the intensities of the two beams change inversely with respect to each other, and the extent of deviation from A or B can be observed in terms of the difference between the intensities of the two beams. As the contrast between the intensities of the two beams is higher in the vicinity of B than A, the optical rotation is measured at B.

Automatic polarimeters find the crossing point B by finding the analyzer angle that gives identical intensities of two beams having a difference of $\theta°$ in their directions of polarization. The beams are generated alternately instead of using a half-shadow mirror system. Accordingly, the reliability of results obtained by this system depends greatly on the stability of the light source.

In measurements of optical rotation, the following precautions are necessary to obtain reliable results.

(1) The quality of the sample tube should be carefully checked, since this may be more important than the quality of the polarimeter itself. The end plates must be exactly at right angles to the light path. Since polarized light generated by surface diffraction inside the tube decreases the accuracy of the results, a large diameter tube with a ground inside surface is desirable. A diameter of more than 3 mm is suitable for a tube 10 cm in length.

(2) Since a strained end plate acts as a polarizer, the application of strong irregular forces to the end plates must be avoided during use of the assembled cell.

(3) To avoid errors due to the above factors, optical rotation should be measured at various rotational positions of the tube and the similarity of the results confirmed.

(4) Since small particles of insoluble materials in the sample solution rotating about the light path tend to generate incidental optical activity in the solution, a check for this should be made. Quick rotation of the sample tube about the light path gives rise to rotation of the solution in the sample tube, so that the above phenomenon can be detected by determining whether the appearance of optical rotation is changed after occasional rotation of the tube about the light path, and whether opposite optical activity is observed on rotation of the tube in the opposite direction.

(5) Although the optical rotation of a colored or turbid sample can be measured with an automatic instrument, the intensity of light from the analyzer is decreased and the change in intensity is greatly decreased near the crossing point B, as shown in Fig. 8.3. Thus, the accuracy of results obtained under such conditions is greatly reduced.

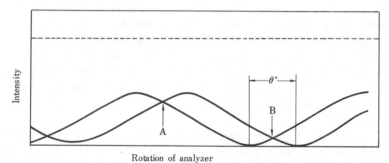

Rotation of analyzer

Fig. 8.3. Representation of the relationship between light intensity and the rotation of the analyzer in a turbid sample.

b. *Special methods for the determination of* [α]

Although the optical method is very effective for analyzing enantiomer ratios, the optical purity of a compound which has a very low value of $[\alpha]_{max}$ cannot be determined without special modification. In addition, the optical purity cannot be calculated if the $[\alpha]_{max}$ value of the compound is unknown. We shall describe here several special methods for the determination of optical purity.

i) *Measurements near maximum absorption*: As a compound which has an absorption maximum shows optical rotatory dispersion, high optical rotation is observed near the absorption maximum (see section 9.2.1.D).

ii) *Introduction of a chromophore*: It may be possible to produce an absorption maximum near the D line in a compound which has an absorption maximum far from the D line by chemical modification (introducing a chromophore). This may not only increase the optical rotation, but can also increase the weight and volume of the sample.

Two examples of chemical modification will be described here. The first is the modification of amino acids by dinitrophenylation (DNP-ation).[2] Silica gel chromatography is commonly used to purify the product. As DNP-amino acids have absorption maxima near the D line, high optical activity can be observed using the D line.[3]

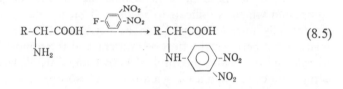

$$ \tag{8.5} $$

DNP-Amino acids are prepared as follows. The sample, containing 100 mg of amino acid, is dissolved in 5 ml of 5% aqueous sodium carbonate. Sixteen ml of ethanol containing 210 mg of 2,4-dinitrofluorobenzene is added to the above solution and the reaction mixture is stirred in the dark for 3 hr. The reaction mixture is then concentrated *in vacuo* and the residue is extracted with water. The water extract is extracted three times with ether, then the water layer is acidified with 6 N HCl and again extracted three times with ether. The ether extracts are combined and washed with water. The ether solution is dried with sodium sulfate and the crude DNP-amino acid is obtained by evaporation of the ether *in vacuo*.

Purification is carried out as follows. The crude DNP-amino acid is dissolved in a small portion of methanol. Chromatography is then carried out on a preparative thin-layer plate of Kiesel gel H using chloroform/methanol/acetic acid (95:5:1). The fraction containing DNP-amino acid is collected and extracted with methanol. Removal of the methanol by evaporation *in vacuo* yields pure DNP-amino acid. For the determination of α, the purified compound is dissolved in 1% sodium carbonate, and the optical rotation determined at 550 nm.

To determine the concentration, the solution used for the determination of rotation is diluted 200 times with 1% sodium carbonate and the content of DNP-amino acid determined by colorimetry at 360 nm.

The second example is the modification of esters of amino acids with methyl acetoacetate to increase the optical rotation and the volume.[4]

$$
\text{R-CHCOOR'} \xrightarrow{\text{CH}_3\text{COCH}_2\text{COOR''}} \text{R-CHCOOR'}
$$

$$
\underset{\text{NH}_2}{\,} \qquad\qquad \underset{\overset{|}{\text{CH}_3}}{\text{N}{=}\text{CCH}_2\text{COOR''}}
\qquad (8.6)
$$

Ethyl alanate (2 g) is dissolved in 10 ml of ether, then 4 ml of ethyl aceto-acetate and 1 g of anhydrous sodium sulfate are added. The reaction mixture is filtered after standing for one night and fractionated under 0.5 mm Hg. The Schiff's base of the alanine ester is collected at 95°/0.5 mm Hg. The $[\alpha]_D$ of L-alanine ethylester-ethylacetoacetate Schiff base is $+138° \pm 4°$.

iii) *Determination of* $[\alpha]_{max}$ *by isotope dilution method*: The isotope dilution method can be used to determine the specific rotation, $[\alpha]_{max}$, of a compound which is difficult to resolve. That is, the specific rotation of an optically pure compound can be determined by examination of the relationship between the isotope content and the optical rotation of the sample and of various mixtures of known amounts of the sample with DL-compound containing a known amount of isotope.

An amount a g of sample (specific rotation α) and an amount b g of DL-compound containing an isotope (isotope content I_0) are mixed, dissolved in the solvent and recrystallized. If the crystals have specific rotation α_i and isotope content I_i, and the optical purity of the sample $(|D - L|)/(D + L)$ to be determined is designated as P, the isotope contents of D and L enantiomers can be calculated from Eqs. 8.7 and 8.8, respectively (the calculation is performed assuming that the D isomer content is higher than that of the L isomer).

$$
I_D = \frac{b \cdot I_0}{b + a(1 + P)} \qquad (8.7)
$$

$$
I_L = \frac{b \cdot I_0}{b + a(1 - P)} \qquad (8.8)
$$

Let the optical purity of the crystals obtained from the mixture be P_i. Then, the isotope content of the crystals, I_i, can be calculated from Eq. 8.9.

$$
I_i = I_D \cdot \frac{1 + P_i}{2} + I_L \cdot \frac{1 - P_i}{2} \qquad (8.9)
$$

Substituting Eqs. 8.7 and 8.8 into Eq. 8.9, we obtain

$$I_i = \frac{(a + b)bI_0}{(a + b)^2 - a^2 P^2} - \frac{aP_iPbI_0}{(a + b)^2 - a^2 P^2}$$

$$= \frac{bI_0}{(a + b)^2 - a^2 P^2}(a + b - aP_iP) \tag{8.10}$$

where $P = [\alpha]/[\alpha]_{max}$, $P_i = [\alpha_i]/[\alpha]_{max}$. Thus, $[\alpha]_{max}$ is given by Eq. 8.11.

$$\left. \begin{aligned} I_i &= bI_0\frac{(a + b)[\alpha]_{max}^2 - a[\alpha][\alpha_i]}{(a + b)^2[\alpha]_{max}^2 - a^2[\alpha]^2} \\[2mm] [\alpha]_{max} &= \sqrt{\frac{I_ia^2[\alpha]^2 - I_0ab[\alpha][\alpha_i]}{I_i(a + b)^2 - I_0b(a + b)}} \end{aligned} \right\} \tag{8.11}$$

From this, Eq. 8.12 can be obtained.

$$P = \frac{[\alpha]}{[\alpha]_{max}} = [\alpha] \times \sqrt{\frac{I_i(a + b)^2 - I_0b(a + b)}{I_ia^2[\alpha]^2 - I_0ab[\alpha][\alpha_i]}} \tag{8.12}$$

The isotope content and the specific rotation of mixtures of the sample and isotope-containing DL-compound are given by $I_i = I_0b/(a + b)$ and $[\alpha]b/(a + b) = [\alpha_1]$, respectively, and putting $Q = I_1/I_0$, we obtain Eq. 8. 13.

$$[\alpha]_{max} = \sqrt{\frac{Q[\alpha_1][\alpha_i] - [\alpha_1]^2}{Q - 1}} \tag{8.13}$$

For example 2-benzyl-4-phenyloxazoline-2-thione was determined to have $[\alpha]_{D_{max}} = 170°$ ($c = 8.3$, benzene) using Eq. 8.12 with 3(α-^{2}H$_1$-benzyl)-4-phenyloxazoline-2-thione as a standard.[5]

c. Biochemical methods

The stereo-differentiating ability of enzymes or bacteria has been applied to the determination of optical purity since early times. However, the enzymes suitable for conventional quantitative determination are limited, including L-amino acid oxidase, L-amino acid transaminase, D-amino acid oxidase and L-amino acid decarboxylase (Table 8.1).[6] L-Amino acid oxidase and D-amino acid oxidase catalyze the oxidation of L- and D-amino acids, respectively, by oxygen to give α-keto acids in the presence of oxygen and a catalytic amount of catalase, and L-Amino acid decarboxylase catalyzes the decarboxylation of L-amino acids, as shown in Eq. 8.14.

TABLE 8.1

The applicability of enzymic analysis for the determination of

Amino acid isomers which can be tested	
L-Isomer[1]	D-Isomer[2]
Alanine	Alanine[3]
β-Aminoalanine	α-Aminoadipic acid
S-Benzylcysteine	Arginine[4]
S-Benzylhomocysteine	Aspartic acid[5]
Butyrine	S-Benzylcysteine
β-Cyclohexylalanine	S-Benzylhomocysteine
α-Cyclohexylglycine	Butyrine
Cystine	β-Cyclohexylalanine
Ethionine	α-Cyclohexylglycine
Heptyline	Cystine
Homoserine	Ethionine
Hydroxyproline	Glutamic acid[5]
Allohydroxyproline	Heptyline
Isoleucine	Histidine
Alloisoleucine	Homoserine
Leucine	Isoleucine
Methionine	Alloisoleucine
Nonyline	Leucine
Norleucine	Lysine[6]
Norvaline	Methionine
Phenylalanine	Nonyline
Phenylglycine	Norleucine
Proline	Norvaline
Serine	Ornithine[7]
Allothreonine	Phenylalanine
Tryptophan	α-Phenylglycine
Tyrosine	Allo-β-phenylserine
Valine	Tryptophan
	Tyrosine
	Valine

[1]   Enzyme employed is hog kidney D-amino acid oxidase
[2]   Unless otherwise noted, the enzyme employed is *Crotalus adamanteus* L-amino acid oxidase
[3]   *Bothrops jararaca* L-amino acid oxidase is used
[4]   *E. coli* 7020 L-arginine decarboxylase is used

enantiomer ratio using certain oxidases and decarboxylases

Amino acid isomers which cannot be tested	
L-Isomer[1]	D-Isomer[2]
α-Aminoadipic acid	β-Aminoalanine
γ-Aminobutyric acid	γ-Aminobutyric acid
Aspartic acid	Hydroxyproline
Glutamic acid	Allohydroxyproline
Histidine	Isovaline
Isovaline	tert-Leucine
tert-Leucine	β-Phenylserine
Lysine	Proline
Ornithine	Serine
β-Phenylserine	Threonine
Allo-β-phenylserine	Allothreonine
Threonine	

[5] *C. welchii* SR-12 decarboxylase is used
[6] *B. cadaveris* 6578 L-lysine decarboxylase is used
[7] *C. septicum* P-III L-ornithine decarboxylase plus added pyridoxal phosphate is used

$$R-\underset{\underset{NH_2}{|}}{CH}COOH + \tfrac{1}{2}O_2 \xrightarrow{\text{Oxidase}} R-\underset{\underset{O}{\|}}{C}-COOH + NH_3$$

$$R-\underset{\underset{NH_2}{|}}{CH}COOH \xrightarrow{\text{Decarboxylase}} R-CH_2NH_2 + CO_2$$

$$(8.14)$$

Oxygen is absorbed in the former reactions, while carbon dioxide is liberated in the latter reaction. Accordingly, the enantiomer ratio of amino acids can be determined by a manometric method using the Warburg apparatus (Fig. 8.4.). The enantiomer ratio of amino acids can also be determined by counting the radioactivity in the products of the enzyme reactions if substrates containing radioisotopes are used. The enantiomer ratio of alanine was determined by using the radioisotope method.[7] That is, D- or L-^{14}C-alanine was converted to pyruvic acid using D-amino acid oxidase (EC 2.61.2) or an L-glutamic acid–pyruvic acid–transaminase system (Eq. 8.15), respectively, and the enantiomer ratio of alanine was obtained by determining the radioactivity in the produced pyruvate.

$$CH_3-\underset{\underset{NH_2}{|}}{CH}-COOH \xrightarrow{\text{HOOC-CH}_2\text{-CH}_2\text{-CO-COOH}} CH_3-\underset{\underset{O}{\|}}{C}-COOH$$

$$+ HOOC-CH_2-CH_2-\underset{\underset{NH_2}{|}}{CH}-COOH \qquad (8.15)$$

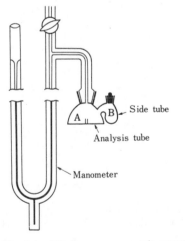

Fig. 8.4.   Warburg manometric apparatus.

However, the simple manometric method is an accurate, conventional method. It is possible to detect 0.1% D- or L-isomer in L- or D-isomer, respectively, with high accuracy. A typical procedure is as follows.

TABLE 8.2

Accuracy in the determination of the enantiomer ratios of amino acids by the enzymatic procedure

Amino acid	Concentration		Enzyme	O_2 uptake or CO_2 evolution (μmole)
	L(μmole)	D(μmole)		
Alanine	0	10,000	D-Amino acid oxidase	0
	1.0	10,000		1.2
	0	1,000		0
	1.0	1,000		1.1
Phenylalanine	1,000	0	L-Amino acid oxidase	0
	1,000	1.0		0.9
	1,000	2.0		1.8
Glutamic acid	1,000	0	Glutamic acid decarboxylase	0
	1,000	1.0		0.9
Aspartic acid	1,000	0	Aspartic acid decarboxylase	0
	1,000	1.0		1.3

One thousand μmole of the sample is placed in 1.5–2.5 ml of buffer solution. In the case of oxidation with D-amino acid oxidase, phosphate buffer (pH 8.1) is used and in the case of L-amino acid oxidase, phosphate or Tris buffer (pH 7.4) containing 25 units of catalase is used. For decarboxylation, acetate buffer (pH 4.9) is used. Next, 0.3–0.4 ml of the enzyme solution is placed in side tube B. For example, D-amino acid oxidase from hog kidney, venom L-amino acid oxidase or L-amino acid decarboxylase is used as saturated solutions. The reaction system is brought to a constant temperature and the enzyme solution is mixed with the sample. In the reaction system shown in Eq. 8.13, the mixture is shaken until adsorption or evolution of gas ceases. It was found that 1 μmole of substrate adsorbed 112 μl of oxygen or released 224 μl of carbon dioxide. Table 8.2 shows the results obtained by Meister.

d. Gas chromatographic method

The determination of enantiomer ratios by gas chromatography was developed independently by three groups in 1965.[9-11] Many improved methods have since been developed, and gas chromatography has become an important method for determination of the enantiomer ratio.

The gas chromatographic method can be used with volatile diastereomers introduced from diastereomers obtained by the reaction of enantiomers with an optically active reagent. The principle of this method is similar to that of usual optical resolution. Designating the enantiomers in the sample and the optically pure reagent (*i.e.* L form) as S_L, S_D and R_L, respectively, diastereomers S_L–R_L and S_D–R_L are produced by reaction of the sample with the reagent R_L. The enantiomer excess (e.e.) of the sample can be calculated from the results of gas chromatography by Eq. 8.16.

$$e.e.(\%) = \frac{|[S_L-R_L] - [S_D-R_L]| \times 100}{[S_L-R_L] + [S_D-R_L]} = \frac{(1 - r) \times 100}{(1 + r)} \quad (8.16)$$

$$d = \frac{[S_D-R_L]}{[S_L-R_L]} = r$$

The enantiomer ratio (r) is proportional to the diastereomer ratio (d).

Gas chromatographic determination is a simple procedure. However, certain precautions are necessary to obtain reliable results, as follows.

(1)   The diastereomers obtained by reaction with the optically active reagent should be introduced into a sufficiently volatile compound for gas chromatography.

(2)   The diastereomers should be stable at the temperature of chromatography. If decomposition takes place, a diastereomer-differentiating reaction will occur and reliable results cannot be obtained.

(3)   Enantiomer differentiation should not take place during the reaction of the sample with the optically active reagent. The reaction should go to completion.

(4)   Racemization of the sample and the optically active reagent, and epimerization of the diastereomers produced should not take place during the preparation of diastereomers or the gas chromatographic procedure.

Since a specific sample does not always satisfy the above conditions, these points should be checked in each case. If an optically pure authentic sample is available, the best conditions can be determined with samples containing known amounts of the optically pure material.

If an optically pure sample is not available, suitable conditions can often be determined using an analogous sample. If neither is available, suitable conditions must be estimated from results obtained under several different sets of conditions in the light of the physical and chemical properties of the sample. Table 8.3 shows representative studies for the determination of enantiomers by gas chromatography. As a conventional optically active reagent, a naturally occurring compound is usually used. However, optically pure compounds obtained from synthetic materials can also be used.

We will describe an improvement of Vitl's method[9a] as an example: it has been used for the determination of the enantiomer ratio of alanine by our group.[9b]

Preparation of alanine menthylester hydrochloride: Fig. 8.5(a) shows the apparatus used for esterification. The sample solution, containing

TABLE 8.3

Some examples of the determination of enantiomer ratios by GLC

Sample	Optically active reagent	Diastereomer	GLC conditions	Ref.
Amino acids	(−)-Menthol	$\begin{array}{c} H \\ R-C-COO-Men \\ NHCCF_3 \\ \parallel \\ O \end{array}$	5% PEGA/Chromosorb W, 0.4 cm, 4 m, 165°C	(9)
$\begin{array}{c} H \\ C_6H_5-C-COOH \\ R \end{array}$ $\left(R = \begin{array}{c} CH_3 \\ C_3H_7 \end{array} \right)$	(−)-Menthol	$\begin{array}{c} H \\ C_6H_5-C-COO-Men \\ CH_3 \end{array}$	10% PEG 20M/Chromosorb W, 175°C	(11)
Amino acid	(+) or (−)-2-Butanol (+) or (−)-2-Octanol	$\begin{array}{c} CH_3 \\ H\; | \\ R-C-COOCH \\ NHCCF_3\;\; R \\ \parallel \\ O \end{array}$	PPG/Capillary column, 150–195°C	(12)
Amines $\left[\begin{array}{c} CH_3-CH-R \\ NH_2 \end{array} \right]$	TFA-Proline chloride	CH_3CHR structure with $NH-C=O$, proline ring, $\overset{\text{O}}{\underset{\parallel}{C}}-CF_3$	0.75% PEGS/0.25% EGSS/ Chromosorb W, 1/8 in 5 ft, 140–180°C	(10)

TABLE 8.3—Continued

Sample	Optically active reagent	Diastereomer	GLC conditions	Ref.
Cyclic amines	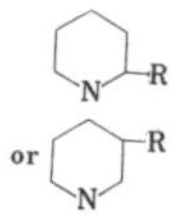 TFA-Proline[†] chloride	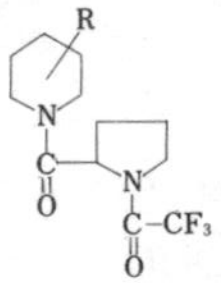	20% HI-EFF-4B/Chromosorb AW, DMCS, 1/4 in 10 ft, 230°C	(13)
Amino acids	Optically active alcohols	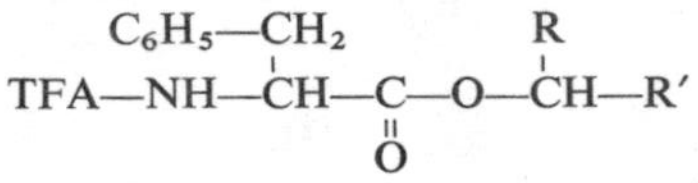 $C_6H_5-CH_2$ TFA—NH—CH—C—O—CH—R′ R,R′ = alkyl or cyclo-alkyl residues 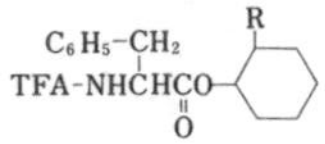 	0.15 DEGS/0.25 EGSS-X/ Chromosorb W, 1/4 in 5 ft, 140–170°C	(14)
Amines	TFA-Proline[†] chloride	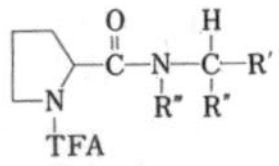 R′ = alkyl R″ = methyl or ethoxycarbonyl R‴ = H or methyl	5% QF-1/Aeropack 36, 1/8 in 5 ft, 200°C or 0.5% EGA/Aeropack 30, 1/8 in 5 ft, 200°C	(14)
1,1,1-Trifluoropropane-2-ol	(−)-O-Methyl-mandelic acid	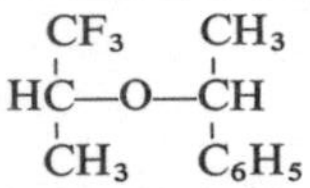	Carbowax 20M/Chromosorb W, 1/4 in 20 ft, 70°C	(15)

† TFA: Trifluoroacetyl

10–20 mg of alanine (dissolved as the hydrochloride), is placed in the bottom of reaction vessel A, then the solution is evaporated to dryness *in vacuo* in such a way that the residue is accumulated at the bottom of the vessel. To the residue, 1–1.5 g of (−)-menthol is added, and the vessel A is connected to apparatus B and C. Part of vessel A is heated in an oil bath at 115–125 °C under dry hydrogen chloride introduced from the top of C for

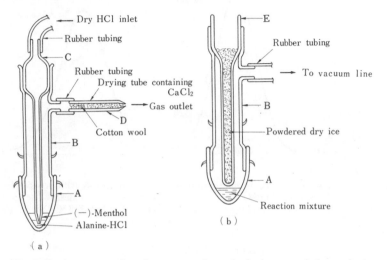

Fig. 8.5. Apparatus for the preparation of alanine menthylester hydrochloride. (a) Esterification reaction. (b) Removal of excess (−)-menthol.

1–1.5 hr. During the esterification, the solid alanine hydrochloride is gradually dispersed. When the reaction is complete, the apparatus C is replaced with E, which is cooled with dry ice as shown in Fig. 8.5(b), and the system is connected to a vacuum line and kept at 10–15 mmHg; heating of A is continued at 115–125 °C. The remaining menthol is evaporated off completely at 120–125 °C in this operation.

Trifluoroacetylation of alanine menthylester hydrochloride: To the menthylester hydrochloride remaining in vessel A, 0.3–0.4 ml of trifluoroacetyl anhydride is added, then the vessel is closed with a glass stopper and kept at room temperature for one night.[†]

Gas chromatography: After dilution of the reaction mixture with 0.2–1.5 ml of dichloromethane, 0.2–0.3 μl of the sample solution is subjected to chromatography under the following conditions: a glass column (3 m × 3 mm) packed with 1.5% neopentyl glycolsuccinate on Chro-

---

† Though fluoroacetylation can be performed at high temperature, racemization takes place above 40 °C.

mosorb AW 60–80 mesh is used at a column temperature of 165–178 °C.[†] The ratio of diastereomers can be determined with an error of less than ±0.5% by weighing the cut portion of chart paper corresponding to the chromatogram peak (Fig. 8.6).

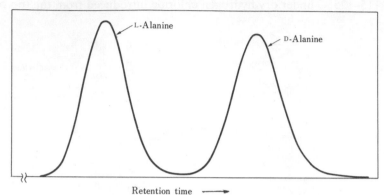

Fig. 8.6.   Gas chromatogram of (−)-menthyl DL-N-trifluoroacetyl alaninate. Column: NPGS (1.5%) on Chromosorb AW (60–80 mesh). The temperatures of the column, sample inlet and detector were 165–178 °C, 230 °C and 250 °C, respectively.

e.   NMR-spectroscopic methods

The procedure for the determination of enantiomer ratios by NMR spectroscopy has been greatly improved recently, and the configuration and enantiomer ratios of many compounds can be determined by this method as data accumulate. Already, NMR spectroscopy is one of the most convenient methods for the determination of enantiomer ratio and configuration. Since the NMR spectrum does not indicate chirality directly, the ratio of enantiomers, S_D and S_L, is determined after conversion to the diastereomeric state, S_D–R_D, etc., in the presence of an optically active reagent or solvent, R. As substances in diastereomeric relation have different free energies of formation, they have different physical properties. Thus, the nuclear shieldings are different and these differences are detectable in the NMR spectrum.

There are three methods for conversion to the diastereomeric state. The first involves conversion of the enantiomers to diastereomers by reaction with an optically active reagent. The second is to use an optically active solvent. The third is to take the NMR spectrum in the presence of an

---

[†]   If the concentration of the stationary phase is increased above 5%, good resolution is not obtained, only an increase in retention time. The same results were obtained with 60–80 mesh support as with 80–100 mesh.

optically active shift reagent. We shall next describe several examples of each method.

Determination of enantiomer ratios as diastereomers: This method was established by Mislow and Mosher to determine enantiomer ratios of alcohols and amines.[16] They determined the ratio from the diastereomer ratio of esters or amides derived from the alcohols or amines by reaction with optically active acids or their derivatives, respectively. This method can also be used for the determination of the enantiomer ratios of acids by using optically active alcohols or amines.

To obtain reliable results, the choice of reagent and reaction conditions is very important. The following conditions are essential to obtain good results.

(1) The reagent should contain a substituent or atom which generates magnetic anisotropy, because a big difference of chemical shift ($\Delta\Delta\delta$) is preferable for accurate analysis.

(2) The reagent should have a simple spectrum. If possible, a reagent having a singlet spectrum should be chosen.

(3) The reagent should react completely with the sample in order to ensure that enantiomer differentiation does not occur.

(4) Racemization should not occur at either of the chiral centers (in the substrate and optically active reagent) during the reaction. Also, the diastereomers produced should not be epimerized under the conditions used for the reaction and determination.

Typical reagents satisfying the above conditions are *O*-methylmandelyl chloride and 2-methoxy-α-trifluoromethylphenylacetyl chloride. However, the latter two conditions mentioned above should be confirmed in each case to ensure that the reagent employed produces sufficient $\Delta\Delta\delta$ for the determination, and that enantiomer differentiation or epimerization does not take place during the preparation of diastereomers by checking that the signal intensities of the enantiomers are the same in authentic racemic substance subjected to the same treatment. As the value of $\Delta\Delta\delta$ depends on the type of solvent employed for the measurement of the NMR spectrum, the choice of solvent and conditions of measurement are also important if good results are to be obtained.

Fig. 8.7 shows the NMR spectrum of the ester of (RS)-C_6H_5CH-(OMe)COCl and (RS)-CF_3CH[C(OH$_3$)$_3$]OH in CF_3–C_6H_5.[17] As can be seen, the signals of all the protons, except for the protons on benzene, and the signals of all the fluorines each separate into two, $RR + SS$ and $SR + RS$, and satisfy the first two of the above conditions. However, the sample was obtained from racemic compound. Thus, the latter two conditions are not satisfied, and the enantiomer ratio cannot be determined with this derivative. Similar results were obtained with the esters of alcohols

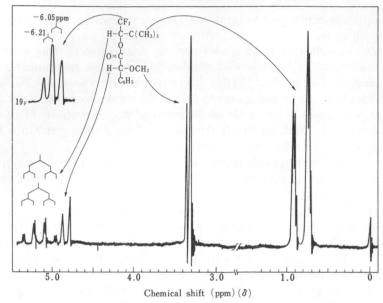

Fig. 8.7. NMR spectrum (60 MHz) of *tert*-butyltrifluoromethylcarbinyl-*O*-methyl mandelate. [From Dale and Mosher (17)].

such as $C_6H_5CH(OH)C(CH)_3$ or $C_2H_5CH(OH)C(CH)_3$. The reason for this is thought to be racemization of the reagent during esterification, and the ratio of diastereomers reaches a constant equilibrium value irrespective of the enantiomer ratio of the sample (i.e., asymmetric transformation occurs).

The enantiomer ratios of several alcohols which can be determined as esters of *O*-methylmandelic acid by NMR spectroscopy, are listed in Table 8.4, together with results obtained by other methods.

TABLE 8.4

Comparison of enantiomer excess values determined by various methods for esters of *O*-methylmandelic acid

$$\underset{\substack{| \\ H}}{C_6H_5}\!-\!\underset{}{\overset{\substack{CH_3O \\ |}}{C}}\!-\!\overset{\substack{O \\ \|}}{C}\!-\!O\!-\!\underset{\substack{| \\ R'}}{\overset{\substack{R \\ |}}{C}}\!-\!H$$

R	R'	Polarimetry	Gas chromatography	NMR
CH_3	C_6H_{11}	2.0 ± 1.0	—	3.0 ± 2.0
CH_3	$n\text{-}C_6H_{13}$	96.1 ± 0.3	—	97.5 ± 2.0
CH_3	CF_3	—	62.3 ± 0.5	62.5 ± 0.3
C_6H_5	CF_3	100.0 ± 0.1	100.0 ± 0.5	100.0 ± 0.5
CH_3	$iso\text{-}C_3H_7$	100.0 ± 1.0	—	100.0 ± 0.5

α-Methoxy-α-trifluorophenylacetic acid (MTPA) can be more widely applied as a reagent than *O*-methylmandelic acid. The diastereomers produced from MTPA not only have a large ΔΔδ value in the proton signal, but also produce an effective fluorine spectrum, yielding very reliable results, since the fluorine spectrum does not overlap with the proton spectrum.[18] Moreover, since MTPA is optically stable, it can be used for the determination of the enantiomer ratio of secondary alcohols which have large steric hindrance. MTPA also gives diastereomers of amides with large ΔΔδ values on reaction with amines. However, there is a possibility that enantiomer-differentiating esterification of alcohols with bulky ligands may take place with MTPA. Accordingly, when the ratio of enantiomers is determined with optically pure MTPA, complete reaction of the sample with MTPA is necessary to prevent enantiomer differentiation. Table 8.5 shows the results obtained by NMR using MTPA in comparison with the results obtained by polarimetry.

TABLE 8.5
Comparison of optical purity with enantiomer excess values for MTPA esters

Sample	Polarimetry	Enantiomer excess (determined by NMR)	
		^{1}H(60 MHz)	^{19}F(100 MHz)
$CH_3CH(OH)C_2H_5$	82.4	82.0	—
$CH_3CH(OH)$-n-C_6H_{11}	96.1	—	96.5
$CH_3CH(OH)$-iso-C_3H_7	95.5	95.0	—
	95.5	95.0	95.3
$CH_3CH(OH)$-$tert$-C_4H_9	7.8	7.5	—
$CH_3CH(OH)CF_3$	75.8	75	—
	97	98	—
$C_6H_5CH(OH)CF_3$	45.2	45.5	—
	45.2	44.9	—
$C_6H_5CH(OH)$-$cyclo$-C_6H_{11}	78.5	77.5	79.0
$C_6H_5CH(OH)$-$tert$-C_4H_9	100	100	—
	100	100	—
$C_6H_5CH(OH)COOCH_3$	67.1	67.0	67.3
$C_6H_5CH(NH_2)CH_3$	100	100	100
	42.2	42.4	—
$C_6H_5CH_2CH(NH_2)CH_3$	100	100	100
	100	100	100
$C_6H_5CH(NH_2)$-α-$C_{10}H_7$	36.4	35.0	36.0

The determination of the ratio of enantiomers produced by deuteration, a case in which the optical rotatory power is very small, can be achieved very conveniently by the NMR method..[19] Fig. 8.8(a, b) shows

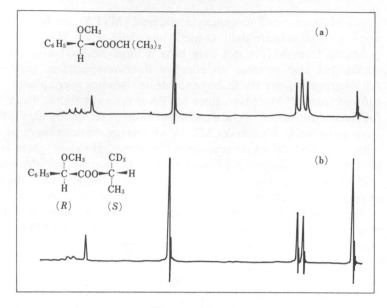

Chemical shift

Fig. 8.8.   NMR spectra of esters of (R)-O-methylmandelic acid: (a) propane-2-ol ester, (b) (S)-propane-2-ol-1-d_3 ester. [From Raban and Mislow (19).]

the NMR spectra of O-methylmandelic acid esters of propane-2-ol and (S)-propane-2-ol-1-d_3, respectively. In Fig. 8.8(b) one of the doublet signals of the methyl groups has disappeared. This indicates that the optical purity of the deuterated alcohol is very high.

Since the signals of the methylene protons of the glycine residues of L-phenylalanylglycine (L-Phe–Gly) and L-leucylglycine (L-Leu–Gly) are not equivalent, it is possible to determine the enantiomer ratio of glycine-2-d_1 with phenylalanyl peptide. The NMR spectrum of glycine-2-d_1 is shown in Fig. 8.9.[20]

Determination using an optically active solvent: The determination of the enantiomer ratio by NMR spectroscopy in an optically active solvent was developed by Pirkle, and represented a unique method for determination of the enantiomer ratio without any chemical treatment, until the shift reagent method was discovered. This method was based on the fact that the fluorine doublet of 2,2,2-trifluoro-1-phenylethanol splits into two pairs of doublets in optically active 2-phenylethylamine.

When the equilibria shown in Fig. 8.10 for the solvation of compound I with compound II hold and the rate of interchange of the compounds in equilibrium is large, then if the chemical shifts of I_L and I_D are designated as δ^{I_L} and δ^{I_D}, and the molar concentrations, $[I_L]$, $[I_L-II_L]$, $[I_L-II_D]$, $[I_D]$, $[I_D-II_D]$ and $[I_D-II_L]$, are designated as p, q, r, p', q' and r', respectively,

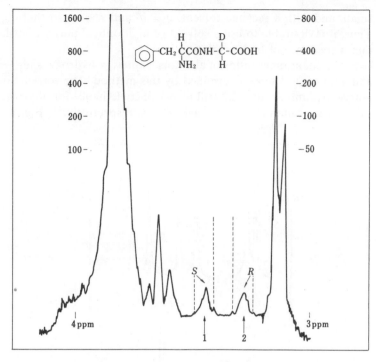

Fig. 8.9. NMR spectrum of (S)-phenylalanyl-(RS)-glycine-2-d_1. Peaks 1 and 2 show the signals due to the methylene protons of (S)-glycine-2-d_1 and (R)-glycine-2-d_1. The dotted lines show the positions of signals due to the methylene protons of (S)-phenylalanylglycine.

the observed chemical shift, δ_{obs}, is given by the equations of Fig. 8.10. If II_L and II_D are present in different amounts, then $q \neq q'$, $r \neq r'$. Thus, δ_{obs}^L and δ_{obs}^D take different values and the signals from I_L and I_D separate. Their intensities are proportional to the concentrations of I_L and I_D, and are independent of the optical purity of the solvent. However, as the optical purity of the solvent decreases, the difference between q and q' becomes

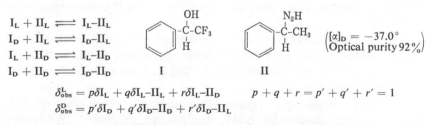

$$I_L + II_L \rightleftharpoons I_L{-}II_L$$
$$I_D + II_L \rightleftharpoons I_D{-}II_L$$
$$I_L + II_D \rightleftharpoons I_L{-}II_D$$
$$I_D + II_D \rightleftharpoons I_D{-}II_D$$

$$\delta_{obs}^L = p\delta I_L + q\delta I_L{-}II_L + r\delta I_L{-}II_D \qquad p + q + r = p' + q' + r' = 1$$
$$\delta_{obs}^D = p'\delta I_D + q'\delta I_D{-}II_D + r'\delta I_D{-}II_L$$

Fig. 8.10. Equilibria and equations for determination of the chemical shift of each enantiomer in a chiral solvent (see the text).

small, and with a racemic solvent, $q = q'$ and $r = r'$, so that $\delta_{obs}^{L} = \delta_{obs}^{D}$. Thus, it is desirable to use a solvent of high optical purity in order to obtain a spectrum of high resolution.

The enantiomer ratios of alcohols, amines, α-hydroxy acids, sulfoxides and amino acids were determined by this method. The solvents used were α-arylethylamine and 2,2,2-trifluoro-1-phenylethanol for alcohols and for sulfoxides, amines and amino acid esters, respectively.[21] Fig. 8.11 shows

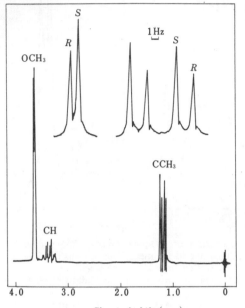

Chemical shift (ppm)

Fig. 8.11.   NMR spectra (60 MHz) of partially resolved methyl (S)-alaninate in (R)-($-$)-2,2,2-trifluoro-1-phenylethanol. [From Pirkle (21)].

the spectrum of partially resolved alanine methylester, (S)-enriched, in optically active 2,2,2-trifluoro-1-phenylethanol. The results indicated 17.8% e.e., in reasonably good correspondence with the theoretical value of 20.0%.

Since $\Delta\Delta\delta$ obtained by this method is smaller than that obtained using a shift reagent, there are limitations in the application of this method for the determination of enantiomer ratios. The method using shift reagents has been preferred recently.

Determination using shift reagents: Lanthanide complexes were discovered as shift reagents for NMR spectroscopy in 1969.[22] Whitesides and Lewis[23] prepared the β-diketone from (+)-camphor and obtained optically active Eu^{3+} and Pr^{3+} complexes of the diketone. He found that the NMR spectrum of racemic amine is not only shifted in the presence of this reagent but also the extents of the shift are different for enantiomers. The difference between the shift values of enantiomers, $\Delta\Delta\delta$, is much larger than that found in the spectra of diastereomers or of enantiomers in optically active solvents. Subsequently, many improved reagents were prepared and NMR spectroscopy using optically active shift reagents has become a widely applied conventional method for the determination of enantiomer ratios.

Although compounds which can chelate with the metal in the shift reagent show $\Delta\Delta\delta$ to some extent, it is generally known that compounds with a strong basic substituent show large values of $\Delta\Delta\delta$ and the ratio of enantiomers can then be determined easily.

$\Delta\Delta\delta$ may arise from diastereomers produced stoichiometrically between the sample and the shift reagent. Alternatively, it may depend on the concentrations of diastereomeric compounds in the equilibrium state. It is generally considered that the former contributes more than the latter to $\Delta\Delta\delta$. However, as the two possible causes are closely related, it seems likely that both contribute. Most of the reagents which have been tried have been prepared from optically active terpenes. Those shown below are known to be superior reagents.

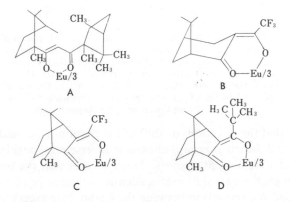

However, since the superiority depends greatly on the nature of the sample, several reagents should be tested to find the best in a particular case. The conditions of determination should also be investigated to obtain good results for each sample. e.g., the ratio of solvent, reagent and sample, the type of solvent, the temperature, impurities in the reagent, etc. As a solvent which competes with the sample for coordination with the reagent is not suitable, nonpolar solvents such as carbon tetrachloride or fluorinated hydrocarbons are generally used. The gradual addition of the shift reagent to the sample solution with examination of $\Delta\Delta\delta$ after each addition is the best way to find the optimum ratio of sample and reagent. Room temperature is usually used for the determination, but good results can be obtained at low temperature with nitro compounds and thiols which have weak affinity for the reagent.

Fig. 8.12 shows the NMR spectra of methyl α-phenyl-α-methylbutyrate in the presence of various amounts of shift reagent C.[24] It is

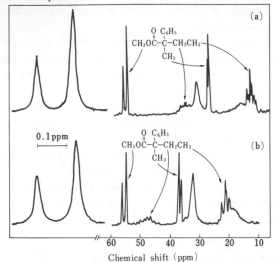

Fig. 8.12. Relationship between the difference of chemical shift of the enantiomers of methyl α-phenyl-α-methylbutyrate and the concentration of the sample in the presence of tris-[3-trifluoromethylhydroxymethylene-(+)-camphorato]-europium (III) (shift reagent C in text). Concentrations were (a) sample 0.9 M, shift reagent 0.5 M (in CCl₄); (b) sample 0.6 M, shift reagent 0.5 M (in CCl₄). The left-hand signals are the α-methyl peaks.
[From Goering and Eikenberry (24)].

noteworthy that the directions of shift of the enantiomeric methyl signal of acylmethyl and 3-methyl groups change in inverse relationship as the concentration of shift reagent changes. This indicates that the proton signals arising from each group shift independently with change of concentration of the reagent. A comparison between the enantiomer excess of the sample as determined by NMR spectroscopy and the optical purity of the same sample is as follows: (a) 25.8 % (polarimetric), 27.7 % (NMR); (b) 25.4 % (polarimetric), 27.3 % (NMR) (see Fig. 8.12).

Fig. 8.13 shows the effect of temperature on the determination of $\Delta\Delta\delta$. It can be seen that a decrease of temperature increases the value of $\Delta\Delta\delta$.[25]

Systematic studies on the relation between the sign of $\Delta\Delta\delta$ and the configuration of the sample have been carried out with amines, amino

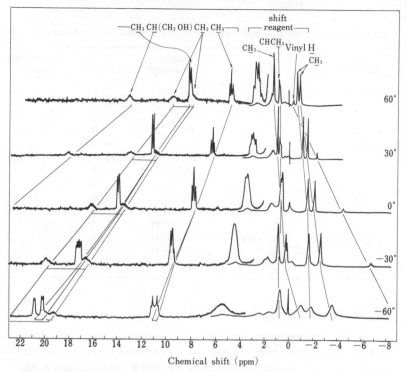

Fig. 8.13. Temperature dependencies of NMR spectra of 2-methyl-1-butanol in the presence of shift reagent A (see the text). These spectra were taken with 0.3 and 0.15 M solutions of 2-methyl-1-butanol and tris-[(+)(+)-dicamphorylmethanato]europium (III) (shift reagent A) in CS_2.
[From McCreary, Lewis, Wernick and Whitesides (25)].

acids and hydroxy acids.[26] Table 8.6 shows $\Delta\Delta\delta$ values for various amino acids with shift reagent C. It can be seen that the signal from the L isomer appears at higher magnetic field than that from the D isomer. However, at present it is difficult to determine configuration accurately by this method, since it has such a short history of development. Nonetheless, the determination of enantiomer ratios by NMR using shift reagents is one of the best methods available. This method will be further developed and should become important for the determination of configuration as well as enantiomer ratios in the near future.

Since results obtained by NMR methods are given as a ratio of signal strengths due to enantiomers, r, the enantiomer excess (e.e.) can be calculated as follows:

$$e.e.(\%) = \frac{1 - r}{1 + r} \times 100 \tag{8.17}$$

TABLE 8.6
$\Delta\Delta\delta$ values of amino acid menthylesters in the presence of shift reagent C (see text)

Amino acid	(L/D ratio)	Signal assignment	$\Delta\Delta\delta$(Hz)	Isomer at higher magnetic field
Alanine	(2/1)	OCH_3	5.4	L
		CH	26.5	L
		CH_3	6.0	D
Phenylalanine	(5/1)	CCH_3	6.0	L
Valine	(2/1)	OCH_3	40.5	L
		CH_3	17.5, 15.0	L, D
Norleucine	(2/1)	OCH_3	6.5	L
Isoleucine	(2/1)	OCH_3	4.5	L
Tryptophan	(2/1)	OCH_3	21.0	L
Aspartic acid	(2/1)	C^1-OCH_3	23.0	L
		C^4-OCH_3	15.5	L
Glutamic acid	(5/1)	C^1-OCH_3	26.0	L
		C^5-OCH_3	5.5	L
Proline	(2/1)	OCH_3	10.5	D

B. Diastereomers

Since diastereomers are different compounds as regards chemical and physical properties, the ratio can be determined by conventional methods of analysis such as chromatography or NMR, as described earlier. When the ratio of diastereomers is determined in terms of the optical purity of derived enantiomers, the determination can be carried out by the methods described in section 8.2.3.A.a.

8.2.4 Relationship between the ratio of stereoisomers produced and degree of stereo-differentiation

We will next consider the relationship between the ratio of stereo-isomers in the product and the degree of stereo-differentiation in a model reaction in which the optical yield of product corresponds to the enantiomer excess and the reaction proceeds through only one type of differentiation process with 100% synthetic yield. In addition, calculation of the optical yield from results obtained by a reaction using optically impure chiral factor will be described.

A. Enantio-differentiation

Taking the chiral factor with 100% optical purity as L*, enantioface- or enantiotopos-differentiating reaction of a substrate S occurs as in the following scheme.

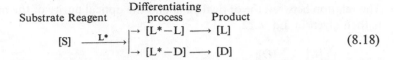

$$
\text{[S]} \xrightarrow{\text{L}^*} \left|
\begin{array}{l}
\rightarrow \text{[L}^*\text{—L]} \longrightarrow \text{[L]} \\
\\
\rightarrow \text{[L}^*\text{—D]} \longrightarrow \text{[D]}
\end{array}
\right. \tag{8.18}
$$

As shown in the scheme, in a reaction with optically pure chiral factor, the proportion of enantiomers produced, [L] and [D], corresponds exactly to the degree of enantio-differentiation. Namely, the optical purity, P(%), of the product corresponds to the optical yield, OY(%), of the product as shown below.[†]

$$
\text{OY}(\%) = \frac{|[L] - [D]|}{[D] + [L]} \times 100 \left(= P(\%) = \text{e.e.}\,(\%) \right) \tag{8.19}
$$

If q_L and q_D represent the mole fractions of L and D isomers in the product, and r is the mole ratio, the relationships among these quantities are given by Eq. 8.20.

$$
\left.
\begin{aligned}
q_L &= \frac{[L]}{[D] + [L]} = \frac{1}{2}\left(1 + \frac{P}{100}\right) = \frac{1}{2}\left(1 + \frac{OY}{100}\right) \\
q_D &= 1 - q_L = \frac{1}{2}\left(1 - \frac{P}{100}\right) = \frac{1}{2}\left(1 - \frac{OY}{100}\right) \\
r &= \frac{q_L}{q_D} = \frac{100 + P}{100 - P} = \frac{100 + OY}{100 - OY}
\end{aligned}
\right\} \tag{8.20}
$$

If a chiral factor with an optical purity P′(%) is used, the material balance of the reaction will be as follows.

		Relative part-icipation of chiral center in reaction	Differen-tiating process	Product	Proportion of product

$$
\text{[S]} \xrightarrow[\text{D}^*]{\text{L}^*} \left|
\begin{array}{c}
\left|\frac{1}{2}\left(1+\frac{P'}{100}\right)\right| \text{[L}^*\text{-S]} \; \frac{1}{2}\left(1+\frac{OY}{100}\right)
\begin{cases}
\xrightarrow{\frac{1}{2}\left(1+\frac{OY}{100}\right)} \text{[L}^*\text{-L]} \xrightarrow{} \text{[L]}_L \; \frac{1}{4}\left(1+\frac{P'}{100}\right)\left(1+\frac{OY}{100}\right) \\
\xrightarrow{\frac{1}{2}\left(1-\frac{OY}{100}\right)} \text{[L}^*\text{-D]} \xrightarrow{} \text{[D]}_L \; \frac{1}{4}\left(1+\frac{P'}{100}\right)\left(1-\frac{OY}{100}\right)
\end{cases} \\[6pt]
\left|\frac{1}{2}\left(1-\frac{P'}{100}\right)\right| \text{[D}^*\text{-S]} \; \frac{1}{2}\left(1+\frac{OY}{100}\right)
\begin{cases}
\xrightarrow{\frac{1}{2}\left(1+\frac{OY}{100}\right)} \text{[D}^*\text{-D]} \xrightarrow{} \text{[D]}_D \; \frac{1}{4}\left(1-\frac{P'}{100}\right)\left(1+\frac{OY}{100}\right) \\
\xrightarrow{\frac{1}{2}\left(1-\frac{OY}{100}\right)} \text{[D}^*\text{-L]} \xrightarrow{} \text{[L]}_D \; \frac{1}{4}\left(1-\frac{P'}{100}\right)\left(1-\frac{OY}{100}\right)
\end{cases}
\end{array}
\right. \tag{8.21}
$$

[†] A new standard, enantiomer yield, will be necessary as a basis for the calculation of enantiomer excess in the future.

The relation between the optical yield and the optical purity of the product is then given by Eq. 8.22.

$$
\begin{aligned}
P(\%) &= \frac{|\sum[L] - \sum[D]|}{\sum[D] + \sum[L]} \times 100 \\
&= \left\{ \frac{1}{4}\left(1 + \frac{P'}{100}\right)\left(1 + \frac{OY}{100}\right) + \frac{1}{4}\left(1 - \frac{P'}{100}\right)\left(1 - \frac{OY}{100}\right) \right. \\
&\quad \left. - \frac{1}{4}\left(1 + \frac{P'}{100}\right)\left(1 - \frac{OY}{100}\right) - \frac{1}{4}\left(1 - \frac{P'}{100}\right)\left(1 + \frac{OY}{100}\right) \right\} \times 100 \\
&= P' \times OY \times \frac{1}{100} \\
\\
OY(\%) &= \frac{P}{P'} \times 100
\end{aligned} \tag{8.22}
$$

As shown in the above calculation, the optical yield of an enantio-differentiating reaction carried out with an optically impure chiral factor can be expressed as $P/P' \times 100$.

In some enantio-differentiating reactions catalyzed by an organo-metallic catalyst with more than 3 optically active ligands, a simple relation such as Eq. 8.22 cannot be expected. For example, in the case of a reaction where the catalyst contains one metal atom, M, and 3 optically active ligands, L_R^* or L_S^*, 4 different catalyst types $(L_R^*)_3M$, $(L_R^*)_2(L_S^*)M$, $(L_R^*)(L_S^*)_2M$ and $(L_S^*)_3M$ can be produced when optically impure ligand is employed. Thus, the effect of the enantiomer contained as an optical impurity on the enantio-differentiation depends on the type (species) of catalyst, and the impurity does not affect the differentiation simply.

The material balance of an enantioface- or enantiotopos-differentiating reaction which gives diastereomers as products is represented by the following scheme, if optically pure reagent, L^*, is used.

$$
\begin{array}{cccc}
 & \text{Differentiating} & & \text{Proportion} \\
\text{Substrate} & \text{process} & \text{Product} & \text{of product}
\end{array}
$$

$$
[S] + [L^*] \xrightarrow{\text{Reagent}}
\begin{cases}
\xrightarrow{\frac{1}{2}\left(1 + \frac{OY}{100}\right)} & [L\text{–}L^*] & \frac{1}{2}\left(1 + \frac{OY}{100}\right) \\
\\
\xrightarrow{\frac{1}{2}\left(1 - \frac{OY}{100}\right)} & [D\text{–}L^*] & \frac{1}{2}\left(1 - \frac{OY}{100}\right)
\end{cases} \tag{8.23}
$$

The optical yield can be calculated from Eq. 8.24.

$$OY(\%) = \frac{|[L-L^*] - [D-L^*]|}{[L-L^*] + [D-L^*]} \times 100 \tag{8.24}$$

That is, in the case of a reaction which gives diastereomers, the optical yield is given by the value of the diastereomer excess. In a reaction which produces diastereomers and in which the optically active reagent racemizes during the reaction (e.g. the Reformatsky reaction with optically active α-bromopropionate) (see p. 90), the optically active reagent acts as a mixture of L* and D*. The degree of differentiation is then given by Eq. 8.25.

$$\frac{OY}{100} = \frac{|[L-L^*] - [D-L^*]|}{[L-L^*] + [D-L^*]} = \frac{|[D-D^*] - [L-D^*]|}{[D-D^*] + [L-D^*]}$$

$$= \frac{|\{[L-L^*] + [D-D^*]\} - \{[D-L^*] + [L-D^*]\}|}{\{[L-L^*] + [D-D^*]\} + \{[D-L^*] + [L-D^*]\}} = \left(\frac{d.e.}{100}\right) \tag{8.25}$$

Thus, the optical purity, P', of a reagent racemized during the reaction does not affect the evaluation of the degree of enantio-differentiation. The optical purity of each diastereomer of the product corresponds to the optical purity of the reagent as actually contributed by the reagent. This may be calculated from Eq. 8.26.

$$\frac{|[L-L^*] - [D-D^*]|}{[L-L^*] + [D-D^*]} = \frac{|[D-L^*] - [L-D^*]|}{[L-D^*] + [D-L^*]} = \frac{|[L^*] - [D^*]|}{[L^*] + [D^*]} = P'/100$$

$$\tag{8.26}$$

B. Diastereo-differentiation

The degree of diastereomer differentiation (DY, %) corresponds to the value of the diastereomer excess (d.e.) and the diastereomer yield (DY, %).

When optically pure substrate is used, the diastereomer yield can be calculated from Eq. 8.27.

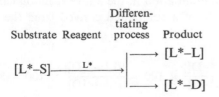

$$DY(\%) = \frac{|[L^*-L] - [L^*-D]|}{[L^*-L] + L^*-D]} \times 100$$

$$= \frac{|[D^*-D] - [D^*-L]|}{[D^*-D] + [D^*-L]} \times 100 \; (=d.e.) \qquad (8.27)$$

The diastereomer yield can be calculated from the ratio of diastereomers, d, which is obtained by gas chromatographic or NMR analysis.

$$DY(\%) = \frac{d - 1}{d + 1} \times 100 \qquad (8.28)$$

If the substrate is not optically pure, the diastereomer yield can be calculated from Eq. 8.29, since the degree of diastereo-differentiation is not affected by the optical purity of the substrate. In other words, a study of diastereoface- or diastereotopos-differentiating reactions can be carried out with a racemic substrate.

$$\frac{DY}{100} = \frac{|[L^*-L] - [L^*-D]|}{[L^*-L] + [L^*-D]} = \frac{|[D^*-D] - [D^*-L]|}{[D^*-D] + [D^*-L]}$$

$$= \frac{|\{[L^*-L] + [D^*-D]\} - \{[L^*-D] + [D^*-L]\}|}{\{[L^*-L] + [D^*-D]\} + \{[L^*-D] + [D^*-L]\}} \qquad (8.29)$$

A reaction scheme for an enantioface- or enantiotopos-differentiating reaction with optically pure substrate, L*–S, is shown below. The degree of differentiation is determined from the optical purity of the compounds L and D derived from the newly produced chiral moiety in the diastereomers produced, L*–L and L*–D.

$$
\begin{array}{cccc}
 & & & \text{Newly produced} \\
\text{Substrate} & \text{Reagent} & \text{Product} & \text{chiral center} \\
 & & \rightarrow [L^*-L] & \longrightarrow [L] \\
[L^*-S] & \longrightarrow & & \\
 & & \rightarrow [L^*-D] & \longrightarrow [D]
\end{array}
\qquad (8.30)
$$

Thus, the degree of diastereo-differentiation in the above reaction can be calculated from the optical purity of a compound derived from the diastereomeric products, using Eq. 8.31.

$$DY(\%) = \frac{|[L^*-L] - [L^*-D]|}{[L^*-L] + [L^*-D]} \times 100 = \frac{|[L] - [D]|}{[L] + [D]} \times 100 \qquad (8.31)$$

If the above reaction is carried out with optically impure substrate (P'), the material balance of the reaction is as shown in the following scheme.

Substrate	Proportion of substrate	Differen-tiating process	Product	Proportion of product

$$[L^*-S]\ \frac{1}{2}\left(1+\frac{P'}{2}\right) \xrightarrow[\downarrow]{\text{Reagent}}
\begin{array}{l}
\boxed{\frac{1}{2}\left(1+\frac{DY}{100}\right)} \xrightarrow{} [L^*-L] \quad \frac{1}{4}\left(1+\frac{P'}{100}\right)\left(1+\frac{DY}{100}\right) \\[2em]
\frac{1}{2}\left(1-\frac{DY}{100}\right) \xrightarrow{} [L^*-D] \quad \frac{1}{4}\left(1+\frac{P'}{100}\right)\left(1-\frac{DY}{100}\right)
\end{array}$$

$$[D^*-S]\ \frac{1}{2}\left(1-\frac{P'}{2}\right) \xrightarrow[\downarrow]{\text{Reagent}}
\begin{array}{l}
\boxed{\frac{1}{2}\left(1+\frac{DY}{100}\right)} \xrightarrow{} [D^*-D] \quad \frac{1}{4}\left(1-\frac{P'}{100}\right)\left(1+\frac{DY}{100}\right) \\[2em]
\frac{1}{2}\left(1-\frac{DY}{100}\right) \xrightarrow{} [D^*-L] \quad \frac{1}{4}\left(1-\frac{P'}{100}\right)\left(1-\frac{DY}{100}\right)
\end{array}$$

$$(8.32)$$

The optical purity, P, of the enantiomers produced from the product can then be calculated from Eq. 8.33.

$$
\begin{aligned}
P(\%) &= \frac{|\sum[L] - \sum[D]|}{\sum[D] + \sum[L]} \times 100 \\[1em]
&= \frac{|\{[L^*-L] + [D^*-L]\} - \{[D^*-D] + [L^*-D]\}|}{\{[L^*-L] + [D^*-L]\} + \{[D^*-D] + [L^*-D]\}} \times 100 \\[1em]
&= P' \times DY \times \frac{1}{100}
\end{aligned}
$$

$$(8.33)$$

That is, the optical purity of the product corresponds to the product of the optical purity of the substrate and the degree of diastereomer yield of the reactions, in the synthesis of optically active enantiomers by diastereoface- or diastereotopos-differentiating reactions.

8.2.5 Distinction between enantio-differentation and diastereo-differentiation

Differentiation is often performed with an optically active substrate and optically active reagent or catalyst, as in Eq. 8.39.[†] The degree of

---

[†] Such a reaction will be called a double-differentiating reaction in this section.

differentiation in double-differentiating reactions can be expressed as follows

[Stereo-differentiating ability] $= f($[enantio-differentiating ability],

[diastereo-differentiating ability]) $\qquad$ (8.34)

However, it is not yet clear whether the two functions operate independently and whether the functions are additive or not.

Several attempts have been made to elucidate the relation between enantio-differentiating ability and diastereo-differentiating ability in double-differentiating reactions. Guetté and Horeau[27] attempted a separation from the viewpoint that the two abilities independently affect the result. That is, in the reaction of the racemic substrate of a diastereo-differentiating reaction with a chiral reagent, they calculated the amount of each of the four enantiomers produced from Eq. 8.35, in which d represents the ratio of diastereomers produced in the reaction and P represents the optical purity of each of the two pairs of enantiomers.

$$\text{Each enantiomer produced } (\%) = \text{d} \times \text{P} \qquad (8.35)$$

Horeau et al.[28] reported a new system for calculating the results of double-differentiating reactions. They assumed that the degree of differentiation in double-differentiating reactions should be related to the sum of the differences of activation free energy between the enantio-differentiating reaction and the diastereo-differentiating reaction. That is, when this difference of activation free energies of the two components of a double-differentiating reaction are designated as $\Delta\Delta G_1^{\neq}$ and $\Delta\Delta G_2^{\neq}$, respectively, and that of the overall double-differentiating reaction is $\Delta\Delta G_3^{\neq}$ they proposed the relation $\Delta\Delta G_3^{\neq} = \Delta\Delta G_1^{\neq} + \Delta\Delta G_2^{\neq}$.

In this case, the degree of differentiation in a double-differentiating reaction should be calculated using Eq. 8.36, where the degree of enantio-differentiation and the degree of diastereo-differentiation are represented by OY and DY, respectively.

$$\text{Degree of double-differentiation}^{\dagger} = \frac{100(\text{OY} + \text{DY})}{100^2 + \text{OY} \times \text{DY}} \qquad (8.36)$$

However, it is difficult to determine values for the degrees of the two differentiations. They obtained such values from Eqs. 8.37 and 8.38, then

---

† In the original paper, the degree of double-differentiation is represented by the optical yield because mandelic acid was obtained as the final product.

attempted to calculate the degree of differentiation for the reaction shown in Eq. 8.39 based on the above results, using Eq. 8.36. However, in order to do this, the differentiating abilities of the reagents in the former two reactions should remain the same, and also the differentiating abilities of the reagents should remain the same in reactions of substrates which have different ester moieties. In other words, the unlikely assumption is required that any kind of reagent should always have a constant differentiating ability regardless of the structure of the substrate.

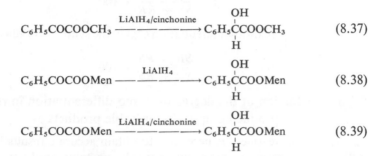

$$C_6H_5COCOOCH_3 \xrightarrow{\text{LiAlH}_4/\text{cinchonine}} C_6H_5\overset{\text{OH}}{\underset{\text{H}}{C}}COOCH_3 \qquad (8.37)$$

$$C_6H_5COCOOMen \xrightarrow{\text{LiAlH}_4} C_6H_5\overset{\text{OH}}{\underset{\text{H}}{C}}COOMen \qquad (8.38)$$

$$C_6H_5COCOOMen \xrightarrow{\text{LiAlH}_4/\text{cinchonine}} C_6H_5\overset{\text{OH}}{\underset{\text{H}}{C}}COOMen \qquad (8.39)$$

In fact, the correspondence between the results obtained and the calculated results was rather poor in their report. This method contains a serious intrinsic defect, as mentioned above.

Since it is not yet known whether the relation between enantio-differentiating ability and diastereo-differentiating ability in double-differentiating reactions is additive or independent, only approximate values can be obtained by calculation at present. Thus, it is practically convenient to determine the approximate extent of the degrees of enantio- and diastereo-differentiation in double-differentiatings reactions by a simple calculation method. For example, in the reaction of a racemic substrate, $RO + SO$, with an optically active reagent or catalyst, C^*, as shown in Fig. 8.14, we can calculate the contribution of the diastereo-differentiating ability of C^* from Eq. 8.40 and also the contribution of enantio-

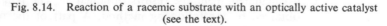

Fig. 8.14.   Reaction of a racemic substrate with an optically active catalyst (see the text).

differentiating ability of C* in the syntheses of each pair of enantiomers from Eqs. 8.41 and 8.42.

Contribution of diastereo-differentiation (%)

$$= \frac{(SS + RR) - (SR + RS)}{(SS + RR) + (SR + RS)} \times 100 \tag{8.40}$$

Contribution of enantio-differentiation (%) in the synthesis of diastereomer

$$A = \frac{SS - RR}{SS + RR} \times 100 \tag{8.41}$$

Contribution of enantio-differentiation (%) in the synthesis of diastereomer

$$B = \frac{SR - RS}{SR + RS} \times 100 \tag{8.42}$$

8.2.6   Evaluation of the degree of stereo-differentiation in reactions producing optically unstable products

A special treatment is necessary to obtain accurate results in stereo-differentiating reactions producing optically unstable products. We shall describe two examples.

A.   Evaluation of enantioface-differentiating ability in cyanhydrin synthesis

Though reactions producing optically unstable compounds are unsuitable for investigation, it is sometimes necessary to study them. The enantioface-differentiating reaction of arylaldehydes with hydrogen cyanide in the presence of an optically active base is a representative case. Since the cyanhydrin produced by this reaction is very optically unstable, it is racemized not only during the reaction but also during purification. Thus, evaluation of the enantioface-differentiating ability of the catalyst must be performed by a special method.

The catalyst is removed from the reaction mixture of benzaldehyde and hydrogen cyanide in chloroform in the presence of an optically active polymer catalyst (see p. 100). Next, the chloroform and hydrogen cyanide are removed *in vacuo*. The residue is dissolved in a specific volume of benzene, and an aliquot of this sample solution is made alkaline with sodium hydroxide. The cyanide liberated from the cyanhydrin is titrated with silver nitrate solution. Another aliquot of the sample solution is used for the determination of optical rotation. The specific rotation of the cyanhydrin is then calculated from the results obtained.[29] The real optical yield can be obtained by the following method. A part of the reaction

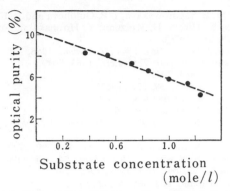

Substrate concentration
(mole/l)

Fig. 8.15. Optical purity of the reaction product in the initial stage of reaction (see the text). [Prelog and Wilhelm (30)].

mixture is removed from the reaction system at specified intervals and the optical yield is measured by the method described above. The optical yield is plotted against the reaction time, as shown in Fig. 8.15 and the real optical yield is determined by extrapolation to zero time.[30]

B. Treatment to prevent racemization

It is sometimes possible to prevent the racemization of a product by conversion of the product to an optically stable substance. For example, since there is a possibility of racemization of the cyclohexene dicarboxylic acid ester shown in Eq. 8.43 during hydrolysis, the ester groups are first reduced with LiAlH$_4$ to the dimethylol derivative, and then the optical purity of this derivative is measured.[31]

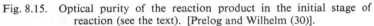

$$(8.43)$$

REFERENCES

1. A. Horeau, *Tetr. Lett.,* **1961,** 3132.
2. H. Ohno and T. Seki, *Tanpakushitsu Kagaku* (Japanese), vol. 4, p. 209, Kyoritsu Shuppan, 1963.
3. K. Yonetani, M. Kawada, T. Toyoshima and T. Shiba, *unpublished.*
4. T. Furuta, T. Harada and Y. Izumi, *unpublished.*
5. H. Gerlach, *Helv. Chim. Acta,* **49,** 2481 (1966).
6. J. P. Greenstein and M. Winitz, *Chemistry of the Amino Acids,* vol. III, p. 1740, Wiley, 1961.
7. M. P. Lambert and F. C. Neuhaus, *J. Bact.,* **110,** 978 (1972).
8. A. Meister, L. Laventow, R. B. Kingsley and J. P. Greenstein, *J. Biol. Chem.,* **192,** 535 (1951).

9. *a.* S. V. Vitl, M. B. Saporowakaya, I. P. Gulohova and V. E. Belikov, *Tetr. Lett.*, **1965**, 2575; *b.* K. Hirota, H. Koizumi, Y. Hironaka and Y. Izumi, *Bull. Chem. Soc. Japan*, **49**, 289 (1976).
10. B. Halpern and J. W. Westley, *Chem. Commun.*, **1965**, 246.
11. J. P. Guetté and A. Horeau, *Tetr. Lett.*, **1965**, 3049.
12. J. Zemlicka, *ibid.*, **1965**, 3057.
13. B. Halpern, *Anal. Chem.*, **39**, 228 (1967).
14. B. Halpern, *ibid.*, **40**, 2046 (1968).
15. S. Mosher, *Chem. Commun.*, **1965**, 614.
16. M. Raban and K. Mislow, *Topics in Stereochemistry*, vol. 2, p. 216, Interscience, 1967.
17. J. A. Dale and H. S. Mosher, *J. Am. Chem. Soc.*, **90**, 2733 (1965).
18. J. A. Dale and H. S. Mosher, *J. Org. Chem.*, **34**, 2543 (1969).
19. M. Raban and K. Mislow, *Tetr. Lett.*, **1966**, 3961.
20. A. Tai, Y. Shimizu and Y. Izumi, *unpublished*.
21. W. H. Pirkle, *J. Am. Chem. Soc.*, **88**, 1937 (1966); *ibid.*, **88**, 4294 (1966); W. H. Pirkle and S. D. Blare, *ibid.*, **89**, 5485 (1967); *ibid.*, **90**, 6250 (1968); *ibid.*, **91**, 5150 (1969).
22. C. C. Hinckley, *ibid.*, **91**, 5160 (1969).
23. G. M. Whitesides and D. W. Lewis, *ibid.*, **92**, 6979 (1970).
24. H. L. Goering and J. N. Eikenberry, *ibid.*, **93**, 5913 (1971).
25. M. D. McCreary, D. W. Lewis, D. L. Wernick and G. M. Whitesides, *ibid.*, **96**, 1038 (1974)
26. K. Ajisaka, M. Kamisaku and M. Kainosho, *Chem. Lett.*, **1972**, 857.
27. J. P. Guetté and A. Horeau, *Bull. Soc. Chim. France*, **1967**, 1747.
28. A. Horeau, H. B. Kagan and J. Vigneron, *ibid.*, **1968**, 3795.
29. S. Tsuboyama, *Bull. Chem. Soc. Japan*, **35**, 1004 (1962).
30. V. Prelog and M. Wilhelm, *Helv. Chim. Acta*, **37**, 1634 (1954).
31. H. M. Walborsky, L. Barash and J. C. Davis, *J. Org. Chem.*, **26**, 4778 (1961).

CHAPTER 9

Basic Principle of Optical Activity

9.1 OPTICAL ROTATION AND THE PROPERTIES OF LIGHT

Before considering optical rotation, some discussion of the properties of light is necessary, and a brief outline of the theory of light will therefore be given.

9.1.1 The wave theory of light

Light is a transverse wave of the alternating electromagnetic field obtained as a solution of Maxwell's equations, and is a type of so-called electromagnetic wave. According to Maxwell's equations, electromagnetic waves are produced by a changing electric field E and magnetic field H induced by each other. The electric field and magnetic fields are given as vectors acting at right angles.

There are spherical waves (emitted spherically from a center) and more complicated waves among electromagnetic waves that propagate in three-dimensional space. However, we will consider the simplest electromagnetic wave, namely a plane wave. The light from an infinitely distant source represents a typical plane wave, and if we take the direction of propagation as the positive z axis, the function can be expressed in terms of its position z and time t. The electric field E and the magnetic field H of a plane wave propagating in an isotropic medium are given by the following expressions from Maxwell's equations (see Appendix A).

$$E_x = f_1(z, t), \qquad H_y = \sqrt{\frac{\varepsilon}{\mu}} f_1(z, t) \tag{9.1}$$

$$E_y = f_2(z, t), \qquad H_x = -\sqrt{\frac{\varepsilon}{\mu}} f_2(z, t) \tag{9.2}$$

$$E_z = H_z = 0 \qquad\qquad (9.3)$$

Here, ε is the dielectric constant, and μ is the magnetic permeability of the medium through which propagation occurs. That is, the electric field E has two independent components; E_x along the positive x axis and E_y along the y axis, but the magnetic field H is given as a dependent function of the electric field, crossing at right angles to it. Moreover, the components in the propagative direction z of the electric and magnetic fields become zero. Thus, it is clear that electromagnetic waves are transverse ones.

Maxwell's equations do not impose any other restriction, except that $f(z, t)$ should be a periodic function. A periodic function can be expressed by the superposition of harmonic oscillators with various frequencies, so even if the discussion of electromagnetic waves is limited to harmonically oscillating waves with fixed frequency v (monochromatic waves), generality can be retained. E_x and E_y, the components of the electric field, are independent of each other, so a plane electromagnetic wave can be described as a sum of the vectors E_x and E_y and becomes a transverse wave vibrating in an arbitrary direction in the plane crossing the z axis at right angles, depending on the ratio of the values of the x and y components.

Based on Eqs. 9.1 through 9.3, the relation between the electric and magnetic fields of a plane electromagnetic wave travelling in the z direction is shown in Fig. 9.1. Since E_x and E_y are independent, even if we suppose

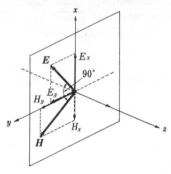

Fig. 9.1.   Electric and magnetic vectors of a plane wave.

that one of them is equal to zero as a special case, the properties of electromagnetic waves can still be well discussed.

Here we will consider a one-dimensional monochromatic travelling wave whose electric field is vibrating in the xz plane ($E_y = 0$) and which is propagating in the z direction with velocity v. Now we consider a wave I which passes a certain point at time t and a wave II which passes at an earlier time t' (Fig. 9.2). Suppose that the distance in the z direction between waves I and II is z_0, then $t - t' = z_0/v$. The amplitude $f(z_0, t)$ at time t and position z_0 is equal to the amplitude $f(0, t')$ at time t' and position zero. Since we give z_0 an arbitrary value, it is possible to replace z_0

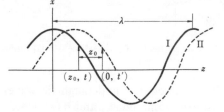

Fig. 9.2. A one-dimensional travelling wave.

with position z in the z direction. In general, the travelling wave in the z direction requires the following relations.

$$f(z, t) = f(0, t') = f\left(t - \frac{z}{v}\right) \qquad (9.4)$$

In the case of a monochromatic wave, $f(z, t)$ becomes a trigonometric function. Let us consider a monochromatic wave having a maximum amplitude A at $z = 0$, and given identical displacements of period T; the equation for the wave becomess

$$f(z, t) = A \cos \frac{2\pi}{T}\left(t - \frac{z}{v}\right) \qquad (9.5)$$

The one-dimensional electromagnetic wave whose electric field amplitude E_x is expressed by Eq. 9.5 can be described as follows (from Eq. 9.1).

$$\left.\begin{array}{l} E_x = A \cos \dfrac{2\pi}{T}\left(t - \dfrac{z}{v}\right) \\[4mm] H_y = \sqrt{\dfrac{\varepsilon}{\mu}}\, A \cos \dfrac{2\pi}{T}\left(t - \dfrac{z}{v}\right) \end{array}\right\} \qquad (9.6)$$

The one-dimensional electromagnetic wave expressed by Eq. 9.6 is shown

† The parameters of wavelength λ, frequency v, etc. are commonly used to express the properties of light: λ is the distance between the nearest positions with the same amplitude, and v is the number of waves rising per second, so the following relations exist among T, v and λ, v: $v = 1/T$, $\lambda = Tv = v/v$.

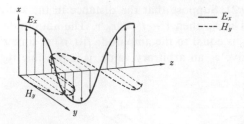

Fig. 9.3.   Electric and magnetic fields of a one-dimensional monochromatic wave.

schematically in Fig. 9.3. Since the electric and magnetic waves cross at right angles to each other with the same phase, and have dependent relations, only one of them need to consider in ordinary cases to discuss the properties of light. Therefore, in this book we will discuss the properties of light in terms of electric waves, except where it is essential to consider both.

9.1.2   Refraction and reflection of light

When light passes the boundary between two different media, as illustrated in Fig. 9.4, a part of it is reflected and the remainder passes into the second medium with a change in the direction of its propagation. The phenomenon of the bending of light at the boundary is called refraction. The degree of refraction, named the refractive index, is designated as n and, as illustrated in Fig. 9.4, it can be defined by using the angle θ which it makes with the normal of the boundary:

$$\frac{\sin \theta_1}{\sin \theta_2} = \frac{n_2}{n_1} \tag{9.7}$$

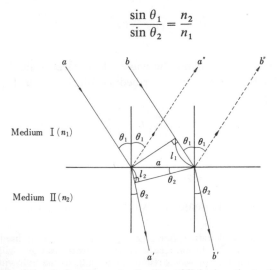

Fig. 9.4.   Refraction and reflection of light (see the text).

where n_1 is the refractive index of medium I, and n_2 is that of medium II. If the velocity of light in medium I is taken as v_1 and that in medium II as v_2, the following relationships can be derived from Fig. 9.4.

$$\frac{l_1}{v_1} = \frac{l_2}{v_2} \tag{9.8}$$

$$l_1 = a \sin \theta_1$$
$$l_2 = a \sin \theta_2 \tag{9.9}$$

Thence we obtain the following relation between the velocity of light in a medium and the refractive index of the medium.

$$\frac{\sin \theta_1}{\sin \theta_2} = \frac{v_1}{v_2} = \frac{n_2}{n_1} \tag{9.10}$$

Since the velocity of light reaches a maximum ($c = 3 \times 10^{10}$ cm/sec) in a vacuum, the refractive index of a vacuum is taken as a standard, and the following relation is obtained from Eq. 9.10.

$$n = \frac{c}{v} \tag{9.11}$$

Refractive index is a physical value which is characteristic of particular materials. The more polarizable substance by electromagnetic waves, the higher the refractive index becomes.

Next we will discuss the behavior of light refracted and reflected at the boundary between medium I and medium II for the case where the refractive index n_1 of medium I is smaller than that of medium II, and a plane electromagnetic wave enters from medium I with an angle of incidence θ_1. Since the electric vector of a plane electromagnetic wave can be separated into two independent vectors acting at right angles, we will consider the component E_S perpendicular to the incident plane and E_P, parallel to the incident plane. Fig. 9.5 illustrates the situation. If the paper is taken as the incident plane, then E_P is in the plane of the paper acting in the direction shown by the arrow →. E_S is perpendicular to the paper, and vectors acting vertically upwards from the paper are shown as ○, while vectors acting downwards are shown as ●. The electric vectors of incident light and light which passes the boundary lie in the same direction, but the electric vector of reflected light lies in the opposite direction (see Appendix B). The equations specifying the relation between incident and reflected light are known as Fresnel's equations, and in the case where the magnetic

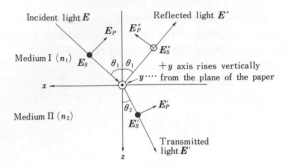

Fig. 9.5. Directions of the electric vectors of incident, reflected and transmitted light (see the text).

permeabilities of medium I and medium II are the same, these equations take the following form (see Appendix C).

$$\frac{E_P''}{E_P} = \frac{\tan(\theta_1 - \theta_2)}{\tan(\theta_1 + \theta_2)} \tag{9.12}$$

$$\frac{E_S''}{E_S} = \frac{\sin(\theta_1 - \theta_2)}{\sin(\theta_1 + \theta_2)} \tag{9.13}$$

If $\theta_1 + \theta_2 = 90°$ then the denominator of Eq. 9.12 becomes infinite. Under these conditions, $E_P''/E_P = 0$ and only E_S'' is reflected. In this case, the vibrational plane of the electric field of the reflected light consists only of the component parallel to the boundary plane. Such light is said to be linearly polarized,[†] and forms a one-dimensional wave, as described in

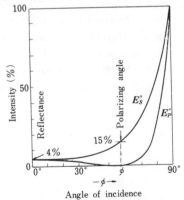

Angle of incidence

Fig. 9.6. Polarization of incident and reflected light at a glass surface.[3] [After Tenkin and White (3).]

[†] The vibration of rectinilearly polarized light occurs in one plane, and it is also known as plane polarized light for this reason. In this book we will use the former term (abbreviated as linear polarization).

section 9.1.1. The angle of incidence which gives complete rectilinear polarization depends on the refractive index of the medium, and is known as Brewster's angle. Fig. 9.6 shows as an example the case where light enters glass ($n = 1.50$) from air ($n =$ approx. 1), when Brewster's angle becomes 57°. In this case, $(E_P''/E_P)^2$ is zero and $(E_S''/E_S)^2$ becomes 0.15, indicating that 15% of the incident light is reflected as linearly polarized light. Light reflected from the boundary at angles of incidence other than Brewster's angle will be incompletely polarized.

9.1.3   Double refraction

In isotropic materials such as liquids, the refractive index is the same in all directions, but in the case of polarizable arrays of atoms with different periodicities in different directions (e.g., certain crystals), the refractive index is anisotropic. When light enters such a crystal, it encounters two different refractive indices, and double refraction occurs. The double image of the sun seen by Malus through Iceland spar was caused by double refraction. The changes of light and shade seen when such a crystal is rotated suggest that double refraction is associated with polarization.

The relation between double refraction and polarization will next be considered for calcite, which shows a large double refraction. The crystal axis of calcite is a threefold proper axis, and perfect crystals are rhombohedral, as shown in Figs. 9.7 and 9.8. If the portion of crystal shown in Fig. 9.7 is cut, double images can be seen through it, since light travels at different velocities in the crystal, emerging as parallel double images, as shown in Fig. 9.8. In calcite the plane of CO_3^{2-} ions lies normal to the

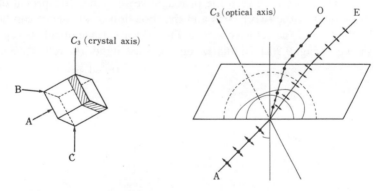

Fig. 9.7.   (*left*)   A crystal of calcite.
Fig. 9.8.   (*right*)   The phenomenon of double refraction in calcite. O, Normal light; E, anomalous light; -●-, light having an electric field vibrating perpendicularly with respect to the paper; ↕ , light having an electric field vibrating in the plane of the paper.

axis of the crystal. Polarization of the electrons in CO_3^{2-} ions occurs more easily in the plane of the ions than perpendicular to this plane. Accordingly, light in which the vibrational plane of the electric vector lies in the plane of the CO_3^{2-} ions "sees" a larger refractive index than light in which the vibrational plane of the electric vector lies perpendicular to the plane of the ions. As a result of this anisotropy of the crystal, light entering it (initially having randomly distributed directions of vibration of the electric field vector) emerges as two linearly polarized beams, one vibrating in the plane of the crystal and the other vibrating perpendicularly to it. In the case of calcite, the refractive index of the former is smaller. If a propagating wave enters the crystal along a crystal axis (direction C in Fig. 9.7), the vibrational planes are not distinguished, and only a single image is observed. Such an axis is called an optical axis of the crystal (in this case the refractive index is a maximum, $n = 1.658$ for calcite). However, light entering perpendicularly to the optical axis of the crystal (direction B in Fig. 9.7) is split into two (in this case, one beam "sees" a refractive index of 1.658, as before, while that for the second beam reaches a minimum, $n = 1.486$), showing maximum double refraction. Light incident between these two positions (e.g., direction A in Fig. 9.7) will also be split into two beams, one with a refractive index of 1.658, and the other with a refractive index intermediate between this and the minimum value.

In general the beam with the unchanging refractive index is referred to as the normal beam, and the other as the anomalous beam, for historical reasons. This terminology does not of course imply any intrinsic abnormality of light, since the effects are simply due to the anisotropy of the crystal.

The relation between the propagative velocities (reciprocal of refractive index) of the two beams and the direction of incidence can be shown by the face of the velocity vector. The vector surface of calcite ($v_E \geqq v_0$) is shown in Fig. 9.9(a). In quartz, on the other hand, the refractive index for

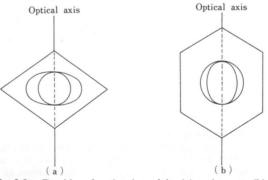

Fig. 9.9. Double refraction by calcite (a) and quartz (b).

the anomalous beam becomes higher than that for the normal beam, as shown in Fig. 9.9(b). Crystals of the former and latter types are sometimes referred to as negative and positive crystals, respectively.

As described above, the double refraction of a plane wave produces two beams of linearly polarized light with mutually perpendicular vibrational planes. Based on the above equations, we will next discuss equipment used to produce linearly polarized light.

9.1.4 Equipment used to generate polarized light

There are three methods for generating polarized light. The first is to use light reflected from the surface of a transparent substance at Brewster's angle. The second is to utilize double refraction by a crystal, and the third is to use a material which is dichroic. Equipment used by Malus to generate polarized light for the first time is shown in Fig. 9.10. The reflected beam is used from a glass plate upon which light is incident at 57°.

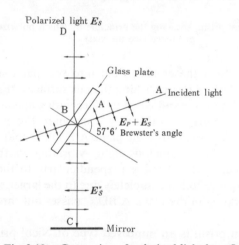

Fig. 9.10. Generation of polarized light by reflection from a glass plate.

Light which enters from direction A at Brewster's angle is partly reflected in direction C, and this light is again reflected by a mirror, passing back through the glass in direction D. The light thus obtained at D is linearly polarized.

In the second method, a prism can be used to separate the two beams produced by double refraction: such a prism is known as a Nicol prism. Fig. 9.11 shows a typical Nicol prism; Fig. 9.11(a) shows a long calcite crystal for use as a Nicol prism. The main section of this crystal ABCD is a parallelogram with $\angle ABD = \angle DCA = 71°$. The planes AEBF and CHDG are cut such that $\angle A'BD = \angle D'CA = 68°$ to form planes

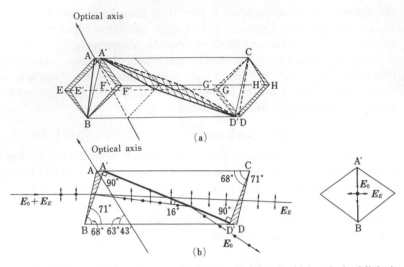

(a)

(b)

Fig. 9.11. A Nicol prism, showing the principle by which polarized light is
produced (see the text).

A′E′BF′ and CG′D′H′. Next the crystal is cut into two parts such that
$\angle$BA′D′ = $\angle$A′D′C = 90°. After polishing the cut surfaces, the section
is replaced in its original position and glued in place with Canada balsam.

A section of a prism prepared in this way is shown in Fig. 9.11(b). A
plane wave which enters the prism from the plane A′E′DI′ is split into two
linearly polarized beams E_E (anomalous light, vibrating in the plane
A′BD′C) and E_0 (normal light, vibrating perpendicularly to the former
beam). At the section, E_0 is reflected completely due to the larger refractive
index, and only E_E, vibrating in the plane A′BD′C passes out through the
plane CG′D′H′.

The Glan-Thomson prism is an improved type of Nicol prism (Fig.
9.12), but the principle is the same. Moreover, a Glan-Thomson-Taylor
prism made of quartz can even be used in the ultraviolet region.

The third method is to use a dichroic material (tourmaline, or synthetic Polaroid films, which are made by adsorption of iodine onto one-

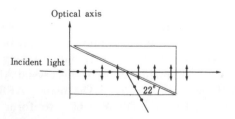

Fig. 9.12. A Glan-Thomson prism.

dimensionally stretched plastic films). If a plane wave enters such a material, light polarized in one of the two mutually perpendicular directions is absorbed strongly. Passage of the remaining light yields a linearly polarized beam (Fig. 9.13). Polarization by this method is not always com-

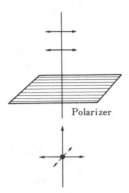

Fig. 9.13.   Generation of polarized light by polarized films. ●, Light having an electric field vibrating perpendicularly to the paper.

plete, and sometimes has a color bias, so this method is used for consumer products such as sunglasses rather than for scientific instruments.

9.1.5   Linearly polarized light and circular polarization

Linearly polarized light is light with its vibrational plane limited to one direction, and is a propagative wave travelling in one plane. There are two approaches to describing the relation between linearly polarized light and circularly polarized light. Kuhn's concept introduces left- and right-circularly polarized light based on linearly polarized light as a propagative wave, while Fresnel's concept regards linearly polarized light as a synthetic wave made up from left- and right-circularly polarized light having the same phase in terms of circular polarization. These two theories are complementary, being in the relation of "which comes first, the chicken or the egg?", so we will consider optical rotation and circular dichroism on the basis of both approaches.

First we will introduce circular polarization based on linearly polarized light. Consider linearly polarized light A having an electric field vector in the xz plane, and B having an electric field vector in the yz plane. The electric vectors are denoted by E_x and E_y (Fig. 9.14(a)), respectively. When the oscilation of E_y is delayed $1/4v$ with E_y, we take the sum of the vectors at various time ($t_1, t_2, t_3, t_4, t_5, t_6$), and, as shown in Fig. 9.14(b). The heads of resulting vector in xy plane ($z = 0$) move on the circle anticlockwise with time. Such light is said to be circularly polarized, and in the case where the rotation is anticlockwise, it is said to be left-circularly polarized.

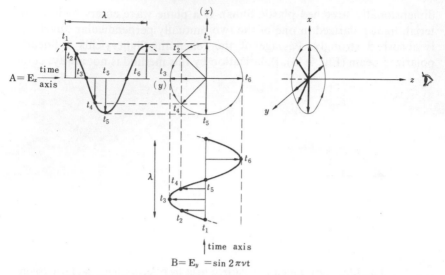

Fig. 9.14. A synthetic wave made up of linearly polarized light A and B with E_x proceeding $2\pi/4(\lambda/4)$ ahead of E_y (left-circular polarization).

As expected, if the linearly polarized light B oscilates $1/4v$ ahead A, the resulting beam is right-circularly polarized. Based on Eq. 9.6, this can be expressed as follows. For beam A, we have

$$E_x = E \cos 2\pi v\left(t - \frac{nz}{c}\right) \tag{9.14}$$

Since E_y is one-quarter of a wavelength ahead of E_x, then

$$E_y = E \cos\left\{2\pi v\left(t - \frac{nz}{c}\right) - \frac{2\pi}{4}\right\} = E \sin 2\pi v\left(t - \frac{nz}{c}\right) \tag{9.15}$$

Now, suppose that i and j are unit vectors in the x and y directions, respectively, then

$$\psi_l = \left\{E \cos 2\pi v\left(t - \frac{nz}{c}\right)\right\}i + \left\{E \sin 2\pi v\left(t - \frac{nz}{c}\right)\right\}j \tag{9.16}$$

However, if E_y is delayed one-quarter wavelength with respect to E_x, we have

$$\psi_r = \left\{E \cos 2\pi v\left(t - \frac{nz}{c}\right)\right\}i - \left\{E \sin 2\pi v\left(t - \frac{nz}{c}\right)\right\}j \tag{9.17}$$

ψ_l and ψ_r correspond to left- and right-circularly polarized light, respectively.

Using the following complex expression (de Maiver Theorem)

$$\left.\begin{array}{l} Ee^{i\varphi} = E\cos\varphi + iE\sin\varphi \\[4pt] \mathrm{Re}(Ee^{i\varphi}) = E\cos\varphi \\[4pt] \mathrm{Re}(iEe^{i\varphi}) = -E\sin\varphi \end{array}\right\} \tag{9.18}$$

where $\varphi = 2\pi v(t - nz/c)$ and $\mathrm{Re}\{\ \}$ expresses the real part of the complex number in $\{\ \}$, i.e., $\mathrm{Re}\{a + bi\} = a$, we can write Eqs. 9.16 and 9.17 in the following simple forms.

$$\left.\begin{array}{l} E_r = \psi_r = \mathrm{Re}\{(i + ij)Ee^{i\varphi}\} \\[4pt] E_l = \psi_l = \mathrm{Re}\{(i - ij)Ee^{i\varphi}\} \end{array}\right\} \tag{9.19}$$

On the other hand, a synthetic wave consisting of left- and right-circularly polarized light in the same phase will clearly correspond to a linearly polarized beam, as shown in Fig. 9.15. That is, by summing

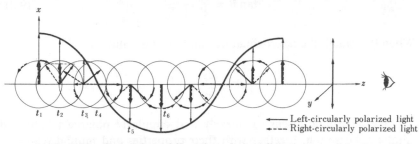

Fig. 9.15. Formation of linearly polarized light by a combination of left- and right-circularly polarized light. Circles represent circular polarization vectors in the xy plane for the time $t_1 \sim t_6$.

Eq. 9.19, we obtain an equation for linearly polarized light propagating in the x direction, as follows.

$$\psi_l + \psi_r = \mathrm{Re}\{2Ee^{i\varphi}\}i = 2(E\cos\varphi)i = 2E_x \tag{9.20}$$

Next, let us consider a synthetic wave made up of left- and right-circularly polarized light with different amplitudes. If $|E_l|$ and $|E_r|$ represent the amplitudes of the two components, respectively, Fig. 9.16 illustrates the case where $|E_l| > |E_r|$. The vector of the synthetic wave rotates anticlockwise in an elliptical fashion (Fig. 9.16). Such behavior is referred to as elliptic polarization. If the ratio of the major axis to the minor axis is taken as $\tan\theta$, and clockwise rotation of vector is difined + direction ($|E_r| > |E_l|$) the characteristics of elliptic polarization can be described in terms of the angle θ, which is called the ellipticity. The following relation exists between the amplitude of circular polarizations and θ.

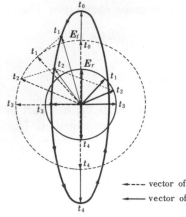

              ◆--- vector of left-circularly polarized light
              ◆— vector of right-circularly polarized light

Fig. 9.16.   The phenomenon of elliptically polarized light (see the text).

$$\tan \theta = \frac{E_r - E_l}{E_r + E_l} \tag{9.21}$$

When θ is small, the approximation $\tan \theta \fallingdotseq \theta$ is valid, i.e.,

$$\theta \fallingdotseq \frac{E_r - E_l}{E_r + E_l} \tag{9.22}$$

We have now discussed linearly and circularly polarized light and elliptic polarization, together with their properties and mutual relations, which are directly connected with the optical properties of optically active materials. Next we will consider the circumstances under which polarization, optical rotation and circular dichroism occur.

9.1.6   The phenomenon of optical rotation

Let us consider how the transmitted light behaves when linearly polarized light passes through a material having different refractive indices for left- and right-circularly polarized light. Fig. 9.17 shows linearly polarized light (vibrating in the xz plane) entering a material at z_0 on the z axis from air, travelling d cm in it, and passing out again into air. If the refractive index of air is n_0, and those of right- and left-circularly polarized light in the medium are n_r and n_l, respectively, then the linearly polarized light $\psi = E_x = 2E\{\cos 2\pi v(t - n_0 z/c)\}i$ entering the medium at z_0 will be split into right- and left-circularly polarized beams which will propagate with different velocities. If they both travel d cm in the medium, from Eqs. 9.16 and 9.17, we have

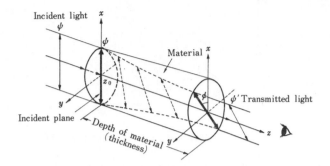

Fig. 9.17. Representation of the phenomenon of optical rotation (for a levorotatory material, $n_r > n_l$). See the text.

$$\psi_l = \left\{E \cos 2\pi v\left(t - \frac{n_0 z_0 + n_l d}{c}\right)\right\}i + \left\{E \sin 2\pi v\left(t - \frac{n_0 z_0 + n_l d}{c}\right)\right\}j$$

$$\psi_r = \left\{E \cos 2\pi v\left(t - \frac{n_0 z_0 + n_r d}{c}\right)\right\}i - \left\{E \sin 2\pi v\left(t - \frac{n_0 z_0 + n_r d}{c}\right)\right\}j$$

$$(9.23)$$

At $z = d$ the light passes out of the medium into air, and the resulting synthetic wave is given by $\psi' = \psi_l + \psi_r$, i.e.,

$$\psi' = \psi_l + \psi_r = 2E \cos 2\pi v\left(t - \frac{n_0}{c}z_0 - \frac{n_r + n_l}{2c}d\right)\left\{\left(\cos \frac{\pi v}{c}d(n_r - n_l)\right)i\right.$$

$$\left. + \left(\sin \frac{\pi v}{c}d(n_r - n_l)\right)j\right\} \qquad (9.24)$$

The expression in the brackets { } in Eq. 9.24 will be constant if d is fixed, and ψ' represents linearly polarized light. The coefficients of i and j show the x and y components of the wave vector, respectively. If $n_l = n_r$, then $\psi' = 2(E \cos \varphi)i = E_x$ and the vibrational planes of incident and transmitted light coincide, but if $n_l \neq n_r$, ψ' has x and y components and the vibrational plane changes as shown in Fig. 9.18. This phenomenon, that the plane of polarization changes and rotates clockwise or anticlockwise, is termed optical rotation, and a material which causes optical rotation is said to be an optically active substance. In other words, optical activity is observed if a material has different refractive indices for left- and right-circularly polarized light.

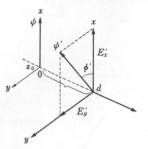

Fig. 9.18. Diagrammatic representation of Fig. 9.17.

Assume that the angle of ψ' from the x direction is ϕ', then

$$\tan \phi' = \frac{E'_y}{E'_x} = \frac{\sin(\pi v d/c)(n_r - n_l)}{\cos(\pi v d/c)(n_r - n_l)} = \tan \frac{\pi v d}{c}(n_r - n_l) \left.\begin{array}{c} \\ \\ \\ \\ \\ \end{array}\right\} \quad (9.25)$$

so that $\qquad \phi' = \dfrac{\pi v d}{c}(n_r - n_l)$

When an observer faces from the $+z$ direction to the $-z$ direction, optical activity such that the plane of polarization is rotated clockwise is defined as $+$, and the angle of its rotation for unit value ($d = 1$ cm), ϕ, is called the angle of optical rotation. From Eq. 9.25, ϕ is as follows.

$$\phi = \frac{\pi v}{c}(n_l - n_r) \qquad\qquad (9.26)$$

From Eq. 9.26, if $n_l > n_r(v_r > v_l)$, ϕ is $+$ and the plane of polarization is rotated clockwise. A material which shows such optical activity is referred to as dextrorotatory. If $n_l < n_r(v_r < v_l)$, ϕ becomes $-$ and the material is levorotatory. Fig. 9.17 shows the passage of linearly polarized light through a levorotatory material.

From Eq. 9.26, ϕ is proportional to d, the distance of the light passes in the material, and to $(n_l - n_r)$, the difference of refractive indices. In general, when the distance d is 10 cm and the angle of rotation is expressed in degrees, we can define the observed rotation α as follows (from Eq. 9.26):

$$\alpha = \frac{180}{\pi} \times 10 \times \phi = 180 \times 10 \times \frac{v}{c}(n_l - n_r) = \frac{1800}{\lambda}(n_l - n_r).$$

The value of α per unit concentration of material per unit length is termed the relative specific rotation, $[\alpha]$; the relation $[\alpha] = \alpha/d[c]$ exists between

α and $[\alpha]$. Where $[c]$ is the concentration of the sample (g/ml), and d is its length (dm).

If the molecular weight of the material is M, $[\alpha] \times M/100$ is defined as the molecular rotation, $[M]$. The relation between $[M]$ and the refractive index is as follows.

$$[M] = \frac{M}{100}[\alpha] = \frac{18 \cdot M}{d[c]\lambda}(n_l - n_r) \tag{9.27}$$

9.1.7 Circular dichroism

In the cases so far discussed, no absorption has occurred in the medium. In this section we will discuss the situation when light is absorbed by the medium. If light is absorbed, its amplitude falls exponentially with the thickness of the medium. Now when linearly polarized light passes through such a medium, the amplitudes of its left- and right-circularly polarized components fall too. When the two components are absorbed equally, the light which emerges from the medium is the same as the initial linearly polarized light, and only its amplitude is reduced, but if one of the circularly polarized components is absorbed preferentially, the transmitted light emerging from the medium will be elliptically polarized, not linearly polarized, as shown in Fig. 9.16. This phenomenon is known as circular dichroism and is described in terms of the ellipticity θ. The intensity I of the light is proportional to the square of the amplitude E. The intensity of light transmitted from a medium obeying Lambert-Beer's law is given by $I = I_0 \exp(-kz)$, where z is the distance passed through the medium and I_0 is the initial intensity. Thus, the amplitudes of left- and right-circularly polarized light passing through such a medium are given by

$$E_l = \sqrt{\frac{I}{I_0}}\exp\left(-\frac{k_l}{2}z\right) \quad E_r = \sqrt{\frac{I}{I_0}}\exp\left(-\frac{k_r}{2}z\right) \tag{9.28}$$

where k is the absorption coefficient. If the molar concentration of the medium is c and the molecular absorption coefficient is ε, then $k = \ln 10 \cdot \varepsilon \cdot c$. Substituting these results into Eq. 9.22, we have

$$\theta = \frac{E_r - E_l}{E_r + E_l} = \frac{\exp(-k_r/2) - \exp(-k_l/2)}{\exp(-k_r/2) + \exp(-k_l/2)} = \frac{\{1 - \exp(k_r - k_l)/2\}}{\{1 + \exp(k_r - k_l)/2\}} \tag{9.29}$$

If $|k_l - k_r| \ll 1$, then $\theta \fallingdotseq (k_l - k_r)/4$. In the case where $k_l > k_r$ the elliptic polarization is clockwise, and this direction is defined as the $+$ direction. Measurements of ellipticity (in degrees) are expressed in terms of specific ellipticity $[\theta']$ based on units of dm for the length of the material and g/ml for concentration. The following relation exists between θ and θ'.

$$[\theta'] = \theta \times \frac{180}{\pi} \times 10 \times \frac{1}{[c]} \quad ([c]: \text{concentration in g/ml}) \quad (9.30)$$

The expression $\theta' \times M/100$ (M = molecular weight) is termed the molecular ellipticity $[\theta]$ and is given by the following equation.

$$[\theta] = \theta' \times \frac{M}{100} = \frac{18M}{4\pi}(k_l - k_r)\frac{1}{[c]} = \frac{18M}{4\pi}\ln 10(\varepsilon_l - \varepsilon_r) \times \frac{c}{[c]}$$

$$= \frac{18 \times 10^3 \times \ln 10}{4\pi}(\varepsilon_l - \varepsilon_r) \quad (9.31)$$

([c], c: concentrations in g/ml and mole/l, respectively)

Eq. 9.31 shows that circular dichroism is a physical phenomenon based on a difference in the absorption coefficients of a material for left- and right-circularly polarized light, and it is thus closely associated with optical rotation, which is based on a difference of refractive indices for left- and right-circularly polarized light. Optical activity and circular dichroism correspond to refraction and absorption of light, respectively, and are thus conversely related. In general, physical values having such a relation can be mathematically treated as having subordinate relations in a complex plane. We can therefore define a complex refractive index of light $\hat{n}$ as follows.

$$\hat{n} = n + ik \quad (9.32)$$

The complex specific rotatory power (Φ) then takes the form

$$\Phi = \frac{\pi v d}{c}(\hat{n}_l - \hat{n}_r) \quad (9.33)$$

on substituting $\hat{n}$ into Eq. 9.26. Since $[M]$ is a function of $(n_l - n_r)$ and $[\theta]$ is a function of $(k_l - k_r)$, we have

$$\Phi = [M] + i[\theta] \quad (9.34)$$

Eq. 9.34 describes the relation between molecular specific rotatory power and molecular ellipticity. It is clear that if one of these quantities is determined experimentally, then the other is implicit. The transformation of the two values can be carried out using the Kroning-Kramer relation. Since this is a standard mathematical method,[1] it will not be described here.

Now if absorption or refraction of light occurs at a particular transition k with resonance vibration v_i in a molecule, the following relations exist between $[M_k(v_i)]$ and $[\theta_k(v_i)]$.

$$[M_k(v)] = \frac{2v^2}{\pi} \int_0^\infty [\theta_k(v_i)] \frac{dv_i}{v_i(v_i^2 - v^2)} \tag{9.35}$$

$$[\theta_k(v)] = \frac{2v^2}{\pi} \int_0^\infty [M_k(v_i)] \frac{dv_i}{v_i(v_i^2 - v^2)} \tag{9.36}$$

Thus, we have so far shown that optical activity and circular dichroism are related phenomena controlled by common factors which in turn depend on the molecular stereostructure of the material itself. In the next section we will consider the features of molecular structure that are involved.

9.2 THEORY OF OPTICAL ROTATION

9.2.1 The spiral model (classical theory)

A. Optical activity and the structure of materials

In 1812 Biot discovered that when linearly polarized light entered quartz from the direction of an optical axis, its plane of polarization was rotated clockwise with respect to the direction of propagation in dextro-quartz and anticlockwise in levo-quartz. This was the first discovery of optical rotation. The rotation of the plane of polarized light corresponds to the facts that in dextro-quartz the SiO_2 chains form a left-handed spiral, while in levo-quartz they form a right-handed spiral. When electromagnetic waves enter such crystals, it is reasonable to assume that the motion of electrons induced in the crystals is dependent on the spiral structure. When quartz is melted it shows no optical activity at all, confirming the view that the crystal structure is involved. Biot also discovered that solutions of sugar and essential oil showed optical activity, even though they are liquid, which clearly indicates that the cause is the configuration of the molecule itself in these cases. Such inherent optical activity of molecules is the object of most current research in this field.

All molecules which show optical activity can be arranged to show a form of spiral structure. Fig. 9.19 shows L-alanine envisaged as a molecular spiral. Such spiral models of molecules can be used with classical electromagnetic theory to account for optical activity.

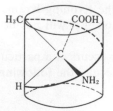

Fig. 9.19.   Structure of L-alanine shown as a spiral.

B.   Electric and magnetic properties of materials

Organic compounds are usually electric insulators, known as dielectrics. Dielectrics are regarded as an assembly of dipoles which produce electric dipole moments in the electric field. There are two kinds of dipoles, a dipole produced by polarity of the molecule itself (i.e., intrinsic) and an induced dipole due to the effect of the external electric field on the nucleus and electrons of a molecule, causing them to shift slightly from the equilibrium positions and polarize into positive and negative charges. When N such dipoles exist per unit volume, the moment P produced is called the polarization. The magnitude of the polarization is proportional to the electric field E in a dielectric $P = \alpha E$ (α = polarizability). Dipole moments produced in this way act in such a way that their electric field opposes the external electric field. In other words, the electric field in a dielectric is determined by the external electric field, but its value is not the same as that which would be produced in a vacuum. Eq. 9.37 shows the relation between the electric field which a charge produces in a vacuum, D (called the electric displacement) and the average electric field which it produces in a dielectric.

$$D = E + 4\pi P = (1 + 4\pi\alpha)E = \varepsilon E \qquad (9.37)$$

E is called the average electric field, since the effective electric field E' at each molecule is not always equal to E. In the case of an isotropic substance such as a liquid, the electric field E' at a given molecule may be expressed approximately as follows.

$$E' = E + \frac{4}{3}\pi P \qquad (9.38)$$

The theory of magnetism of a molecule is similar to that of electric polarization as described above in many respects: if the magnetic induction is B, the magnetic field in the material is H and the intensity of magnetization due to magnetic dipole moments existing in the material is I, then as for the electric field, $I = \kappa H$ (κ = magnetic susceptibility) and we can introduce the following relationship.

$$B = H + 4\pi I = (1 + 4\pi\kappa)H \tag{9.39}$$

In general, the magnetic permeability of molecules is small, and the effective magnetic field at each molecule is almost the same as the averaged one, so we can approximate as follows in this case.

$$H' = H \tag{9.40}$$

As described previously, light is a rapidly vibrating electromagnetic field, and thus in a dielectric substance it will become polarized. Permanent dipole moments of a molecule cannot follow the rapid vibration of the electromagnetic field, but an electron, having small mass, can. As a result, light interacts with electrons to produce rapidly changing electric and magnetic dipole moments. According to classical electromagnetism, electromagnetic waves are radiated by vibrating electromagnetic dipole moments, so these moments, produced by the incident light, radiate electromagnetic waves with the same frequency, and these are observed as transmitted light.

The phenomenon of optical rotation occurs because the vibrational plane of linearly polarized light is changed between the incident and transmitted light, so we can consider the causes of optical rotation by relating the vibrational planes of the electric and magnetic dipole moments induced by incident light to the motions of electrons within the molecular structure. In a previous section, it was shown that optically active molecules can be considered as having spiral structure, and by assuming that electrons are constrained to move in such a spiral, we can develop equations describing optical rotation phenomena.

C. Theory of optical rotation based on the spiral model

We will consider linearly polarized light transmitted through a medium composed of molecules which can be considered to form right spirals. Suppose that the z axis is the direction of propagation, and the vibrational plane of the electric field is the xz plane, so that the magnetic field lies in the xy plane. As shown in Fig. 9.20(a), the axis of the molecular spiral is placed in the polarized field such that it coincides with the direction of the magnetic field.

An electric dipole moment $\alpha E'$ in the x direction is induced proportionately to the local electric field E' which operates on the spiral molecule.

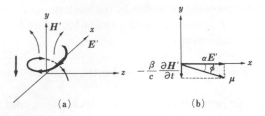

Fig. 9.20. Electric dipole moment, μ, caused by an electron moving in a spiral molecule (see the text).

At the same time, changes of the magnetic field passing through the plane of the spiral produce an electromotive force on the spiral (Faraday's law), and an electron will move along the arrow shown in Fig. 9.20(a). When an electron is shifted from its equilibrium position in this way, an electric dipole moment is produced parallel to the axis of the spiral and acting in the opposite direction to the magnetic field (components in the other direction (xz) are cancelled because the current represented by the electron moves along the loop) and takes a value proportional to the time changes of the magnetic field, being given by $-(\beta/c)(\partial H'/\partial t)$.

Thus, the electric dipole moment produced by a spiral molecule such as that shown in Fig. 9.20b can be expressed as follows (see Appendix D).

$$\mu = \alpha E' - \frac{\beta}{c}\frac{\partial H'}{\partial t} \qquad (9.41)$$

The vector direction of this dipole moment μ is shown in Fig. 9.20(b), and is not the same as that of the electric field of the incident light. Since the vibrational plane of light radiated by an oscillating dipole moment will be the same as that of the dipole moment, it is intuitively clear that the vibrational plane of transmitted light will be rotated clockwise by an angle ϕ with respect to the incident light. The value of β in Eq. 9.41 is determined by the form of the spiral (of radius r and pitch S), the frequency of incident light ν and the characteristic frequency of the molecule ν_0. Taking account of the fact that the axial directions of the spiral molecules are arbitrary and do not all take the y direction, the value of β will be reduced by one-third, and we have

$$\beta = \frac{1}{3}\sum_j \frac{e^2}{m}\cdot\frac{r^2 S}{4\pi^2 r^2 + S^2}\cdot\frac{1}{4\pi(\nu_{0j}^2 - \nu^2)} \qquad (9.42)$$

There are also two types of induced magnetic dipole moments; one is proportional to the alternating magnetic field, and the other is caused by

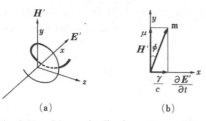

Fig. 9.21. Magnetic dipole moment, M, caused by an electron moving in a spiral molecule (see the text).

the alternating electric field. As shown in Fig. 9.21, we can consider a dextro spiral such that its axis coincides with the vibrational plane of the electric field. The magnetic dipole moment produced by the local magnetic field H' becomes $\kappa H'$ in the direction of the magnetic field. At the same time, changes in the electric field, which passes through the plane containing the spiral, move an electron along it. The current, which is produced by electronic transition, produces a magnetic dipole moment in the direction of the electric field E' in the same way as a solenoid, and its value is proportional to the time changes of the electric field, being given by $(\gamma/c)(\partial E'/\partial t)$. The magnetic dipole moment **m** produced by such a spiral can be expressed by Eq. 9.43, and takes the direction shown in Fig. 9.21(b).

$$m = \kappa H' + \frac{\gamma}{c}\frac{\partial E'}{\partial t} = \kappa H' + \frac{\beta}{c}\frac{\partial E'}{\partial t} \qquad (9.43)$$

The value of γ becomes equal to β (see Appendix E), so the vibrational plane of the magnetic dipole moment rotates clockwise by ϕ with respect to that of the magnetic field of incident light, which naturally produces the same results as for the electric dipole moment, so that the transmitted light is linearly polarized light with its vibrational plane rotated clockwise by an angle ϕ. In the case of a molecule forming a left spiral, the discussion is similar, and the transmitted light is rotated anticlockwise by an angle ϕ with respect to the incident light. Thus, it is qualitatively clear that each molecule rotates the vibrational plane of incident polarized light.

Based on Eqs. 9.41 and 9.43 we will next discuss phenomena relating to the whole system. μ and **m** are the dipole moments produced by each molecule. Taking the number of molecules per unit volume as N, the electric dipole moment P and the magnetic dipole moment I induced by light in the whole system can be expressed as follows.

$$P = N\mu, \quad I = N\mathbf{m} \qquad (9.44)$$

In addition, the relations between the average electromagnetic field in a system and the local electromagnetic field in a molecule are given in Eqs. 9.38 through 9.40, and by combining these equations with Eqs. 9.41 and 9.43 (see Appendix F), we obtain

$$D = \varepsilon'E - \xi\frac{\partial H}{\partial t} \tag{9.45}$$

$$B = \mu H + \xi\frac{\partial E}{\partial t} \tag{9.46}$$

where ε', μ and ξ are constants and take the following forms.

$$\varepsilon' = \text{dielectric constant, } \varepsilon' = \frac{3 + 8\pi N\alpha}{3 - 4\pi N\alpha} \tag{9.47}$$

$$\mu = \text{magnetic permeability, } \mu = 1 + 4\pi N\kappa \tag{9.48}$$

$$\xi = 4\pi N \cdot \frac{\beta}{c} \cdot \frac{\varepsilon' + 2}{3} \tag{9.49}$$

Here, the dielectric constant ε' is based on the local electric field in a molecule, and by summing this with α we obtain the Clausius-Mosotti equation, $(\varepsilon' - 1)/(\varepsilon' + 2) = 4\pi N\alpha/3$. Among D, B, E and H in Eqs. 9.45 and 9.46, the following relation holds, satisfying Maxwell's electromagnetic wave equations.

$$\left.\begin{array}{ll} \nabla \cdot D = 0 & \nabla \times E = -\dfrac{1}{c}\dfrac{\partial B}{\partial t} \\[2mm] \nabla \cdot B = 0 & \nabla \times H = \dfrac{1}{c}\dfrac{\partial D}{\partial t} \end{array}\right\} \tag{9.50}$$

The incident light is linearly polarized, so each electric field can be divided into right- and left-circularly polarized beams, expressed by Eqs. 9.51 and 9.52.

$$E_r = \text{Re}\{(i + ij)Ee^{i\varphi}\} \tag{9.51}$$

$$E_l = \text{Re}\{(i - ij)Ee^{i\varphi}\} \tag{9.52}$$

Using these equations, we can obtain the conditions required to satisfy Eq. 9.50 and Eqs. 9.45 and 9.46, yielding Eq. 9.53 (see Appendix G).

$$\begin{array}{l} n_r = \varepsilon'^{1/2} - 2\pi\nu\xi \\[2mm] n_l = \varepsilon'^{1/2} + 2\pi\nu\xi \end{array} \tag{9.53}$$

Eq. 9.42 shows that in a molecule having spiral structure, the pitch S does not become zero, so ξ also does not become zero. That is to say, the refractive indices of right- and left-circularly polarized light passing through such a molecule will be different, as shown in Eq. 9.53.

The angle of polarization of linearly polarized light composed of right- and left-circularly polarized light can be expressed as follows.

$$\phi = \pi v \frac{n_l - n_r}{c} \tag{9.54}$$

Substituting Eq. 9.53 into 9.54, we have

$$\phi = \frac{4\pi^2 v^2 \xi}{c} \tag{9.55}$$

Substituting in Eq. 9.49, and rewriting ε' according to the relation $\varepsilon' = (n_r + n_l)^2/4 = n^2$, we obtain

$$\phi = \frac{4\pi^2 v^2}{c}\left(4\pi N \frac{\beta}{c} \cdot \frac{\varepsilon' + 2}{3}\right) = \frac{4\pi^2 v^2}{c}\left(4\pi N \frac{\beta}{c} \cdot \frac{n^2 + 2}{3}\right) \tag{9.56}$$

The value of β is positive in a right spiral and negative in a left one, so the plane of polarization rotates clockwise in a right spiral and anti-clockwise in a left one.

Our consideration of optical activity has so far been based on an intuitive spiral model, but actual molecules are not completely described by such a model, which provides only qualitative results, and for real molecules treatment in an exact form by means of quantum mechanical analysis is necessary.

D. Optical rotatory dispersion

In this section, we will consider the relation between the frequency of incident light and optical rotation. A change in the refractive index with change in frequency of light is known as dispersion, and the relation between ϕ, a value based on the difference of refractive index, and the frequency of incident light is known as optical rotatory dispersion. It is described by Eq. 9.57, obtained by substituting Eq. 9.42 into Eq. 9.55.

$$\begin{aligned} \phi &= \sum_j \frac{\alpha_j v^2}{v_{0j}^2 - v^2} \\ \alpha_j &= \frac{1}{3} \cdot \frac{e^2}{m} \cdot \frac{r^2 S}{4\pi^2 r^2 + S^2} \cdot \frac{4\pi N}{c} \cdot \frac{\varepsilon + 2}{3} \end{aligned} \tag{9.57}$$

In a field where the eigenvibration v_{0j} of an electron and the frequency of light are different, Eq. 9.57 is continuous and is known as normal dispersion, but at $v_0 = v$, i.e., in a field where the eigenvibration of an electron is close to the frequency of the light (absorption region), the value of Eq. 9.57 becomes infinite. This is because there is no damping term for the motion of the electron. Inserting such a term, Γ, we obtain

$$\phi = \mathrm{Re}\left\{\sum_j \frac{\alpha_j v^2}{v_{0j}^2 - v^2 + iv\Gamma}\right\} \tag{9.58}$$

Let us illustrate the significance of Eq. 9.58 by setting $j = 1$; this produces the simplest form of optical rotatory dispersion spectrum, as shown in Fig. 9.22. The optical rotatory dispersion spectra of real mate-

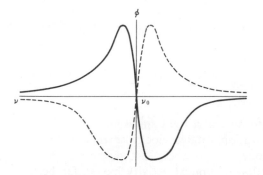

Fig. 9.22.   Illustration of the results of Eq. 9.58 (see the text).

rials arise from many v_{0j}, and show quite complex forms in many cases. Such spectra can be measured for many chemical compounds and are widely used for structure analysis of materials based on experimental rules.[2] However, this lies outside the scope of this book, and here we are simply attempting to show how an optical rotatory dispersion spectrum arises, based on the spiral model.

9.2.2   Quantum mechanical treatment

Quantum theory was applied by Rosenfeld in 1928 to account for the phenomenon of optical rotation. The procedure is essentially the same as that outlined in the previous section. Thus, we have to obtain expressions for the electric and magnetic dipole moments arising due to interaction between a material and electromagnetic waves. Using quantum theory, we can do this by using a wave function which expresses the states of electrons in molecules of the material, so that the results obtained are directly related to the molecular structure of the material.

According to quantum mechanics, if the wave function for a molecule is represented by Ψ and the operators of the electric and magnetic dipole moments μ and **m** are represented by D and M, then μ and **m** are given by Eqs. 9.59 and 9.60.

$$\mu = \text{Re}\left(\int \Psi^* D\Psi \, d\tau \right) \tag{9.59}$$

$$\mathbf{m} = \text{Re}\left(\int \Psi^* M\Psi \, d\tau \right) \tag{9.60}$$

Thus, we have to obtain equations for the operators D and M and the wave function Ψ representing the states of an electron in a molecule upon which light is incident. Using these in conjunction with Eqs. 9.59 and 9.60, we will be able to account for the optical activity of molecules.

A. The quantum mechanical operators D and M

These operators can easily be obtained from classical electromagnetic theory. Let us consider the j-th electron in a system having charge e and potential vector r_j from the center of the molecule. The electric dipole moment becomes er_j and the total moment of the system is given by the sum of the moments of each electron, so that

$$D = \sum er_j \tag{9.61}$$

Magnetic dipole moment occurs when an electron moves about a molecule. According to electromagnetic theory, if we set the velocity of the electron as v and its distance from the center of the molecule as r, then the magnetic dipole moment **m** becomes

$$|\mathbf{m}| = -\frac{erv}{2c} = \frac{-erp}{2cm} \tag{9.62}$$

where $p = mv$, and m is the mass of an electron.

Considering the j-th electron in the system, and taking the potential vector from the center of the molecule to be r_j and the momentum of the electron p_j, the vector of magnetic moment appears vertical to the plane of motion of the electron, and its value is given by the vector product of r_j and p_j. Its direction is opposite to that of the angular momentum vector given by the vector product. Hence,

$$\mathbf{m}_j = \frac{e}{2mc}(r_j \times p_j) \tag{9.63}$$

A quantum mechanical operator is given by the transformation $p_j \rightarrow -i\hbar\nabla_j$ ($\hbar = h/2\pi$, h = planck's constant). The total magnetic moment operator of a system is given by summing over each electron, i.e.,

$$M = \sum_j \left\{ -\frac{ie\hbar}{2mc}(r_j \times \nabla_j) \right\} \tag{9.64}$$

B. Wave function of a molecule in oscillating electromagnetic waves

A wave function ψ expressing the states of an electron in a molecule under the influence of a changing electromagnetic wave such as light can be obtained by solving the time-dependent wave equation shown below.

$$i\hbar\frac{\partial\Psi}{\partial t} = H\Psi \tag{9.65}$$

However, this equation cannot be solved directly, so approximate solutions must be obtained by means of perturbation theory. Assume that H^0 is the time-independent stationary Hamiltonian and H' is the time-dependent perturbed Hamiltonian, and put $H = H^0 + H'$. Now suppose that the time-independent wave equation (Eq. 9.66) has been solved.

$$H^0\Psi^0 = i\hbar\frac{\partial\Psi^0}{\partial t} \tag{9.66}$$

The eigenfunction ψ^0 with eigenvalue E_n satisfying Eq. 9.66 is given by the following equation.

$$\Psi_n^0(q, t) = u_n(q)\exp\left(-i\frac{E_n}{\hbar}t\right) \tag{9.67}$$

where u_n is an eigenfunction of an eigenvalue E_n which depends only on the position coordinate q. By using the time-independent eigenfunction u, we can obtain an approximate solution of the time-dependent wave equation (Eq. 9.65) by perturbation theory, yielding the wave function shown below (see Appendix H).

$$\Psi(q, t) = u_n\exp\left(-i\frac{E_n}{\hbar}t\right) + \sum_m c_m(t)u_m\exp\left(-i\frac{E_m}{\hbar t}\right) \tag{9.68}$$

$c_m(t)$ is a time-dependent coefficient and is given by the following equation,

$$c_m(t) = \frac{1}{i\hbar} \int_0^t \langle u_m^*|H'|u_n\rangle \cdot \exp\left(i\frac{E_{mn}}{\hbar}t\right)dt \qquad (9.69)$$

where $E_{mn} = E_m - E_n$, and $\langle u_m^*|H'|u_n\rangle = \int u_m^* H' u_n d\tau$.

In order to complete Eq. 9.68, we must obtain the perturbed Hamiltonian H', and then calculate $c_m(t)$ from Eq. 9.69.

C. Introduction of the perturbed Hamiltonian

When a molecule is placed in a rapidly oscillating electric and magnetic field, we may consider that only electrons, which have small mass, are affected. The force to which electrons in the fields are subjected is as follows.

$$F = m\frac{d^2r}{dt^2} = e\left\{E + \frac{1}{c}(v \times H)\right\} \qquad (9.70)$$

where m is the mass of the electron, and e and v are its charge and velocity. The electric and magnetic fields can be expressed by Eq. 9.71 on the basis of electromagnetic theory (see Appendix I).

$$\left. \begin{aligned} E + \frac{1}{c}\frac{\partial A}{\partial t} &= -\nabla\varphi \\ H &= \nabla \times A \end{aligned} \right\} \qquad (9.71)$$

where A is the vector potential of the electromagnetic waves and φ is their scalar potential. Substituting Eq. 9.71 into Eq. 9.70, the equation of motion of the electron becomes

$$m\frac{d^2r}{dt^2} = -\frac{e}{c}\cdot\frac{\partial A}{\partial t} - e\nabla\varphi + \frac{e}{c}\{v \times (\nabla \times A)\} \qquad (9.72)$$

The Hamiltonian function (expressing the total energy of the system in terms of coordinates and momentum) satisfying this equation of motion is given by Eq. 9.73 on the basis of classical mechanics.

$$H = \frac{1}{2m}\left(p - \frac{e}{c}A\right)^2 + e\varphi \qquad (9.73)$$

Here p is a value known as general momentum. In this case, there is the following relation between p and ordinal momentum mv.

$$p = mv + \frac{e}{c} A \tag{9.74}$$

The quantum mechanical operator H from which we can obtain the total energy of the system is obtained by changing the general momentum in Eq. 9.73 into the form of an operation. Since $p = -i\hbar\nabla$ the Hamiltonian becomes

$$H = \frac{1}{2m}\left(-i\hbar\nabla - \frac{e}{c}A \right)^2 + e\varphi \tag{9.75}$$

Expanding Eq. 9.75, we obtain

$$H = \frac{1}{2m}\left(-\hbar^2\nabla^2 + i\hbar\frac{e}{c}\nabla\cdot A + i\hbar\frac{e}{c}A\cdot\nabla + \frac{e^2}{c^2}|A|^2 \right) + e\varphi \tag{9.76}$$

Operating the wave function ψ on the second term of Eq. 9.76 and setting $\nabla\cdot A = 0$,

$$\frac{i\hbar e}{c}\nabla\cdot(A\Psi) = \frac{i\hbar e}{c}(\Psi\nabla\cdot A + A\cdot\nabla\Psi) = \frac{i\hbar e}{c}A\cdot\nabla\Psi \tag{9.77}$$

In the electric and magnetic fields caused by light, $\varphi = 0$ and $\nabla\cdot A = 0$ (Appendix I). Now, suppose that the perturbation of a system by light is small, then based on $|A^2| \fallingdotseq 0$ we can rearrange Eq. 9.76, and, taking the sum over all electrons in the system and adding the term for the potential energy of electrons, V, the total Hamiltonian operator of the system becomes

$$H = \sum_j \left[\frac{1}{2m}\left(-\hbar^2\nabla_j^2 + 2i\hbar\frac{e}{c}A\cdot\nabla_j \right) + V \right] \tag{9.78}$$

This equation can be separated into time-dependent H' and time-independent H^0 parts.

$$H' = \sum_j \frac{i\hbar e}{mc}A\cdot\nabla_j \tag{9.79}$$

$$H^0 = \sum_j \frac{-\hbar^2}{2m}\nabla_j^2 + V \tag{9.80}$$

$$(H = H^0 + H')$$

Here H^0 is the unperturbed Hamiltonian and H' is the perturbed Hamiltonian that we require.

Now, suppose the vector potential at one point in the molecule to be A_0, then the vector potential A of an electron which is situated at a distance from that point can be expressed as follows (Taylor expansion).

$$A = A_0 + r_j(\nabla \cdot A)_0 + \cdots \tag{9.81}$$

Substituting Eq. 9.81 into the second term of Eq. 9.79, we obtain Eq. 9.82.[†]

$$H' = \sum_j \frac{i\hbar e}{mc}\left\{ A_0\nabla_j + \frac{1}{2}(\nabla \times A)_0 \cdot (r_j \times \nabla_j) + Q \cdots \right\} \tag{9.82}$$

Terms of third order and higher in Eq. 9.82 do not relate to the dipole moment, so for the purposes of the present chapter, we can omit these terms. H' then becomes

$$H' = \sum_j \frac{i\hbar e}{mc}\left\{ A_0\nabla_j + \frac{1}{2}(\nabla \times A)_0(r_j \times \nabla_j) \right\} \tag{9.83}$$

D. Derivation of $c_m(t)$ based on electromagnetic theory

The Hamiltonian H' was obtained in Eq. 9.69, and by substituting the vector potential A_0 of electromagnetic waves into H', we can obtain $c_m(t)$.

First, we rewrite $\langle u_m^* | H' | u_n \rangle$, then

$$\langle u_m^* | H' | u_n \rangle = \frac{i\hbar}{mc} \langle u_m^* | e \sum \nabla_j | u_n \rangle \cdot A_0 + \left\langle u_m^* \left| \frac{i\hbar e}{2mc} \sum r_j \times \nabla_j \right| u_n \right\rangle \cdot (\nabla \times A)_0 \tag{9.84}$$

Rearranging this, we obtain Eq. 9.85 (see Appendix J).

$$\langle u_m^* | H' | u_n \rangle = -\left\{ \frac{iE_{mn}}{c\hbar} \langle u_m^* | D | u_n \rangle \cdot A_0 + \langle u_m^* | M | u_n \rangle \cdot (\nabla \times A)_0 \right\} \tag{9.85}$$

where $D = \sum_j er_j \quad M = \sum_j -(ie\hbar/2mc)(r_j \times \nabla_j)$.

[†] Substitution of Eq. 9.81 into Eq. 9.79 up to the second term yields

$$H' = \sum \frac{i\hbar e}{mc}[A_0\nabla + \{(r_j\nabla)(A_0\nabla_j)\}]$$

Applying the vector formula $(V_1 \cdot V_2) \cdot (V_3 \cdot V_4) = \frac{1}{2}(V_2 \times V_3) \cdot (V_1 \times V_4) + \frac{1}{2}\{(V_1 \cdot V_2) \cdot (V_3 \cdot V_4) + (V_1 \cdot V_3)(V_2 \cdot V_4)\}$ to the above equation, and abbreviating third-order and higher-order terms as Q, we obtain Eq. 9.82.

A_0 in Eq. 9.85 is an electromagnetic wave. We assume the time-independent part of this to be A_0^0, and we then have Eq. 9.86.

$$A_0 = \frac{1}{2} A_0^0 \left\{ \exp\left(i\frac{E}{\hbar} t \right) + \exp\left(-i\frac{E}{\hbar} t \right) \right\} \tag{9.86}$$

By substituting Eq. 9.86 into Eq. 9.85, we obtain Eq. 9.87.

$$\langle u_m^* | H' | u_n \rangle = -\frac{1}{2} \left\{ \frac{iE_{mn}}{c\hbar} D_{mn} \cdot A_0^0 + M_{mn} (\nabla \times A^0)_0 \right\}$$

$$\times \left\{ \exp\left(i\frac{E}{\hbar} t \right) + \exp\left(-i\frac{E}{\hbar} t \right) \right\} \tag{9.87}$$

where $\langle u_m^* | D | u_n \rangle = D_{mn}$, $\langle u_m^* | M | u_n \rangle = M_{mn}$.

Using Eq. 9.87, we integrate Eq. 9.69, and $c_m(t)$ is given by Eq. 9.88.

$$c_m(t) = -\frac{1}{2i\hbar} \int_0^t \left\{ \frac{iE_{mn}}{c\hbar} \cdot D_{mn} A_0^0 + M_{mn} \cdot (\nabla \times A^0)_0 \right\} \left\{ \exp\left(\frac{i(E_{mn} + E)t}{\hbar} \right) \right.$$

$$\left. + \exp\left(\frac{i(E_{mn} - E)t}{\hbar} \right) \right\} dt = \frac{1}{2} \left\{ \frac{iE_{mn}}{c\hbar} \cdot D_{mn} \cdot A_0^0 + M_{mn} \cdot (\nabla \times A^0)_0 \right\}$$

$$\times \left(\frac{\exp\{i(E_{mn} + E)t/\hbar\} - 1}{E_{mn} + E} + \frac{\exp\{i(E_{mn} - E)t/\hbar\} - 1}{E_{mn} - E} \right) \tag{9.88}$$

Here we define $c_m(t = 0) = 0$ because the constant of integration is independent of time.

Substituting $c_m(t)$ in Eq. 9.88 into Eq. 9.68, we obtain the perturbed wave function, $\psi(q,t)$.

E. Derivation of electric and magnetic dipole moments

Since we have obtained the perturbed wave function, μ and $\mathbf{m}$ can be obtained from Eqs. 9.59 and 9.60. Substituting Eq. 9.68 into these equations, we obtain (see Appendix K).

$$\mu = \text{Re}(\langle \Psi^* | D | \Psi \rangle) = \text{Re} \left\{ D_{nn} + 2\sum c_m(t) D_{nm} \cdot \exp\left(-i\frac{E_{mn}}{\hbar} t \right) \right\} \tag{9.89}$$

$$\mathbf{m} = \text{Re}(\langle \Psi^* | M | \Psi \rangle) = \text{Re} \left\{ M_{nn} + 2\sum c_m(t) M_{nm} \cdot \exp\left(-i\frac{E_{mn}}{\hbar} t \right) \right\} \tag{9.90}$$

The first terms on the right-hand sides of these equations correspond to the permanent moments of the molecule in an unperturbed state, and the second terms of the moments produced by perturbation; these latter are connected with optical activity. Substituting $c_m(t)$ from Eq. 9.88 into the second terms of the above equations and expanding, we obtain Eq. 9.91 from Eq. 9.89.

$$
\begin{aligned}
\mu = \mathrm{Re}\Bigg\{ &\sum D_{nm}\cdot D_{mn}\cdot\left[\frac{iE_{mn}^2}{ch(E_{mn}^2 - E^2)}\,A_0^0\left[\exp\left(i\frac{E}{\hbar}t\right) + \exp\left(-i\frac{E}{\hbar}t\right)\right]\right.\\
&-\frac{iE_{mn}\cdot E}{ch(E_{mn}^2 - E^2)}\,A_0^0\left[\exp\left(i\frac{E}{\hbar}t\right) - \exp\left(-i\frac{E}{\hbar}t\right)\right]\Bigg]\\
&+\sum D_{nm}M_{mn}\left[\frac{E_{mn}}{E_{mn}^2 - E^2}(\nabla \times A^0)_0\left[\exp\left(i\frac{E}{\hbar}t\right) + \exp\left(-i\frac{E}{\hbar}t\right)\right]\right.\\
&-\frac{E}{E_{mn}^2 - E^2}(\nabla \times A^0)_0\left[\exp\left(i\frac{E}{\hbar}t\right) - \exp\left(-i\frac{E}{\hbar}t\right)\right]\Bigg]\Bigg\} \quad (9.91)
\end{aligned}
$$

Using $E_0 = -1/c\cdot\partial A_0/\partial t$, $H_0 = \nabla \times A_0$ this equation can be rewritten (see Appendix L) to give Eq. 9.92.

$$
\begin{aligned}
\mu = 2\mathrm{Re}\Bigg\{ &\sum \frac{E_{mn}}{E_{mn}^2 - E^2} D_{nm}\cdot D_{mn}\cdot E_0 + \sum \frac{i\hbar E_{mn}^2}{(E_{mn}^2 - E^2)E^2} D_{nm}\cdot D_{mn}\cdot\frac{dE_0}{dt}\\
&+ \sum \frac{E_{mn}}{E_{mn}^2 - E^2} D_{nm}\cdot M_{mn}\cdot H_0 + \sum \frac{i\hbar}{E_{mn}^2 - E^2} D_{nm}\cdot M_{mn}\cdot\frac{dH_0}{dt}\Bigg\} \quad (9.92)
\end{aligned}
$$

By a similar procedure, Eq. 9.93 can be obtained from Eq. 9.90.

$$
\begin{aligned}
m = 2\mathrm{Re}\Bigg\{ &\sum \frac{E_{mn}}{E_{mn}^2 - E^2} M_{nm}\cdot D_{mn}\cdot E_0 + \sum \frac{i\hbar E_{mn}^2}{(E_{mn}^2 - E^2)E^2} M_{nm}\cdot D_{mn}\cdot\frac{dE_0}{dt}\\
&+ \sum \frac{E_{mn}}{E_{mn}^2 - E^2} M_{nm}\cdot M_{mn}\cdot H_0 + \sum \frac{i\hbar}{E_{mn}^2 - E^2} M_{nm}\cdot M_{mn}\frac{dH_0}{dt}\Bigg\} \quad (9.93)
\end{aligned}
$$

The moments given by Eqs. 9.92 and 9.93 are vectors. However, the observable value is that measured from the direction of the incident electromagnetic waves. We may consider that the molecule can take any direction with respect to this direction, so the observed moment will be the mean of moments in all directions, and will thus be one-third of the values given in Eqs. 9.92 and 9.93. Meanwhile, we need to use the local electric and magnetic fields E', H' in a molecule in place of E_0, H_0.

On the basis of the above discussion, we can take the real parts of Eqs. 9.92 and 9.93; rearranging them and using the relations $E_{mn} = h\nu_{mn}$,

$E = hv$ we obtain the following equations (see Appendix M).

$$\boldsymbol{\mu} = \alpha_n \boldsymbol{E'} + \gamma_n \boldsymbol{H'} - \frac{\beta_n}{c} \cdot \frac{\partial \boldsymbol{H'}}{\partial t}$$

$$\mathbf{m} = \kappa_n \boldsymbol{H'} + \gamma_n \boldsymbol{E'} + \frac{\beta_n}{c} \cdot \frac{\partial \boldsymbol{E'}}{\partial t}$$

(9.94)

$$\alpha_n = \frac{2}{3h} \sum \frac{v_{mn} |D_{nm}|^2}{v_{mn}^2 - v^2}, \quad \kappa_n = \frac{2}{3h} \sum \frac{v_{mn} |M_{nm}|^2}{v_{mn}^2 - v^2}$$

$$\gamma_n = \frac{2}{3h} \sum \frac{v_{mn} \, \mathrm{Re}\{D_{nm} \cdot M_{mn}\}}{v_{mn}^2 - v^2}, \quad \beta_n = \frac{c}{3\pi h} \sum \frac{\mathrm{Im}\{D_{nm} \cdot M_{mn}\}}{v_{mn}^2 - v^2}$$

(9.95)

where $E_{mn}/h = v_{mn}$, $E/h = v$, $\mathrm{Im}(a + bi) = b$.[†]

Equations 9.94 and 9.95 describe a molecule in the state u_n before interaction with light, and this can therefore be regarded as a ground state. Assuming the probability of its existence to be ρ_n then $\alpha = \sum \rho_n \alpha_n$, $\beta = \sum \rho_n \beta_n$, $\kappa = \sum \rho_n \kappa_n$. If we suppose that γ_n is small and has only a secondary influence on the optical rotation, then the observed electric and magnetic dipole moments become

$$\boldsymbol{\mu} = \alpha \boldsymbol{E'} - \frac{\beta}{c} \cdot \frac{\partial \boldsymbol{H'}}{\partial t}$$

$$\mathbf{m} = \kappa \boldsymbol{H'} + \frac{\beta}{c} \cdot \frac{\partial \boldsymbol{E'}}{\partial t}$$

(9.96)

Eq. 9.96, which was derived from a quantum mechanical approach, corresponds quite well with the intuitive results based on classical theory. The angle of optical rotation ϕ can be determined from Eq. 9.96 as previously described (Eq. 9.56), and using the expression for β in Eq. 9.95, we obtain

$$\phi = \frac{16\pi^2}{3ch} N \left(\frac{n^2 + 2}{3} \right) \sum_n \rho_n \sum_m \frac{v^2 R_{mn}}{v_{mn}^2 - v^2}$$

$$R_{mn} = \mathrm{Im}(\langle u_n^* | D | u_m \rangle \langle u_m^* | M | u_n \rangle) = \mathrm{Im}(D_{nm} \cdot M_{mn})$$

(9.97)

The molecular rotation $[M]$ is obtained from ϕ using the relation in section 9.1.6, as

$$[M] = \frac{32\pi}{ch} M \cdot N(n^2 + 2) \sum \rho_n \sum \frac{v^2 R_{mn}}{v_{mn}^2 - v^2}$$

(9.98)

[†]   Im { } expresses the imaginal part of the complex number in { }.

Thus, we may say that the optical activity of a molecule is represented by R_{mn} and molecules for which $R_{mn} \neq 0$ will exhibit optical activity.

F.   Circular dichroism

Circular dichroism is closely related to optical activity, and occurs when light is absorbed resonantly at $v_{mn} \fallingdotseq v$. The absorption of light is proportional to the transition probability $c_m^*(t)c_m(t)$ when a molecule is transferred from the state n to state m by the absorption of light. By comparing $c_m^*(t)c_m(t)$ for left- and right-circularly polarized light, circular dichroism can be accounted for theoretically. Now $c_m(t)$ is given by Eq. 9.88, but around $E_m \simeq E(v_{mn} \fallingdotseq v)$ we find that [exp $\{i(E_{mn} - E/\hbar)t\} - 1$]/ $(E_{mn} - E) \gg$ [exp $\{i(E_{mn} + E/\hbar)t\} - 1$]/$(E_{mn} + E)$ and thus the latter term may be neglected, so we have

$$c_m(t) = \frac{1}{2}\left[\left\{ \frac{iE_{mn}}{c\hbar} \cdot D_{mn} \cdot A_0^0 + M_{mn}(\nabla \times A^0)_0 \right\} \left\{ \frac{\exp[i(E_{mn} - E/\hbar)t] - 1}{E_{mn} - E} \right\} \right]$$

$$(9.99)$$

On the other hand, right- and left-circularly polarized light A_r, A_l propagating in the z direction, can be described as follows.

$$(A_0)_r = \text{Re}\{A(i \pm ij)e^{i\varphi}\} \qquad \varphi = \frac{E}{\hbar}\left(t - \frac{z}{c}\right) \tag{9.100}$$

From this $(\nabla \times A)_{0r}$ becomes as follows.

$$(\nabla \times A)_{0r} = \frac{-E}{c\hbar} \text{Re}\{A(\pm i + ij)e^{i\varphi}\} \tag{9.101}$$

Substituting the potential vectors from Eqs. 9.100 and 9.101 into Eq. 9.99, we have

$$c_m(t)_r = \frac{1}{2}\left(\frac{iE_{mn}}{c\hbar} \cdot D_{mn} \mp \frac{E}{c\hbar} M_{mn} \right) \text{Re}\left\{ A(i \pm ij) \exp\left(-i\frac{E}{c\hbar} \cdot z \right) \right\}$$

$$\times \left\{ \frac{\exp[i(E_{mn} - E/\hbar)t] - 1}{E_{mn} - E} \right\} \tag{9.102}$$

Assuming that molecules are randomly distributed with respect to the z axis we can calculate $c_m^*(t)c_m(t)$ from the average, obtaining Eq. 9.103 (see Appendix N).

$$\mathrm{Re}\{c_m^*(t)\cdot c_m(t)\}_r$$
$$= \frac{E_{mn}^2 A^2}{3c^2\hbar^4}\left\{|D_{mn}|^2 \mp 2\frac{E}{E_{mn}}\,\mathrm{Im}(D_{nm}\cdot M_{mn})\right\}\frac{\sin^2\{(E_{mn}-E)\cdot t/2\hbar\}}{4\{(E_{mn}-E)/2\hbar\}^2} \quad (9.103)$$

Now we take the absorption coefficients for right- and left-circularly polarized light to be ε_r, ε_l and take the average. We have $\varepsilon_r \propto \{c_m^*(t)c_m(t)\}_r$, $\varepsilon_l \propto \{c_m^*(t)c_m(t)\}_l$ and $\varepsilon = (\varepsilon_r + \varepsilon_l)/2$. Thus, the anisotropy factor $g = (\varepsilon_l - \varepsilon_r)/\varepsilon$ becomes

$$g = \frac{\varepsilon_l - \varepsilon_r}{\varepsilon} = 4\frac{E}{E_{mn}}\cdot\frac{\mathrm{Im}(D_{nm}\cdot M_{mn})}{|D_{mn}|^2} = \frac{4v}{v_{mn}}\cdot\frac{R_{mn}}{S_{mn}} \quad (9.104)$$

where $S_{mn} = |D_{mn}|^2$, $R_{mn} = \mathrm{Im}(D_{nm}\cdot M_{mn})$.

Supposing that $v_{mn} \simeq v$, Eq. 9.104 can be approximated as follows.

$$\frac{\varepsilon_l - \varepsilon_r}{\varepsilon} = \frac{4R_{mn}}{S_{mn}} \quad (9.105)$$

Absorption of light occurs widely in real molecules, and the following relation exists between S_{mn} and the absorption spectrum.

$$S_{mn} = \frac{3\times 10^3 \ln 10\cdot ch}{8N_0\pi^3}\int_0^\infty \frac{\varepsilon(v)}{v}\,dv \quad (N_0 = \text{Avogadro's number}) \quad (9.106)$$

Taking $\varepsilon_l - \varepsilon_r = \Delta\varepsilon$, a corresponding relation can be obtained between R_{mn} and circular dichroism.

$$R_{mn} = \frac{3\times 10^3 \ln 10\cdot ch}{32N_0\pi^3}\int_0^\infty \frac{\Delta\varepsilon(v)}{v}\,dv \quad (9.107)$$

The ellipticity $[\theta] = (18\times 10^3 \times \ln 10/4\pi)\Delta\varepsilon$ (Eq. 9.31) can be introduced into this equation to give the following general equation for circular dichroism spectra.

$$R_{mn} = \frac{ch}{48\pi^2 N_0}\int_0^\infty \frac{[\theta(v)]}{v}\,dv \quad (9.108)$$

G. Conclusion

The states of electrons in a molecule interacting with electromagnetic waves can be calculated approximately by means of perturbation theory, and from these equations, values of molecular optical rotation $[M]$ and molecular ellipticity $[\theta]$ have been obtained. To summarize, these relations are as follows.

$$
\left.
\begin{aligned}
[M] &= \frac{32MN}{ch}(n^2 + 2)\sum\rho_n\sum\frac{v^2 R_{mn}}{v_{mn}^2 - v^2} \\
&= \frac{32\pi N_0}{ch}(n^2 + 2)\sum\rho_n\sum\frac{v^2 R_{mn}}{v_{mn}^2 - v^2} \\
R_{mn} &= \frac{ch}{48\pi^2 N_0}\int_0^\infty \frac{[\theta(v)]}{v}\,dv
\end{aligned}
\right\}
\qquad (9.109)
$$

We must now consider the physical value R_{mn} which determines $[\theta]$ and $[M]$. Previously we stated that these quantities are not independent, being related by the Kroning-Kramer relation, and in fact they are both determined by R_{mn}. In the case where R_{mn} is zero for a molecule, both $[\theta]$ and $[M]$ become zero, and no optical activity appears, while for $R_{mn} \neq 0$ both quantities take finite values, and optical activity is seen. R_{mn} is referred to as the optical rotatory power.

Theoretically, by calculating R_{mn} for each molecule we can obtain values of $[M]$ and $[\theta]$, but in fact this is difficult, except in some special cases where extensive approximations can validly be made. For detailed descriptions of such calculations, the reference books listed later should be consulted. In this chapter, we will simply seek ways to decide qualitatively whether R_{mn} is zero or not, i.e., whether the molecule is optically active or not.

9.3 MOLECULAR SYMMETRY AND OPTICAL ACTIVITY

9.3.1 Symmetry operations and the conservation of scalars

As described in the previous section, the condition for a molecule to show optical activity is that $R_{mn} \neq 0$. R_{mn} is expressed as a scalar product of $\mathscr{D}$ and $\mathscr{M}$ as shown in Eq. 9.110.

$$
R_{mn} = \mathrm{Im}(\langle u_m^*|D|u_n\rangle\langle u_n^*|M|u_m\rangle) = \mathrm{Im}(\mathscr{D}\cdot\mathscr{M}) \qquad (9.110)
$$

$$
\left.
\begin{aligned}
\langle u_m^*|D|u_n\rangle &= \mathscr{D} \\
\langle u_n^*|M|u_m\rangle &= \mathscr{M}
\end{aligned}
\right\}
\qquad (9.111)
$$

Since a scalar does not have direction, its value must remain unchanged under symmetry operations, i.e., Eq. 9.112 must hold under any symmetry operation, $\mathscr{A}$. In other words, $\mathscr{D} \cdot \mathscr{M}$ must always satisfy Eq. 9.112. This is called the law of conservation of scalars.

$$\mathscr{A}(\mathscr{D}) \cdot \mathscr{A}(\mathscr{M}) = \mathscr{D} \cdot \mathscr{M} \qquad (9.112)$$

Here, both $\mathscr{D}$ and $\mathscr{M}$ are made up of eigenfunctions and quantum mechanical operators, and $\mathscr{A}(\mathscr{D})$ and $\mathscr{A}(\mathscr{M})$ are given by Eqs. 9.113 and 9.114.

$$\mathscr{A}(\mathscr{D}) = \langle u_m^*(\mathscr{A}r) | e\mathscr{A}r | u_n(\mathscr{A}r) \rangle \qquad (9.113)$$

$$\mathscr{A}(\mathscr{M}) = \left\langle u_n^*(\mathscr{A}r) \left| -\frac{e\hbar}{2mc} \mathscr{A}r \times \mathscr{A}\nabla \right| u_m(\mathscr{A}r) \right\rangle \qquad (9.114)$$

These equations show that if $\mathscr{A}(\mathscr{D}) \cdot \mathscr{A}(\mathscr{M}) \neq \mathscr{D} \cdot \mathscr{M}$, the only value of $\mathscr{D} \cdot \mathscr{M}$ which satisfies Eq. 9.112 is zero. Thus, once the point group to which a molecule belongs is decided, we can determine whether $\mathscr{D} \cdot \mathscr{M}$ is zero or not for every symmetry operation of that point group. If, in all the symmetry operations of the point group to which a molecule belongs the value of $\mathscr{D} \cdot \mathscr{M}$ is constant, then $\mathscr{D} \cdot \mathscr{M} \neq 0$ and the molecule has optical activity. However, if even a single symmetry operation changes the value of $\mathscr{D} \cdot \mathscr{M}$ then $\mathscr{D} \cdot \mathscr{M} = 0$ and the molecule belonging to the point group does not have optical activity. Therefore, even without calculating the value of R_{mn} directly, we can qualitatively determine the relation between molecular structure and optical activity by carrying out symmetry operations on R_{mn} and applying the law of conservation of scalars.

As described in Chapter 2, section 2.2, there are two kinds of operations constituting point groups: rotation (C_n) and improper rotation (S_n), so by obtaining the transformation formulae for eigenfunctions and operators under these operations, it is possible to calculate Eqs. 9.113 and 9.114.

9.3.2 Transformation of eigenfunctions by symmetry operations

Consider the case that the eigenfunction u corresponding to the eigenvalue E is nondegenerate. Here we suppose that u is normalized and orthogonalized. The relation between u and E is given by the wave equation shown below.

$$H \cdot u = E \cdot u \qquad (9.115)$$

When Eq. 9.115 is operated on by a symmetry operation $\mathscr{A}$, the energy of the molecule does not change, so Eq. 9.116 holds.

$$\mathscr{A} \cdot H \cdot u = H(\mathscr{A} \cdot u) = E(\mathscr{A} \cdot u) \qquad (9.116)$$

On the other hand, since u is a normalized orthogonal, $\mathscr{A}u$ operated on by $\mathscr{A}$ is also a normalized orthogonal, and Eq. 9.117 holds good.

$$\int (\mathscr{A}u^*)(\mathscr{A}u)\,d\tau = 1 \tag{9.117}$$

u is transformed to either $+u$ or $-u$ by each symmetry operation. Now suppose that $\sigma = \pm 1$, then we have

$$\mathscr{A}u = \sigma u \tag{9.118}$$

In the case that an eigenfunction is nondegenerate, the symmetry operations of the molecular point group can be operated on this function, and it will take the value $+1$ or -1.

Next we will consider the case of an eigenfunction with degeneracy s. In this case not only each eigenfunction $\{u^{(1)}, u^{(2)}, \cdots, u^{(i)}, \cdots, u^{(s)}\}$ but also an arbitrary linear combination $u^{(i)'}$ of these functions is an eigenfunction.

$$u^{(i)'} = \sum_{j=1}^{s} c_{ij} u^{(j)} \tag{9.119}$$

When the s new eigenfunctions $\{u^{(1)'}, u^{(2)'}, \cdots, u^{(i)'}, \cdots, u^{(s)'}\}$ formed by linear combination satisfy the orthonormal condition Eq. 9.120, Eq. 9.119 becomes a transformation formula from the orthonormal system $\{u^{(i)}\}$ to another orthonormal system $\{u^{(i)'}\}$.

$$\int u^{(i)'} u^{(j)'}\,d\tau = \delta_{ij} \tag{9.120}$$

where $\delta_{ij} = 1$ $(i = j)$ $\delta_{ij} = 0$ $(i \neq j')$.

Substituting Eq. 9.119 into Eq. 9.120, the coefficient c_{ij} must satisfy the following relation.

$$\sum_{j=1}^{s} c_{ij} \cdot c_{ij'} = \delta_{jj'} \tag{9.121}$$

Now even if we carry out a symmetry operation $\mathscr{A}$ on the eigenfunction $\{u^{(i)}\}$ the eigenvalue is fixed, so the new eigenfunction $\{\mathscr{A}u^{(i)}\}$ becomes a linear combination of $\{u^{(i)}\}$.

$$\mathscr{A} \cdot u^{(i)} = \sum_{j=1}^{s} a_{ij} u^{(j)} \tag{9.122}$$

Comparing Eq. 9.122 with Eq. 9.119, Eq. 9.122 can be regarded as a transformation formula for $\{u^{(i)}\}$ to $\{\mathscr{A}u^{(i)}\}$. The coefficient a_{ij} must be chosen to satisfy the orthonormal condition (Eq. 9.120), so that $\sum\limits_{j=1}^{s} a_{ij}a_{ij'} = \delta_{jj'}$. Eq. 9.123 expresses Eq. 9.122 in matrix form.

$$\mathscr{A}\begin{pmatrix} u^{(1)} \\ u^{(2)} \\ \vdots \\ u^{(i)} \\ \vdots \\ u^{(s)} \end{pmatrix} = \begin{pmatrix} a_{11} \cdots\cdots\cdots\cdots\cdots a_{1s} \\ \vdots \quad\quad \vdots \quad\quad \vdots \\ a_{i1} \cdots\cdots a_{ij}\cdots\cdots \\ \vdots \quad\quad \vdots \quad\quad \vdots \\ a_{s1} \cdots\cdots\cdots\cdots\cdots a_{ss} \end{pmatrix}\begin{pmatrix} u^{(1)} \\ u^{(2)} \\ \vdots \\ u^{(i)} \\ \vdots \\ u^{(s)} \end{pmatrix} \qquad (9.123)$$

That is, when a symmetry operation is carried out on a normalized orthogonal eigenfunction with degree of degeneracy s, the result can be expressed as an $(s \times s)$ matrix $((a_{ij}))$. Once the point group to which a molecule belongs is decided, a pair of matrix expressions of this type exist for the symmetry operations belonging to it, forming expressions of a group with an unchanging eigenvalue. In the case of nondegeneracy, the value of the expression is ± 1, but this can be regarded as a (1×1) matrix. Such matrix expressions cannot be reduced any further, and are known as irreducible representations.

In summary, if the symmetry operations constituting the point group to which the molecule belongs are carried out, the transformation can be expressed by a matrix (± 1) of order one in the case that the degree of degeneracy of the eigenfunction is one, and by a matrix $((a_{ij}))$ of order s in the case that the degree of degeneracy is s.[†]

9.3.3 Transformation of operators by symmetry operations

The operators D and M are expressed as follows

$$D = e \sum r_j = e \sum_j (x_j\mathbf{i} + y_j\mathbf{j} + z_j\mathbf{k}) \qquad (9.124)$$

$$M = -\frac{e\hbar i}{2mc} \sum_j \left\{ \left(z_j\frac{\partial}{\partial y_j} - y_j\frac{\partial}{\partial z_j} \right)\mathbf{i} + \left(x_j\frac{\partial}{\partial z_j} - z_j\frac{\partial}{\partial x_j} \right)\mathbf{j} \right.$$

$$\left. + \left(y_j\frac{\partial}{\partial x_j} - x_j\frac{\partial}{\partial y_j} \right)\mathbf{k} \right\} \qquad (9.125)$$

When C_n operation is carried out about the z axis, Eq. 2.3 (Chapter 2)

[†] A description of matrix expressions corresponding to the symmetry operations of each point group in terms of a character is known as a character table, and is widely used to describe the properties of a point group.

holds between the coordinates (x, y, z) and (x', y', z'). Substituting this into Eqs. 9.124 and 9.125, the following relations are obtained between D and D'.

$$
\left.
\begin{aligned}
D'_x &= D_x \cos\frac{2\pi}{n} - D_y \sin\frac{2\pi}{n} \\[2mm]
D'_y &= D_x \sin\frac{2\pi}{d} + D_y \cos\frac{2\pi}{n} \\[2mm]
D'_z &= D_z
\end{aligned}
\right\}
\tag{9.126}
$$

and between M and M'.

$$
\left.
\begin{aligned}
M'_x &= M_x \cos\frac{2\pi}{n} - M_y \sin\frac{2\pi}{n} \\[2mm]
M'_y &= M_x \sin\frac{2\pi}{n} + M_y \cos\frac{2\pi}{n} \\[2mm]
M'_z &= M_z
\end{aligned}
\right\}
\tag{9.127}
$$

In the case of S_n operation about the z axis, Eq. 2.4 holds between (x, y, z) and (x', y', z'). Thus, we have

$$
\left.
\begin{aligned}
D'_x &= D_x \cos\frac{2\pi}{n} - D_y \sin\frac{2\pi}{n} \\[2mm]
D'_y &= D_x \sin\frac{2\pi}{n} + D_y \cos\frac{2\pi}{n} \\[2mm]
D'_z &= -D_z
\end{aligned}
\right\}
\tag{9.128}
$$

and

$$
\left.
\begin{aligned}
M'_x &= -M_x \cos\frac{2\pi}{n} + M_y \sin\frac{2\pi}{n} \\[2mm]
M'_y &= -M_z \sin\frac{2\pi}{n} - M_y \cos\frac{2\pi}{n} \\[2mm]
M'_z &= M_z
\end{aligned}
\right\}
\tag{9.129}
$$

9.3.4 Transformation of $\mathscr{D}$ and $\mathscr{M}$ by symmetry operations

Now that we have considered the transformations of eigenfunctions and operators under symmetry operations, we can obtain $\mathscr{A}(\mathscr{D}) \cdot \mathscr{A}(\mathscr{M})$.

First we will consider the case that the symmetry factors of the molecular point group consist only of rotation, $\mathscr{A} = C_n$, and that the eigenfunctions u_m and u_n are nondegenerate. Separating the x, y, z components of the scalar product $\mathscr{D} \cdot \mathscr{M}$ and calculating each of them, we have

$$\mathscr{D} \cdot \mathscr{M} = \mathscr{D}_x \mathscr{M}_x + \mathscr{D}_y \mathscr{M}_y + \mathscr{D}_z \mathscr{M}_z$$

$$C_n(\mathscr{D}) \cdot C_n(\mathscr{M}) = (\mathscr{D}' \cdot \mathscr{M}') = \mathscr{D}'_x \mathscr{M}'_x + \mathscr{D}'_y \mathscr{M}'_y + \mathscr{D}'_z \mathscr{M}'_z \quad (9.130)$$

$$\mathscr{D}'_x \mathscr{M}'_x = \langle C_n u_m^* | D'_x | C_n u_n \rangle \langle C_n u_n^* | M'_x | C_n u_m \rangle$$

$$= \left\langle \sigma u_m^* \left| D_x \cos \frac{2\pi}{n} - D_y \sin \frac{2\pi}{n} \right| \sigma u_n \right\rangle$$

$$\times \left\langle \sigma u_n^* \left| M_x \cos \frac{2\pi}{n} - M_y \sin \frac{2\pi}{n} \right| \sigma u_m \right\rangle$$

$$= \left\{ \langle u_m^* | D_x | u_n \rangle \cos \frac{2\pi}{n} - \langle u_m^* | D_y | u_n \rangle \sin \frac{2\pi}{n} \right\} \left\{ \langle u_n^* | M_x | u_m \rangle \cos \frac{2\pi}{n} \right.$$

$$\left. - \langle u_n^* | M_y | u_m \rangle \sin \frac{2\pi}{n} \right\} \quad \because \sigma^2 = 1$$

$$= \mathscr{D}_x \mathscr{M}_x \cos^2 \frac{2\pi}{n} + \mathscr{D}_y \mathscr{M}_y \sin^2 \frac{2\pi}{n} - (\mathscr{D}_x \mathscr{M}_y + \mathscr{D}_y \mathscr{M}_x) \sin \frac{2\pi}{n} \cos \frac{2\pi}{n}$$

$$(9.131)$$

Similarly,

$$\mathscr{D}'_y \mathscr{M}'_y = \mathscr{D}_x \mathscr{M}_x \sin^2 \frac{2\pi}{n} + \mathscr{D}_y \mathscr{M}_y \cos^2 \frac{2\pi}{n} + (\mathscr{D}_x \mathscr{M}_y + \mathscr{D}_y \mathscr{M}_x) \sin \frac{2\pi}{n} \cos \frac{2\pi}{n}$$

$$(9.132)$$

$$\mathscr{D}'_z \mathscr{M}'_z = \mathscr{D}_z \mathscr{M}_z \quad (9.133)$$

Summarizing Eqs. 9.131, 9.132 and 9.133, we obtain

$$C_n(\mathscr{D}) \cdot C_n(\mathscr{M}) = \mathscr{D}'_x \mathscr{M}'_x + \mathscr{D}'_y \mathscr{M}'_y + \mathscr{D}'_z \mathscr{M}'_z$$

$$= \mathscr{D}_x \mathscr{M}_x + \mathscr{D}_y \mathscr{M}_y + \mathscr{D}_z \mathscr{M}_z = \mathscr{D} \cdot \mathscr{M} \quad (9.134)$$

This equation holds good for any rotational operation and $\mathscr{D} \cdot \mathscr{M}$ takes a finite value, i.e., $\mathscr{D} \cdot \mathscr{M} \neq 0$. For C_n operation the law of conservation of scalars holds for any value of $\mathscr{D} \cdot \mathscr{M}$, so R_{mn} for a molecule belonging to a point group containing only C_n operations may take any value except zero, and a molecule belonging to such a point group will therefore show optical activity. Point groups of this type are C_n, D_n, (O, T), etc. (see Chapter 2, Table 2.3); in these point groups, mirror images of molecules cannot be superposed.

Next we will consider the case of improper rotations S_n when u_m is not degenerate, but u_n has degeneracy of order s. There will exist s eigenfunctions $u_n^{(1)}$, $u_n^{(2)} \cdots u_n^{(s)}$ having the same eigenvalue, and $\mathscr{D}$ and $\mathscr{M}$ will be given by the following equation.

$$\mathscr{D} \cdot \mathscr{M} = \sum_i^s \langle u_m^*|D|u_n^{(i)}\rangle \langle u_n^{(i)*}|M|u_m\rangle \qquad (9.135)$$

As before, the vectors are separated into x, y, z components.

$$\mathscr{D}_x \mathscr{M}_x = \sum_i^s \langle u_m^*|D_x|u_n^{(i)}\rangle \langle u_n^{(i)*}|M_x|u_m\rangle$$

$$\mathscr{D} \cdot \mathscr{M} = \mathscr{D}_x \mathscr{M}_x + \mathscr{D}_y \mathscr{M}_y + \mathscr{D}_z \mathscr{M}_z \qquad (9.136)$$

$$S_n(\mathscr{D}) \cdot S_n(\mathscr{M}) = \mathscr{D}' \cdot \mathscr{M}' = \mathscr{D}_x' \mathscr{M}_x' + \mathscr{D}_y' \mathscr{M}_y' + \mathscr{D}_z' \mathscr{M}_z' \qquad (9.137)$$

Substituting Eqs. 9.122, 9.128 and 9.129 into Eq. 9.135, we obtain

$$\mathscr{D}_x' \mathscr{M}_x' = \sum_i^s \left\{ \left\langle \sigma u_m^* \left| D_x \cos \frac{2\pi}{n} - D_y \sin \frac{2\pi}{n} \right| \sum_j^s a_{ij} u_n^{(j)} \right\rangle \right.$$

$$\times \left. \left\langle \sum_j^s a_{ij} u_n^{(j)*} \left| - M_x \cos \frac{2\pi}{n} + M_y \sin \frac{2\pi}{n} \right| \sigma u_m \right\rangle \right\}$$

$$= \sum_i^s \left\{ \sigma \sum_j^s a_{ij} \left(\langle u_m^*|D_x|u_n^{(j)}\rangle \cos \frac{2\pi}{n} - \langle u_m^*|D_y|u_n^{(j)}\rangle \sin \frac{2\pi}{n} \right) \right.$$

$$\times \left. \sigma \sum_j^s a_{ij} \left(\langle u_n^{(j)*}|M_x|u_m\rangle \left(- \cos \frac{2\pi}{n} \right) + \langle u_n^{(j)*}|M_y|u_m\rangle \sin \frac{2\pi}{n} \right) \right\}$$

$$(9.138)$$

Using $\sigma^2 = 1$, $\sum_j^s \sum_{j'}^s a_{ij} a_{ij'} = \delta_{jj'}$, Eq. 9.138 becomes

$$\mathscr{D}_x' \mathscr{M}_x' = - \mathscr{D}_x \mathscr{M}_x \cos^2 \frac{2\pi}{n} - D_y \mathscr{M}_y \sin^2 \frac{2\pi}{n}$$

$$+ (D_x \mathscr{M}_y + \mathscr{D}_y \mathscr{M}_x) \sin \frac{2\pi}{n} \cos \frac{2\pi}{n} \qquad (9.139)$$

Similarly,

$$\mathscr{D}'_y \mathscr{M}'_y = - \mathscr{D}_x \mathscr{M}_x \sin^2 \frac{2\pi}{n} - \mathscr{D}_y \mathscr{M}_y \cos^2 \frac{2\pi}{n}$$

$$- (\mathscr{D}_x \mathscr{M}_y + \mathscr{D}_y \mathscr{M}_x) \sin \frac{2\pi}{n} \cos \frac{2\pi}{n} \tag{9.140}$$

$$\mathscr{D}'_z \mathscr{M}'_z = - \mathscr{D}_z \mathscr{M}_z \tag{9.141}$$

$$\mathscr{D}' \cdot \mathscr{M}' = \mathscr{D}'_x \mathscr{M}'_x + \mathscr{D}'_y \mathscr{M}'_y + \mathscr{D}'_z \mathscr{M}'_z \tag{9.142}$$

Substituting Eqs. 9.139, 9.140 and 9.141 into Eq. 9.142, we obtain

$$S_n(\mathscr{D}) \cdot S_n(\mathscr{M}) = \mathscr{D}' \cdot \mathscr{M}' = - \mathscr{D} \cdot \mathscr{M} \tag{9.143}$$

If the condition $S_n(\mathscr{D}) S_n(\mathscr{M}) = \mathscr{D} \cdot \mathscr{M}$ is to hold good, then $\mathscr{D} \cdot \mathscr{M} = 0$. That is, R_{mn} must take a value of zero, based on the law of conservation of scalars.

Thus, we have shown that a molecule having an improper axis as a symmetry operation must have $R_{mn} = 0$, and thus cannot be optically active. Molecules having optical activity are limited to those having only proper axes as symmetry operations. This supports the empirical stereochemical approach of Chapter 2, section 2.2, providing a physical basis for the empirical rule.

REFERENCES

1. D. J. Caldwell and H. Eyring, *The Theory of Optical Activity*, Wiley, 1971.
2. F. Ciardelli and P. Salvadori, *Fundamental Aspects and Recent Developments in Optical Rotatory Dispersion and Circular Dichroism*, Heyden, 1973.
3. F. A. Tenkin and H. E. White, *Fundamental of Optics*, McGrow-Hill, 1957.

APPENDIX A

Maxwell's equations for a plane wave

Maxwell's equations in a uniform material become as follows

$$\nabla \cdot B = 0 \tag{A.1}$$

$$\nabla \cdot D = 0 \tag{A.2}$$

$$\nabla \times H = \frac{1}{c}\frac{\partial D}{\partial t} \tag{A.3}$$

$$\nabla \times E = -\frac{1}{c}\frac{\partial B}{\partial t} \tag{A.4}$$

In addition

$$D = (1 + 4\pi\alpha)E = \varepsilon E \tag{A.5}$$

$$B = (1 + 4\pi\kappa)H = \mu H \tag{A.6}$$

where ε is the dielectric constant and μ is the magnetic permeability (see 9.2.1B). Substituting Eq. A.5 in to Eq. A.3 and Eq. A.6 into Eq. A.4, we obtain

$$\nabla \times H = \frac{\varepsilon}{c}\frac{\partial E}{\partial t} \tag{A.7}$$

$$\nabla \times E = -\frac{\mu}{c}\frac{\partial H}{\partial t} \tag{A.8}$$

Differentiating Eq. A.7 with respect to time, t,

$$\frac{\partial}{\partial t}(\nabla \times H) = \frac{\varepsilon}{c}\frac{\partial^2 E}{\partial t^2} \tag{A.9}$$

Using Eq. A.8, the left-hand side of this equation can be rewritten as follows.

$$\frac{\partial}{\partial t}(\nabla \times H) = \nabla \times \frac{\partial H}{\partial t} = \nabla \times \left(-\frac{c}{\mu}\cdot\nabla \times E\right) \tag{A.10}$$

From standard vector formula,

$$\nabla \times (\nabla \times C) = \nabla(\nabla \cdot C) - (\nabla \cdot \nabla)C \qquad (A.11)$$

Rewriting Eq. A.9 with Eqs. A.2, A.10 and A.11, we obtain

$$\frac{\partial}{\partial t}(\nabla \times H) = \frac{c}{\mu}\nabla^2 E = \frac{\varepsilon}{c}\frac{\partial^2 E}{\partial t^2} \qquad (A.12)$$

Rearranging, we have

$$\frac{\partial^2 E}{\partial t^2} = \frac{c^2}{\varepsilon\mu} \cdot \nabla^2 E \qquad (A.13)$$

Similarly, from Eq. A.8, we obtain

$$\frac{\partial^2 H}{\partial t^2} = \frac{c^2}{\varepsilon\mu}\nabla^2 H \qquad (A.14)$$

and thus E and H satisfy the same vector equation, so that we need only consider the solution of one of them. Separating the vector equation for the electric field (Eq. A.13) into components and rewriting, we find

$$\frac{\partial^2 E_x}{\partial t^2} = \frac{c^2}{\varepsilon\mu}\nabla^2 E_x, \quad \frac{\partial^2 E_y}{\partial t^2} = \frac{c^2}{\varepsilon\mu}\nabla^2 E_y, \quad \frac{\partial^2 E_z}{\partial t^2} = \frac{c^2}{\varepsilon\mu}\nabla^2 E_z \qquad (A.15)$$

From the conditions for plane waves, E and H become functions of z and t, being independent of x and y. In this case, Eqs. A.1 and A.2 become

$$\nabla \cdot D = \varepsilon\nabla \cdot E = \varepsilon\frac{\partial E_z}{\partial z} = 0 \quad \nabla \cdot B = \mu\nabla \cdot H = \mu\frac{\partial H_z}{\partial z} = 0 \qquad (A.16)$$

E_z becomes an arbitrary constant which does not depend on z, and if we suppose it to be zero, the results do not lose generality. Thus,

$$E_z = H_z = 0 \qquad (A.17)$$

Next we apply the conditions for plane waves to Eqs. A.7 and A.8.

$$\left.\begin{aligned}-\frac{\partial H_y}{\partial z} &= \frac{\varepsilon}{c} \cdot \frac{dE_x}{dt} \\[2mm] \frac{\partial H_x}{\partial z} &= \frac{\varepsilon}{c} \cdot \frac{dE_y}{dt}\end{aligned}\right\} \qquad (A.18)$$

$$\frac{\partial E_y}{\partial z} = \frac{\mu}{c} \cdot \frac{dH_x}{dt} \Bigg\} $$

$$-\frac{\partial E_x}{\partial z} = \frac{\mu}{c} \cdot \frac{dH_y}{dt} \Bigg\}$$

(A.19)

Under these conditions, Eq. A.15 becomes as follows.

$$\frac{\partial^2 E_x}{\partial t^2} = \frac{c^2}{\varepsilon\mu} \cdot \frac{\partial^2 E_x}{\partial z^2}, \quad \frac{d^2 E_y}{dt} = \frac{c^2}{\varepsilon\mu} \cdot \frac{\partial^2 E_y}{\partial z^2}$$

(A.20)

Eq. A.20 is called a wave function, and its solution takes the form of a periodic function $f(t, z)$. Since this does not determine the form of the wave, E_x and E_y are generally expressed in the following forms.

$$E_x = f_1\left(t - \frac{z}{v}\right) + g_1\left(t + \frac{z}{v}\right)$$

(A.21)

$$E_y = f_2\left(t - \frac{z}{v}\right) + g_2\left(t + \frac{z}{v}\right)$$

(A.22)

Here, f and g are arbitrary functions: f expresses a wave propagating in the $+z$ direction, and g a wave propagating in the $-z$ direction. We now suppose that E_x is a propagative wave travelling in the $+z$ direction, so that

$$E_x = f_1\left(t - \frac{z}{v}\right)$$

(A.23)

Substituting this into Eqs. A.18 and A.19, we have

$$-\frac{\partial H_y}{\partial z} = \frac{\varepsilon}{c} \cdot f_1'\left(t - \frac{z}{v}\right)$$

(A.24)

Integrating Eq. A.24 with respect to z,

$$H_y = \frac{\varepsilon v}{c} \cdot f_1\left(t - \frac{z}{v}\right) = \frac{\varepsilon v}{c} E_x$$

(A.25)

Using the expression $v = c/\sqrt{\varepsilon\mu}$, we obtain

$$H_y = \sqrt{\frac{\varepsilon}{\mu}} f_1\left(t - \frac{z}{v}\right) = \sqrt{\frac{\varepsilon}{\mu}} E_x$$

(A.26)

Similarly, for E_y we have

$$H_x = -\sqrt{\frac{\varepsilon}{\mu}} f_2\left(t - \frac{z}{v}\right) = -\sqrt{\frac{\varepsilon}{\mu}} E_y \tag{A.27}$$

Rearranging Eqs. A.17, A.26 and A.27, Maxwell's equations for plane waves give us the following relations between the electric and magnetic fields.

$$\left. \begin{array}{c} \sqrt{\mu} H_y = \sqrt{\varepsilon} E_x \\ -\sqrt{\mu} H_x = \sqrt{\varepsilon} E_y \\ H_z = E_z = 0 \end{array} \right\} \tag{A.28}$$

The selection of the periodic function f is unrestricted, but a trigonometric function is used in many cases for the simplest waves. For a monochromatic wave, Eq. 9.5 in section 9.1.1 is applicable.

Appendix B

Relationship between incident and reflected light

When light meets the boundary of two media, I and II, the tangential components of E and H must be continuous, since no current exists at the boundary, and the fundamental equations of electromagnetism show that $\nabla \times E = 0$ and $\nabla \times H = 0$ for the electric and magnetic fields. Now in the case where incident light enters the boundary vertically according to Maxwell's equations,

$$\frac{E_x}{H_y} = \sqrt{\frac{\mu_1}{\varepsilon_1}} \tag{B.1}$$

and for transmitted light

$$\frac{E'_x}{H'_y} = \sqrt{\frac{\mu_2}{\varepsilon_2}} \tag{B.2}$$

The reflected wave propagates in the opposite direction to the incident wave, i.e., the phase is reversed, so that

$$\frac{E_x''}{H_y''} = -\sqrt{\frac{\mu_1}{\varepsilon_1}} \tag{B.3}$$

Applying the boundary condition to E_x, E_x' and E_x'' or E_x and its derivatives,

$$E_x + E_x'' = E_x' \quad \sqrt{\frac{\varepsilon_1}{\mu_1}} (E_x - E_x'') = \sqrt{\frac{\varepsilon_2}{\mu_2}} E_x' \tag{B.4}$$

From Eq. B.4, E_x' can be cancelled out.

$$\frac{E_x''}{E_x} = \frac{\sqrt{\varepsilon_1/\mu_1} - \sqrt{\varepsilon_2/\mu_2}}{\sqrt{\varepsilon_1/\mu_1} + \sqrt{\varepsilon_2/\mu_2}} \tag{B.5}$$

In the same way, the relation between H_y and H_y'' is obtained as

$$\frac{H_y''}{H_y} = \frac{\sqrt{\varepsilon_2/\mu_2} - \sqrt{\varepsilon_1/\mu_1}}{\sqrt{\varepsilon_2/\mu_2} + \sqrt{\varepsilon_1/\mu_1}} \tag{B.6}$$

Taking the refractive indices of media I and II to be n_1 and n_2, and assuming $n_1 < n_2$, then $\sqrt{\varepsilon_1\mu_1} < \sqrt{\varepsilon_2\mu_2}$ and $\varepsilon_2 > \varepsilon_1$, $\mu_1 = \mu_2 = 1$, since the media are nonmagnetic. Applying this relation to Eq. B.5 and Eq. B.6,

$$E_x''/E_x < 0, \qquad H_y''/H_y > 0 \tag{B.7}$$

The direction of the electric field vector is reversed between the incident and reflected waves, whereas the direction of the magnetic field vector is the same in both cases. This relation also holds good for E_y and $-H_x$. The relation between the vector directions of incident and reflected light given in Eq. B.7 holds even if the angle of incidence is changed. Fig. B.1 shows the incident and reflected light diagrammatically in terms of the vectors involved.

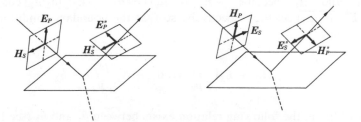

Fig. B.1. Electric and magnetic field vectors of incident and reflected light.

APPENDIX   C

Fresnel's equations

Fresnel's equations can be obtained from the boundary conditions of the tangential components of E and H in the case that the angle of incidence is not zero. The incident light is separated into components E_P, parallel to the incident plane, and E_S, vertical to the incident plane, and the relation between the reflected and transmitted light is as shown in Appendix B, Fig. B.1. Taking the paper as the incident plane, the relationship can be expressed as shown in Fig. C.1.

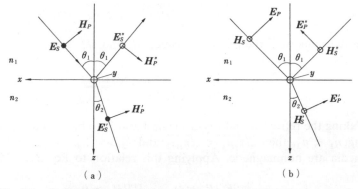

Fig. C.1.   Relationships between incident and reflected light taking the plane of the paper as the incident plane. ●, Vectors acting down into the plane of the paper; ○, vectors acting vertically upwards from the paper.

First we will consider E_S and its derivatives. Since these are vertical to the incident plane, as for the tangential component,

$$E_S - E_S'' = E_S' \tag{C.1}$$

On the other hand, from Fig. C.1(a), the tangential components of H_P, H_P' and H_P'' become $-E_S\sqrt{(\varepsilon_1/\mu_1)} \cos \theta_1$, $-E_S'\sqrt{(\varepsilon_2/\mu_2)} \cos \theta_2$ and $-E_S''\sqrt{(\varepsilon_1/\mu_1)} \cos \theta_1$, respectively, so that the boundary condition for H_P becomes

$$(E_S + E_S'')\sqrt{\frac{\varepsilon_1}{\mu_1}} \cos \theta_1 = E_S'\sqrt{\frac{\varepsilon_2}{\mu_2}} \cos \theta_2 \tag{C.2}$$

In addition, the following relation exists between θ_1 and θ_2 (see Eq. 9.10, section 9.1.1).

$$\frac{\sin \theta_1}{\sin \theta_2} = \frac{n_2}{n_1} = \sqrt{\frac{\varepsilon_2 \mu_2}{\varepsilon_1 \mu_1}} \tag{C.3}$$

Using Eq. C.3, Eq. C.2 can be rewritten as follows.

$$E_S + E_S'' = E_S' \frac{\mu_1 \tan \theta_1}{\mu_2 \tan \theta_2} \tag{C.4}$$

Multiplying Eq. C.1 by $(\mu_1/\mu_2)(\tan \theta_1/\tan \theta_2)$ and subtracting Eq. C.4, we obtain

$$E_S \left(\frac{\mu_1 \tan \theta_1}{\mu_2 \tan \theta_2} - 1 \right) = E_S'' \left(\frac{\mu_1 \tan \theta_1}{\mu_2 \tan \theta_2} + 1 \right) \tag{C.5}$$

Since media I and II are paramagnetic, we assume $\mu_1 = \mu_2 = 1$, so

$$\frac{E_S''}{E_S} = \frac{\tan \theta_1 - \tan \theta_2}{\tan \theta_1 + \tan \theta_2} = \frac{\sin (\theta_1 - \theta_2)}{\sin (\theta_1 + \theta_2)} \tag{C.6}$$

Hence we obtain Eq. 9.13, section 9.1.2.

On the other hand, from Fig. C.1(b) $H_S = \sqrt{(\varepsilon_1/\mu_1)} E_P$, $H_S' = \sqrt{(\varepsilon_2/\mu_2)} E_P'$, $H_S'' = \sqrt{(\varepsilon_1/\mu_1)} E_P''$ are vertical to the incident plane, so they each express tangential components. The tangential components of E_P, E_P' and E_P'' become $(E_P)_x = E_P \cos \theta_1$, $(E_P')_x = E_P' \cos \theta_2$ and $(E_P'')_x = -E_P'' \cos \theta_1$. The boundary conditions for E are given by

$$(E_P - E_P'') \cos \theta_1 = E_P' \cos \theta_2 \tag{C.7}$$

and for H by

$$(E_P + E_P'') \sqrt{\frac{\varepsilon_1}{\mu_1}} = E_P' \sqrt{\frac{\varepsilon_2}{\mu_2}} \tag{C.8}$$

Multiplying Eq. C.7 by $(\mu_1/\mu_2)(\sin \theta_1/\sin \theta_2)$ and Eq. C.8 by $(\cos \theta_2/\cos \theta_1)$ and subtracting, we have

$$E_P \left(\frac{\mu_1 \sin \theta_1}{\mu_2 \sin \theta_2} - \frac{\cos \theta_2}{\cos \theta_1} \right) = E_P'' \left(\frac{\mu_1 \sin \theta_1}{\mu_2 \sin \theta_2} + \frac{\sin \theta_2}{\cos \theta_1} \right) \tag{C.9}$$

We now suppose that $\mu_1 = \mu_2 = 1$ and Eq. C.9 becomes

$$\frac{E_P''}{E_P} = \frac{\sin \theta_1 \cos \theta_1 - \sin \theta_2 \cos \theta_2}{\sin \theta_1 \cos \theta_1 + \sin \theta_2 \cos \theta_2} = \frac{\tan (\theta_1 - \theta_2)}{\tan (\theta_1 + \theta_2)} \qquad \text{(C.10)}$$

Eq. C.10 corresponds to Eq. 9.12 in section 9.1.2, and Eqs. C.10 and C.6 are known as Fresnel's equations.

APPENDIX D

The spiral model: Electric moment induced by an alternating magnetic field.

If the magnetic field changes along a spiral axis, then the electric field produced will be as follows, based on Maxwell's equations.

$$\int_S (\nabla \times E) \, dS = \int_c E \, dc = -\frac{1}{c} \int \frac{\partial H}{\partial t} \, dS \qquad \text{(D.1)}$$

Taking the radius of the spiral as r, the electric field produced by a magnetic field passing the area covered by the spiral becomes

$$2\pi r E = -\frac{\pi r^2}{c} \frac{\partial H}{\partial t} \qquad \text{(D.2)}$$

The force produced by this electric field on an electron moving in the spiral is as follows (see Fig. D.1).

$$F = eE \cos \theta = \frac{-\pi r^2 e}{c\sqrt{4\pi^2 r^2 + S^2}} \frac{\partial H}{\partial t} \qquad \text{(D.3)}$$

where $$H = H_0 \cos \omega t$$

Next we obtain an equation of motion for an electron on the spiral, considering the dynamical stability and inertial force,

$$m\ddot{q} + m\omega_0^2 q - \frac{\pi r^2 e}{c\sqrt{4\pi^2 r^2 + S^2}} \omega H_0 \sin \omega t = 0 \qquad \text{(D.4)}$$

where m is the mass of an electron and $\omega_0 = 2\pi v_0$, $\omega = 2\pi v$.
Solving this equation, we have

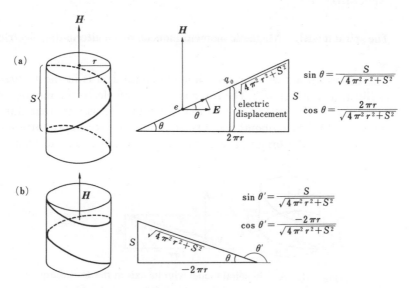

Fig. D.1. (a) Right spiral model with its axis in the *H* direction, and its development. (b) Left spiral model and its development.

$$q_0 = \frac{-\pi r^2 e}{m \cdot c \sqrt{4\pi^2 r^2 + S^2}(\omega_0^2 - \omega^2)} \frac{\partial H}{\partial t} \tag{D.5}$$

The electric dipole moment μ' is cancelled out except in the *H* direction, so the induced moment in the *H* direction becomes $eq_0 \sin \theta$, and from Eq. D.5 we have

$$\mu' = eq_0 \sin \theta = -\frac{\pi r^2 e^2 \cdot S}{mc(4\pi^2 r^2 + S^2)(\omega_0^2 - \omega^2)} \frac{\partial H}{\partial t} \tag{D.6}$$

Since the molecular direction is arbitrary, by taking the average $\beta = 1/3 \cdot eq_0 \sin \theta$ and summing over all the electrons, we obtain Eq. 9.42 in section 9.2.1. Substituting this into Eq. 9.41, we obtain the value of μ.

$$\mu = \alpha E - \sum_j \frac{\pi r^2 S e^2}{c \cdot 3m(4\pi^2 r^2 + S^2)(\omega_{0j}^2 - \omega^2)} \frac{\partial H}{\partial t} \tag{D.7}$$

Since $\omega = 2\pi \nu$, $\omega_0 = 2\pi \nu_0$, this can be substituted into Eq. D.7, yielding Eq. 9.42. In a left spiral, as shown in Fig. D.1(b), $\theta' = \pi - \theta$ and the value of $\cos \theta'$ becomes negative, while the sign of β is reversed.

APPENDIX E

The spiral model: Magnetic moment induced by an alternating electric field

Let us set a right spiral so that the spiral axis is in the E direction, as illustrated in Fig. E.1. The force acting on an electron e moving on the spiral due to the electric field is $e \cdot E \sin \theta$ and we assume that this force is balanced by the dynamic stability and inertial force in proportion to the distance from the origin. The equation of motion of an electron on the

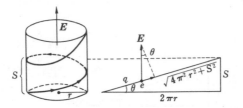

Fig. E.1.   A right spiral model with its axis in the E direction.

spiral is then given by Eq. E.1, and the electron transition velocity is given by Eq. E.3.

$$m\ddot{q} + m\omega_0 q + \frac{SeE}{\sqrt{4\pi^2 r^2 + S^2}} = 0 \tag{E.1}$$

Now supposing that $E = E_0 \sin \omega t$,

$$q = \frac{-eSE_0 \sin \omega t}{m\sqrt{4\pi^2 r^2 + S^2}(\omega_0^2 - \omega^2)} \tag{E.2}$$

so that

$$\dot{q} = \frac{-eS\omega E_0 \cos \omega t}{m\sqrt{4\pi^2 r^2 + S^2}(\omega_0^2 - \omega^2)} = \frac{-eS}{m\sqrt{4\pi^2 r^2 + S^2}(\omega_0^2 - \omega^2)} \frac{\partial E}{\partial t} \tag{E.3}$$

When a current i passes in a loop of radius r, the magnetic dipole moment $\mathbf{m}'$ is given by Eq. E.4 on the basis of electromagnetic theory (the Biot-Savort law). Here, the loop current produced when an electron moves on the ring with velocity v becomes $-ev/2\pi r$. Thus,

$$\mathbf{m}' = \frac{1}{c}\pi r^2 i = -\frac{erv}{2c} \tag{E.4}$$

When an electron moves on a spiral with velocity $\dot{q}$ this corresponds to a velocity $\dot{q}\cos\theta$ on a ring, and substituting this into Eq. E.4, we find

$$\mathbf{m}' = -\frac{er}{2c}\cdot\frac{-e\cdot S}{m\sqrt{4\pi^2 r^2 + S^2(\omega_0^2 - \omega^2)}}\cdot\frac{2\pi r}{\sqrt{4\pi^2 r^2 + S^2}}\cdot\frac{\partial E}{\partial t}$$

$$= \frac{\pi r^2 S e^2}{mc(4\pi^2 r^2 + S^2)(\omega_0^2 - \omega^2)}\frac{\partial E}{\partial t} \tag{E.5}$$

As was done for the electric dipole moment, the average of the magnetic dipole moments over every direction is taken, and thus $\frac{1}{3}\mathbf{m}'$ corresponds to the dipole moment induced by the electric field. Summing over all electrons, Eq. 9.43 in section 9.2.1 becomes as follows

$$\mathbf{m} = \kappa H' + \sum_j \frac{\pi r^2 S e^2}{3cm(4\pi^2 r^2 + S^2)(\omega_{0j}^2 - \omega^2)}\frac{\partial E}{\partial t} \tag{E.6}$$

It can be seen that $\gamma = \beta$.

For a left spiral, the sign of γ is reversed, as in the case of electric dipole moment. By rewriting the coefficients of the second term of Eq. E.6 using the relations $\omega = 2\pi\nu$, $\omega_{0j} = 2\pi\nu_{0j}$, Eq. 9.42 in the text is obtained.

APPENDIX F

Derivation of Eqs. 9.45 and 9.46

Eqs. 9.38 and 9.39 in section 9.2.1 describe the relation between P and I given by Eqs. 9.41, 9.43 and 9.44, and the external electric and magnetic fields E and H. Rearranging these,

$$\left.\begin{array}{l} \mu = \alpha E' - \dfrac{\beta}{c}\dfrac{\partial H'}{\partial t} \\[2ex] \mathbf{m} = \kappa H' + \dfrac{\beta}{c}\dfrac{\partial E'}{\partial t} \end{array}\right\} \tag{F.1}$$

$$\left.\begin{array}{l} P = N\mu \\[1ex] I = N\mathbf{m} \end{array}\right\} \tag{F.2}$$

$$\left.\begin{array}{l} D = E + 4\pi P \\ B = H + 4\pi I \end{array}\right\} \qquad \text{(F.3)}$$

$$\left.\begin{array}{l} E' = E + \tfrac{4}{3}\pi P \\ H' = H \end{array}\right\} \qquad \text{(F.4)}$$

Substituting Eq. F.2 into Eq. F.3 yields

$$\left.\begin{array}{l} D = E + 4\pi N\mu \\ B = H + 4\pi Nm \end{array}\right\} \qquad \text{(F.5)}$$

Substituting Eq. F.4 into Eq. F.1 gives

$$\mu = \alpha\left(E + \frac{4}{3}\pi P\right) - \frac{\beta}{c}\frac{\partial H}{\partial t} \qquad \text{(F.6)}$$

$$m = \kappa H + \frac{\beta}{c}\frac{\partial}{\partial t}\left(E + \frac{4}{3}\pi P\right) \qquad \text{(F.7)}$$

Next, Eq. F.6 can be rewritten, utilizing Eq. F.2, as follows

$$\mu = \alpha\left(E + \frac{4}{3}\pi N\mu\right) - \frac{\beta}{c}\frac{\partial H}{\partial t} \qquad \text{(F.8)}$$

Rearrangement gives

$$\mu = \left(\alpha E - \frac{\beta}{c}\frac{\partial H}{\partial t}\right)\bigg/\left(1 - \alpha\frac{4\pi}{3}N\right) \qquad \text{(F.9)}$$

Substituting Eq. F.9 into Eq. F.7, we can obtain **m** (omitting the β^2 term).

$$m = \kappa H + \frac{\beta}{c}\cdot\frac{\partial}{\partial t}\left(E + \frac{4}{3}\pi N\frac{\alpha E}{1 - (4\pi/3)\cdot\alpha N}\right)$$

$$= \kappa H + \frac{\beta}{c}\left(1 + \frac{4}{3}\pi N\alpha \times \frac{1}{1 - (4\pi/3)\alpha N}\right)\frac{\partial E}{\partial t} = \kappa H + \frac{\beta}{c}\left(\frac{3}{3 - 4\pi\alpha N}\right)\frac{\partial E}{\partial t} \qquad \text{(F.10)}$$

Substituting Eqs. F.9 and F.10 into Eq. F.5, we have

$$D = E + 4\pi N \times \left(\frac{3}{3 - 4\pi\alpha N}\right)\left(\alpha E - \frac{\beta}{c}\frac{\partial H}{\partial t}\right)$$

$$= \left(\frac{3 + 8\pi\alpha N}{3 - 4\pi\alpha N}\right)E - \left(\frac{12\pi N}{3 - 4\pi\alpha N} \times \frac{\beta}{c}\right)\frac{\partial H}{\partial t} \tag{F.11}$$

$$B = H + 4\pi N\left(\kappa H + \frac{3}{3 - 4\pi\alpha N} \times \frac{\beta}{c}\right)\frac{\partial E}{\partial t}$$

$$= (1 + 4\pi N\kappa)H + \left(\frac{12\pi N}{3 - 4\pi\alpha N} \times \frac{\beta}{c}\right)\frac{\partial E}{\partial t} \tag{F.12}$$

By noting that $\varepsilon' = (3 + 8\pi\alpha N)/(3 - 4\pi\alpha N)$, $\mu = 1 + 4\pi N\kappa$, $\xi = (12\pi N)/(3 - 4\pi\alpha N) \times (\beta/c) = \{4\pi N(\varepsilon' + 2)\}/3 \times (\beta/c)$, we can obtain Eqs. 9.45 and 9.46 of section 9.2.1.

APPENDIX G

Derivation of Eq. 9.53

Substituting Eq. 9.51 into Eq. 9.50 (section 9.2.1), we have

$$\nabla \times E_r = \text{Re}\begin{vmatrix} i & j & k \\ \partial/\partial x & \partial/\partial y & \partial/\partial z \\ Ee^{i\varphi} & iEe^{i\varphi} & 0 \end{vmatrix} = \text{Re}\left\{\left(-iE\frac{\partial e^{i\varphi}}{\partial z}\right)i + \left(E\frac{\partial e^{i\varphi}}{\partial z}\right)j\right\} \tag{G.1}$$

Now $\varphi = 2\pi v(t - nz/c)$, so that

$$\nabla \times E_r = \text{Re}\left\{\frac{2\pi vn}{c}(-i - ij)Ee^{i\varphi}\right\} = -\frac{1}{c}\frac{\partial B}{\partial t} \tag{G.2}$$

Integrating with respect to time, t,

$$B = n\,\text{Re}\{(-ii + j)Ee^{i\varphi}\} \tag{G.3}$$

On the other hand, we can substitute Eq. 9.51 into Eq. 9.46, and supposing that $\mu = 1$ (a good approximation in a nonmagnetic substance), we have

$$B = H + \xi\frac{\partial}{\partial t} \cdot \text{Re}\{(i + ij)Ee^{i\varphi}\} = H + 2\pi v\xi\,\text{Re}\{(ii - j)Ee^{i\varphi}\} \tag{G.4}$$

Substituting Eq. G.3 into Eq. G.4,

$$H = (n + 2\pi v\xi)\,\mathrm{Re}\{(-ii + j)Ee^{i\varphi}\} \tag{G.5}$$

and we may replace H with E. Moreover, by substituting Eq. G.5 into Eq. 9.50, we obtain

$$\nabla \times H = (n + 2\pi v\xi)\,\mathrm{Re}\begin{vmatrix} i & j & k \\ \partial/\partial x & \partial/\partial y & \partial/\partial z \\ -iEe^{i\varphi} & Ee^{i\varphi} & 0 \end{vmatrix}$$

$$= (n + 2\pi v\xi)\frac{2\pi vn}{c}\,\mathrm{Re}\{(ii - j)Ee^{i\varphi}\} = \frac{1}{c}\frac{\partial D}{\partial t} \tag{G.6}$$

Integrating with respect to time, t, we have

$$D = n(n + 2\pi v\xi)\,\mathrm{Re}\{(i + ij)Ee^{i\varphi}\} \tag{G.7}$$

Substituting Eq. 9.51 into this expression, we obtain

$$D = n(n + 2\pi v\xi)E \tag{G.8}$$

Next we substitute Eq. G.5 into Eq. 9.45, so that

$$D = \varepsilon'E - \xi\frac{\partial}{\partial t}[(n + 2\pi v\xi)\,\mathrm{Re}\{(-ii + j)Ee^{i\varphi}\}]$$

$$= \varepsilon'E - 2\pi v\xi(n + 2\pi v\xi)\,\mathrm{Re}\{(i + ij)Ee^{i\varphi}\} \tag{G.9}$$

Substituting Eq. G.7 into this,

$$D = \varepsilon'E - \frac{2\pi v\xi}{n}D \tag{G.10}$$

$$D = \frac{n\varepsilon'}{n + 2\pi v\xi}E \tag{G.11}$$

Eqs. G.8 and G.11 must be equal, so we have

$$\left.\begin{aligned} n(n + 2\pi v\xi) &= \frac{n\varepsilon'}{n + 2\pi v\xi} \\ n_r &= \varepsilon'^{1/2} - 2\pi v\xi \end{aligned}\right\} \tag{G.12}$$

In the same way as for left-handed circular polarization, we obtain

$$n_l = \varepsilon'^{1/2} + 2\pi v\xi \tag{G.13}$$

Eqs. G.12 and G.13 correspond to Eq. 9.53 in the text.

APPENDIX H

Perturbed wave equation

If we assume that H^0 is the stationary Hamiltonian and H' is the time-dependent perturbed Hamiltonian $(H^0 + H' = H)$, then the perturbed wave function becomes as follows.

$$i\hbar \frac{\partial \Psi(t)}{\partial t} = (H^0 + H')\Psi(t) \tag{H.1}$$

If we suppose that the solution for the stationary wave equation is known, then

$$i\hbar \frac{\partial \Psi^0}{\partial t} = H^0\Psi^0 = E\Psi^0 \tag{H.2}$$

The eigenfunction Ψ_n^0, giving the eigenvalue E_n, is given by Eq. H.3.

$$\Psi_n^0 = u_n \exp\left(-i\frac{E_n}{\hbar}t\right) \tag{H.3}$$

where u_n indicates a time-independent wave function. Here we will approximate the time-dependent wave function as a sum of eigenfunctions of H^0, so that

$$\Psi(t) = \sum_n c_n(t)u_n \exp\left(-i\frac{E_n}{\hbar}t\right) \tag{H.4}$$

where $c_n(t)$ is a time-dependent coefficient. Substituting Eq. H.4 into Eq. H.1,

$$i\hbar \frac{\partial \Psi(t)}{\partial t} = i\hbar \sum \left\{ \frac{dc_n(t)}{dt} u_n \exp\left(-i\frac{E_n}{\hbar}t \right) - \frac{iE_n}{\hbar} c_n(t)u_n \exp\left(-i\frac{E_n}{\hbar}t \right) \right\}$$

(H.5)

$$(H^0 + H')\Psi(t) = \sum \left\{ H^0 u_n c_n(t) \exp\left(-i\frac{E_n}{\hbar}t \right) + H' u_n c_n(t) \exp\left(-i\frac{E_n}{\hbar}t \right) \right\}$$

(H.6)

and rearranging, with $H^0 u_n = E_n u_n$

$$i\hbar \sum \frac{dc_n(t)}{dt} u_n \exp\left(-i\frac{E_n}{\hbar}t \right) = \sum c_n(t) H' u_n \exp\left(-i\frac{E_n}{\hbar}t \right) \quad \text{(H.7)}$$

Multiplying both sides of Eq. H.7 by $\Psi_m^{0*} = u_m^* \exp\{i(E_m/\hbar)t\}$ and integrating both sides, we obtain

$$i\hbar \sum \frac{dc_n(t)}{dt} \int u_m^* \cdot u_n \exp\left(i\frac{E_m - E_n}{\hbar}t \right) d\tau$$

$$= \sum c_n(t) \int u_m^* H' u_n \exp\left(i\frac{E_m - E_n}{\hbar}t \right) d\tau \quad \text{(H.8)}$$

Here $\int u_m^* u_n \exp\{i(E_m - E_n/\hbar)t\}d\tau$ is equal to 1 if $n = m$, and to 0 if $n \neq m$, so Eq. H.8 becomes

$$\frac{dc_m(t)}{dt} = \frac{1}{i\hbar} \sum c_n(t) \cdot \langle u_m^* | H' | u_n \rangle \exp\left(i\frac{E_{mn}}{\hbar}t \right) \quad \text{(H.9)}$$

Next we consider that $c_n(t)$ can be expanded as a power series in λ by estimating H' as a small perturbed term. Using an arbitrary parameter λ (later we will put $\lambda = 1$), we have

$$c_n = c_n^{(0)} + \lambda c_n^{(1)} + \lambda^2 c_n^{(2)} + \ldots \quad \text{(H.10)}$$

$$H' = \lambda H' \quad \text{(H.11)}$$

Substituting into Eq. H.9 and setting the coefficients of each power of λ as equal, we have

$$\frac{dc_m^0}{dt} = 0 \quad \text{(H.12)}$$

$$\frac{dc_m^{(1)}}{dt} = \frac{1}{i\hbar} \sum_n \langle u_m^*|H'|u_n\rangle c_n^{(0)} \exp\left(i\frac{E_{mn}}{\hbar}t\right) \tag{H.13}$$

$$\frac{dc_m^{(2)}}{dt} = \frac{1}{i\hbar} \sum_n \langle u_m^*|H'|u_n\rangle c_n^{(1)} \exp\left(i\frac{E_{mn}}{\hbar}t\right) \tag{H.14}$$

Assume that a perturbation H' is added from time 0 to time t, then $\Psi(0) = u_n$, so that

$$c_n^{(0)} = 1, \quad c_m^{(0)} = 0 \; (m \neq n) \tag{H.15}$$

If we calculate c_n approximately by taking the first term (i.e., using only the first-order perturbed term), then by substituting Eq. H.10 into Eq. H.4 and applying the condition shown in Eq. H.15, the perturbed wave function takes the form shown in Eq. H.16 (provided we put $\lambda = 1$ and neglect $c_n^{(1)}$, on the assumption that $c_n^{(1)}$ is too small compared with $c_n^{(0)}$ so that $c_n^0 + c_n^{(1)} \doteq 1$.

$$\Psi(t) = u_n \exp\left(-i\frac{E_n}{\hbar}t\right) + \sum_{n \neq m} c_m^{(1)}(t)u_m \exp\left(-i\frac{E_m}{\hbar}t\right) \tag{H.16}$$

$c_m^{(1)}(t)$ in Eq. H.16 is obtained by the integration of Eq. H.13 with respect to time, t, as follows.

$$c_m^{(1)}(t) = \frac{1}{i\hbar} \int \langle u_m^*|H'|u_n\rangle \exp\left(i\frac{E_{mn}}{\hbar}t\right)dt \tag{H.17}$$

APPENDIX I

Vector treatment of electromagnetic waves

If a vector a satisfies the relation $\nabla \cdot a = 0$ then the vector a can be expressed in terms of a vector A according to the equation $a = \nabla \times A$. If another vector b satisfies the relation $\nabla \times b = 0$, the vector b can be expressed as $b = -\nabla \cdot \varphi$ using a scalar φ.

In the electromagnetic wave equation (Eq. 9.50) for a vacuum, $B = H$, $D = E$ hold. Thus we have

$$\nabla \cdot H = 0 \tag{I.1}$$

$$\nabla \cdot E = 0 \tag{I.2}$$

$$\nabla \times E = -\frac{1}{c}\frac{\partial H}{\partial t} \tag{I.3}$$

$$\nabla \times H = \frac{1}{c}\frac{\partial E}{\partial t} \tag{I.4}$$

Taking H as a, H can be expressed as

$$H = \nabla \times A \tag{I.5}$$

Substituting Eq. I.3 into Eq. I.5, we have

$$\nabla \times \left(E + \frac{1}{c}\frac{\partial A}{\partial t}\right) = 0 \tag{I.6}$$

If we assume that $b = \{E + (1/c)(\partial A/\partial t)\}$ then we obtain

$$E + \frac{1}{c}\frac{\partial A}{\partial t} = -\nabla\varphi \tag{I.7}$$

Eqs. I.7 and I.5 correspond to Eq. 9.71 in the text. Thus, the electromagnetic wave equation has been rewritten in terms of A and φ. However the values of A and φ are not unequivocally determined without an additional condition (if we suppose $A' = A + \nabla \cdot \lambda$, $\varphi' = \varphi - (1/c)(\partial\lambda/\partial t)$, A' and φ' are arbitrarily determined by changes in λ, and each pair of A' and φ' also satisfy Eq. I.5 and Eq. I.7). Thus, we can take φ and A, satisfying the following relation, as an addition condition (the so-called Lorentz condition).

$$\frac{1}{c}\frac{\partial\varphi}{\partial t} + \nabla \cdot A = 0 \tag{I.8}$$

Substitution of Eq. I.4 into Eqs. I.5 and I.7, followed by rearrangement with Eq. I.8 using a vector formula (see Appendix A), yields

$$\left(\frac{1}{c^2}\frac{\partial^2}{\partial t^2} - \nabla^2\right)\Phi = 0 \tag{I.9}$$

Here Φ can be arbitrarily chosen from A, φ, H, E.

Eq. I.9 corresponds to the equation derived in Appendix A, if we set

$\varepsilon = \mu = 1$. The conditions expressed by Eq. I.9 (the wave function of light satisfies these conditions), taken with Eq. I.8, mean that if we chose $\varphi = 0$, then the corresponding vector A becomes $\nabla \cdot A = 0$.

APPENDIX J

Derivation of Eq. 9.85

We can rewrite $\langle u_m^* | H | u_n \rangle$ shown in Eq. 9.84 as follows.
The first term of this equation can be rewritten using the exchange law of operators, $H \sum r_j - \sum r_j H = -(\hbar^2/m) \nabla_j$.

$$\frac{-\hbar^2}{m} \langle u_m^* | e \sum \nabla_j | u_n \rangle = \langle u_m^* | H e \sum r_j | u_n \rangle - \langle u_m^* | e \sum r_j H | u_n \rangle \quad (J.1)$$

Using the relations $u_m^* H = H u_m^* = E_m u_m^*$, $H u_n = E_n u_n$,

$$\frac{-\hbar^2}{m} \langle u_m^* | e \sum \nabla_j | u_n \rangle = E_m \langle u_m^* | e \sum r_j | u_n \rangle - E_n \langle u_m^* | e \sum r_j | u_n \rangle$$

$$= E_{mn} \langle u_m^* | e \sum r_j | u_n \rangle \quad (J.2)$$

Substituting Eq. J.2 into Eq. 9.84 and using $e \sum r_j = D$, $-(i\hbar e/2mc) \sum r_j \times \nabla_j = M$ we can rewrite $\langle u_m^* | H' | u_n \rangle$ as follows

$$\langle u_m^* | H' | u_n \rangle = - \left\{ \frac{i E_{mn}}{c\hbar} \langle u_m^* | D | u_n \rangle A_0 + \langle u_m^* | M | u_n \rangle (\nabla \times A)_0 \right\} \quad (J.3)$$

This corresponds to Eq. 9.85 in section 9.2.2.

APPENDIX K

Derivation of Eqs. 9.89 and 9.90

From Eq. 9.68 in section 9.2.2 we have

$$\Psi(t) = u_n \exp\left(-i\frac{E_n}{\hbar}t\right) + \sum c_m(t)u_m \exp\left(-i\frac{E_m}{\hbar}t\right) \qquad \text{(K.1)}$$

$$\Psi^*(t) = u_n^* \exp\left(i\frac{E_n}{\hbar}t\right) + \sum c_m^*(t)u_m^* \exp\left(i\frac{E_m}{\hbar}t\right) \qquad \text{(K.2)}$$

Substituting these into Eqs. 9.59 and 9.60, we have

$$\text{Re}\left\{\langle\Psi^*(t)|D|\Psi(t)\rangle\right\} = \text{Re}\left\{\langle u_m^*|D|u_n\rangle + \sum c_m^*(t)\langle u_m^*|D|u_n\rangle \exp\left(i\frac{E_{mn}}{\hbar}t\right)\right.$$

$$\left. + \sum c_m(t)\langle u_n^*|D|u_m\rangle \exp\left(-i\frac{E_{mn}}{\hbar}t\right) + \cdots\right\} \qquad \text{(K.3)}$$

where $c_m(t)^2$ is omitted, since c_m is small. Now the second term of Eq. K.3 becomes as follows (since D is a Hermitic operator, $D_{mn} = D_{nm}^*$).

$$\text{Re}\left\{\sum c_m^*(t)\langle u_m^*|D|u_n\rangle \exp\left(i\frac{E_{mn}}{\hbar}t\right)\right\}$$

$$= \text{Re}\left\{\left[\sum c_m^*(t)\langle u_m^*|D|u_n\rangle \exp\left(i\frac{E_{mn}}{\hbar}t\right)\right]^*\right\}$$

$$= \text{Re}\left\{\sum c_m(t)\langle u_n^*|D|u_m\rangle \exp\left(-i\frac{E_{mn}}{\hbar}t\right)\right\} \qquad \text{(K.4)}$$

Substituting Eq. K.4 into Eq. K.3,

$$\text{Re}\{\langle\Psi^*|D|\Psi\rangle\} = \text{Re}\left\{\langle u_n^*|D|u_n\rangle + 2\sum c_m(t)\langle u_n^*|D|u_m\rangle \exp\left(-i\frac{E_{mn}}{\hbar}t\right)\right\} \qquad \text{(K.5)}$$

and similarly

$$\mathrm{Re}\{\langle\Psi^*|M|\Psi\rangle\} = \mathrm{Re}\left\{\langle u_n^*|M|u_n\rangle + 2\sum c_m(t)\langle u_n^*|M|u_m\rangle \exp\left(-i\frac{E_{mn}}{\hbar}t\right)\right\}$$

$$\text{(K.6)}$$

APPENDIX   L

Derivation of Eq. 9.92

Since the following relations are given,

$$A_0 = \frac{1}{2}A_0^0\left[\exp\left(i\frac{E}{\hbar}t\right) + \exp\left(-i\frac{E}{\hbar}t\right)\right] \tag{L.1}$$

$$E_0 = -\frac{1}{c}\frac{\partial A_0}{\partial t} \tag{L.2}$$

we have

$$E_0 = \frac{-iE}{2c\hbar}A_0^0\left[\exp\left(i\frac{E}{\hbar}t\right) - \exp\left(-i\frac{E}{\hbar}t\right)\right] \tag{L.3}$$

$$\frac{dE_0}{dt} = \frac{E^2}{2c\hbar^2}A_0^0\left[\exp\left(i\frac{E}{\hbar}t\right) + \exp\left(-i\frac{E}{\hbar}t\right)\right] \tag{L.4}$$

On the other hand, $H_0 = (\nabla \times A)_0$. This operation is carried out only on potential vectors, so it has no connection with the time-variable terms. that is

$$H_0 = (\nabla \times A)_0 = (\nabla \times A^0)_0\frac{1}{2}\left\{\exp\left(i\frac{E}{\hbar}t\right) + \exp\left(-i\frac{E}{\hbar}t\right)\right\} \tag{L.5}$$

$$\frac{dH_0}{dt} = \frac{\partial}{\partial t}(\nabla \times A)_0 = \frac{iE}{2\hbar}(\nabla \times A^0)_0\left[\exp\left(i\frac{E}{\hbar}t\right) - \exp\left(-i\frac{E}{\hbar}t\right)\right] \tag{L.6}$$

The first term of Eq. 9.91 in section 9.2.2 corresponds to Eq. L.4, the second term to Eq. L.3, the third term to Eq. L.5 and the fourth term to Eq. L.6, so by arranging the coefficients of each term and using E_0 and H_0 we can rewrite Eq. 9.91 to obtain Eq. 9.92.

APPENDIX   M

Derivation of Eq. 9.94

Since the coefficients of the first term in Eq. 9.92 are all real, we have

$$\text{Re}\left\{\sum\frac{E_{mn}}{E_{mn}^2 - E^2}D_{nm}D_{mn}\right\}E_0 = \sum\frac{v_{mn}}{h(v_{mn}^2 - v^2)}|D_{nm}|^2 E_0 \qquad (M.1)$$

The coefficient of the second term in { } becomes purely imaginary, so that

$$\text{Re}\left\{\sum\frac{i\hbar E_{mn}^2}{E_{mn}^2 - E^2)E^2}D_{nm}D_{mn}\right\}\frac{\partial E_0}{\partial t} = 0 \qquad (M.2)$$

The coefficient of the third term in { } is a complex number, so

$$\text{Re}\left\{\sum\frac{E_{mn}}{E_{mn}^2 - E^2}D_{nm}M_{mn}\right\}H_0 = \text{Re}\left\{\sum\frac{v_{mn}}{h(v_{mn}^2 - v^2)}D_{nm}\cdot M_{mn}\right\}H_0 \quad (M.3)$$

Noting that $\text{Re}\{i(a + ib)\} = -\text{Im}\{(a + ib)\} = -b$ and since the coefficient of the fourth term in { } is also a complex number, we have

$$\text{Re}\left\{\sum\frac{i\hbar}{E_{mn}^2 - E^2}\cdot D_{nm}M_{mn}\right\}\frac{\partial H_0}{\partial t} = -\text{Im}\left\{\sum\frac{1}{2\pi h(v_{mn}^2 - v^2)}\cdot D_{nm}\cdot M_{mn}\right\}\frac{\partial H_0}{\partial t}$$

$$(M.4)$$

From these equations, taking the average over all molecular directions and using the local electromagnetic fields, we find

$$\mu = \frac{2}{3h}\sum\frac{v_{mn}}{v_{mn}^2 - v^2}\cdot|D_{nm}|^2\cdot E' + \frac{2}{3h}\text{Re}\left\{\sum\frac{v_{mn}}{v_{mn}^2 - v^2}\cdot D_{nm}\cdot M_{mn}\right\}H'$$

$$-\frac{1}{3\pi h}\text{Im}\left\{\sum\frac{1}{v_{mn}^2 - v^2}\cdot D_{nm}\cdot M_{mn}\right\}\frac{\partial H'}{\partial t} \qquad (M.5)$$

Now supposing that

$$\alpha_n = \frac{2}{3h}\sum\frac{v_{mn}|D_{mn}|^2}{v_{mn}^2 - v^2}, \qquad \gamma_n = \frac{2}{3h}\sum\frac{v_{mn}\cdot\text{Re}\{D_{nm}\cdot M_{mn}\}}{v_{mn}^2 - v^2},$$

$$\beta_n = \frac{c}{3\pi h}\sum\frac{\text{Im}\{D_{nm}\cdot M_{mn}\}}{v_{mn}^2 - v^2} \qquad (M.6)$$

μ can be expressed by Eq. M.7, which is equivalent to Eq. 9.94 in section 9.2.2.

$$\mu = \alpha_n E' + \gamma_n H' - \frac{\beta_n}{c}\frac{\partial H'}{\partial t} \tag{M.7}$$

The first term of Eq. 9.93 becomes a complex number, and may be expressed as follows (see Appendix K).

$$\text{Re}\left\{\sum \frac{E_{mn}}{E_{mn}^2 - E^2} \cdot M_{nm} \cdot D_{mn}\right\} E_0 = \frac{1}{h}\text{Re}\left\{\sum \frac{v_{mn}}{v_{mn}^2 - v^2} \cdot M_{nm} \cdot D_{mn}\right\} E_0$$

$$= \frac{1}{h}\text{Re}\left\{\sum \frac{v_{mn}^2}{v_{mn}^2 - v^2} \cdot D_{nm}^* \cdot M_{nm}^*\right\} E_0 = \frac{1}{h}\text{Re}\left\{\sum \frac{v_{mn}}{v_{mn}^2 - v^2} \cdot D_{nm} \cdot M_{mn}\right\} E_0 \tag{M.8}$$

The second term can be divided into two, as follows.

$$\text{Re}\left\{\sum \frac{i\hbar E_{mn}^2}{(E_{mn}^2 - E^2)E^2} \cdot M_{nm} \cdot D_{mn}\right\}\frac{dE_0}{dt}$$

$$= \text{Re}\left\{\frac{i\hbar}{E^2}\sum M_{nm} \cdot D_{mn} + \sum \frac{i\hbar}{E_{mn}^2 - E^2} \cdot M_{nm} \cdot D_{mn}\right\}\frac{dE_0}{dt} \tag{M.9}$$

Since the first term in Eq. M.9 is only the sum of products of matrix elements, we take the sum of them and this term becomes purely imaginary (the real part is zero because $(i\hbar/E^2)\sum M_{nm} \cdot D_{mn} = (i\hbar/E^2)(MD)_{nn}$, $(MD)_{nn} = \int u_m^* |MD| u_n d\tau$. The second term becomes

$$\sum \frac{1}{2\pi h(v_{mn}^2 - v^2)}\,\text{Re}\,\{i \cdot M_{nm} \cdot D_{mn}\} \tag{M.10}$$

and

$$\text{Re}\,\{i \cdot M_{nm} \cdot D_{mn}\} = \text{Re}\,\{-i(D_{mn}^* \cdot M_{nm}^*)\} = \text{Re}\,\{-i(D_{nm} \cdot M_{mn})\}$$

$$= \text{Im}(D_{nm} \cdot M_{mn}) \tag{M.11}$$

so this term becomes as follows

$$\frac{1}{2\pi h}\sum \frac{\text{Im}\{D_{nm} \cdot M_{mn}\}}{v_{mn}^2 - v^2} \cdot \frac{dE_0}{dt} \tag{M.12}$$

Since the third term of Eq. 9.93 consists of real numbers,

$$\text{Re}\left\{\sum \frac{E_{mn}}{E_{mn}^2 - E^2} \cdot M_{nm} \cdot M_{mn}\right\} H_0 = \frac{1}{h} \sum \frac{v_{mn}}{v_{mn}^2 - v^2} |M_{nm}|^2 \cdot H_0 \quad (\text{M.13})$$

The coefficient in { } of the fourth term is purely imaginary, and Re { } becomes zero. Taking the average of Eq. 9.78 over all molecular directions and using the local electromagnetic fields, we can rewrite as follows (this is equivalent to Eqs. 9.79 and 9.80 in section 9.2.2).

$$\mathbf{m} = \frac{2}{3h} \sum \frac{v_{mn}}{v_{mn}^2 - v^2} |M_{nm}|^2 \cdot H' + \frac{2}{3h} \sum \frac{v_{mn} \, \text{Re}\{D_{nm} \cdot M_{mn}\}}{v_{mn}^2 - v^2} E'$$

$$+ \frac{1}{3\pi h} \sum \frac{\text{Im}\{D_{nm} \cdot M_{mn}\}}{v_{mn}^2 - v^2} \frac{\partial E'}{\partial t} = \kappa_n H' + \gamma_n E' + \frac{\beta_n}{c} \cdot \frac{\partial E'}{\partial t} \quad (\text{M.14})$$

where

$$\kappa_n = \frac{2}{3h} \sum \frac{v_{mn}}{v_{mn}^2 - v^2} |M_{nm}|^2 \quad (\text{M.15})$$

This is equivalent to Eq. 9.94 in section 9.2.2.

APPENDIX N

Value of $\text{Re}\{c_m^*(t)c_m(t)\}_{\stackrel{r}{l}}$ (Eq. 9.103)

Taking the following equations

$$c_m(t)_{\stackrel{r}{l}} = \frac{1}{2}\left(\frac{iE_{mn}}{ch} \cdot D_{mn} \mp \frac{E}{ch} M_{mn}\right) \text{Re}\left\{A(i \pm ij) \exp\left(-i\frac{E}{ch}z\right)\right\}$$

$$\times \left\{\frac{\exp\left(i\dfrac{E_{mn} - E}{h}t\right) - 1}{E_{mn} - E}\right\} \quad (\text{N.1})$$

$$c_m^*(t)_{\stackrel{r}{l}} = \frac{1}{2}\left(\frac{-iE_{mn}}{ch} \cdot D_{mn}^* \mp \frac{E}{ch} M_{mn}^*\right) \text{Re}\left\{A(i \mp ij) \exp\left(i\frac{E}{ch}z\right)\right\}$$

$$\times \left\{\frac{\exp\left(-i\dfrac{E_{mn} - E}{h}t\right) - 1}{E_{mn} - E}\right\} \quad (\text{N.2})$$

$$\text{Re }\{c_m^*(t)\cdot c_m(t)\}_r = \frac{1}{4}\text{Re}\left\{\left(\frac{E_{mn}^2}{c^2\hbar^4}|D_{mn}|^2 \pm \frac{iE_{mn}E}{c^2\hbar^4}\cdot D_{mn}^*\cdot M_{mn}\right.\right.$$

$$\mp \frac{iE_{mn}E}{c^2\hbar^4}\cdot M_{mn}^*\cdot D_{mn} + \left.\frac{E^2}{c^2\hbar^4}|M_{mn}|^2\right)A^2(i^2+j^2)\right\}$$

$$\times\left\{\frac{-\left\{\exp\left(i\frac{E_{mn}-E}{2\hbar}t\right) - \exp\left(-i\frac{E_{mn}-E}{2\hbar}t\right)\right\}^2}{(E_{mn}-E)^2/\hbar^2}\right\}$$

$$\text{(N.3)}$$

$$\text{Re}\,(iD_{mn}^*\cdot M_{mn}) = -\text{Im}(D_{nm}\cdot M_{mn}) \qquad \because \quad D_{mn}^* = D_{nm}$$

$$\text{Re}\,(iM_{mn}^*\cdot D_{mn}) = -\text{Im}(M_{nm}\cdot D_{mn}) = \text{Im}(D_{nm}\cdot M_{mn})$$

we apply the expression

$$\left[\exp\left\{\frac{i(E_{mn}-E)t}{2\hbar}\right\} - \exp\left\{\frac{-i(E_{mn}-E)t}{2\hbar}\right\}\right]^2$$

$$= \left\{2i\sin\frac{(E_{mn}-E)t}{2\hbar}\right\}^2 = -4\sin^2\frac{(E_{mn}-E)t}{2\hbar} \qquad \text{(N.4)}$$

and omitting $|M_{mn}|^2$ since it is small, we obtain

$$\text{Re}\{c_m^*(t)c_m(t)\}_r = \frac{E_{mn}^2}{c^2\hbar^4}\left\{|D_{mn}|^2 \mp 2\frac{E}{E_{mn}}\cdot\text{Im}(D_{nm}\cdot M_{mn})\right\}$$

$$\times \frac{\sin^2\{(E_{mn}-E)t/2\hbar\}}{\{(E_{mn}-E)/2\hbar\}^2}\{A^2(i^2+j^2)\} \qquad \text{(N.5)}$$

Now, supposing that a molecule may take any direction about the z axis, and taking the average, the following equation holds good.

$$\text{Re }\{c_m^*(t)c_m(t)\}_r = \frac{E_{mn}^2 A^2}{3c^2\hbar^4}\left\{|D_{mn}|^2 \mp 2\frac{E}{E_{mn}}\text{Im}(D_{nm}\cdot M_{mn})\right\}$$

$$\times \frac{\sin^2\{(E_{mn}-E)t/2\hbar\}}{\{(E_{mn}-E)/2\hbar\}^2} \qquad \text{(N.6)}$$

Author Index

Numbers in parentheses are reference numbers and indicate that an author's work is referred to although his name is not cited in the text. Numbers in italic show the page on which the complete reference is listed.

Subject Index

A, B

absolute asymmetric snythesis, **158**, **181**
achiral, **15**
—molecule, 23, 37, 193
aconitase, **156**
activation energy, **180**, 204, 207
acylase, 157
alcohol dehydrogenase, **157**
L-amino acid decarboxylase, 223
L-amino acid oxidase, 157, 223
L-amino acid transaminase, 223
anisotropy factor, 159, **288**
anomalous beam, **260**
anti, 62
Arrhenius plots, 205
assignment table, **35**
asymmetric (=dissymmetric), **9**
asymmetric reaction
classification of, 75
historical definition of, 174
new definition and classification of, 75
—transformation, **75**, **179**
atrop isomer, 8, 59
axial chirality,
specification of, 59
axis
improper—, *see* improper axis
proper—, *see* proper axis

bijection, 49
Brewster's angle, **259**, 261

C

calcite, 259
camphorsulfonic acid, 123
Cartesian coodinate, **16**
catalyst, 3, 151
MRNi type, 105
recemization, 169
silk palladium type, 102
Wilkinson type, 4, 94, 208, 209
central chirality,
specification of, 59
chiral, **9**
normenclature for, 57

chiral factor, 82
—molecule, **21**, 23, 37
chirality, **7**, **9**, **19**
axial—, *see* axial chirality
central—, *see* central chirality
planar—, *see* planar chirality
—function, **194**
circular dichroism, 159, **287**
circularly polarized light, **158**, 168, **263**
circularly polarized γ-ray, 158
cis, 62
—addition, 72
clathrate compound, 154
clinal, 63
condensation polymerization, **151**
configuration, description of
by D,L-system, 58
by *dl*-system, 57
by *k*,*s*-system, 58
by (+), (−)-system, 57, **59**
configurational diastereoface, **68**
—diastereo-differentiating reaction, **80**
—enantiomer-differentiating reaction, **158**
—plane, **66**
—stereoisomer, 7
conformational diastereo-differentiating
reaction, **80**
—diastereoface, **68**
—enantiomer-differentiating reaction, **160**
—plane, **66**
—stereoisomer, 8
nomenclature for, 62
conjugate class, 51
Cornforth's rule, **186**
coset, 23, **50**, 193
Cram's rule, **185**
Curtin-Hammett principle, **207**

D

diastereo-differentiating ability, **80**, **131**,
192, 248
diastereo-differentiating reaction
definition of, **76**
effect of differentiating ability of
reagent or catalyst on, 131

DATE DUE